INTRODUCTION TO
MODERN VIROLOGY
BASIC MICROBIOLOGY
SERIES

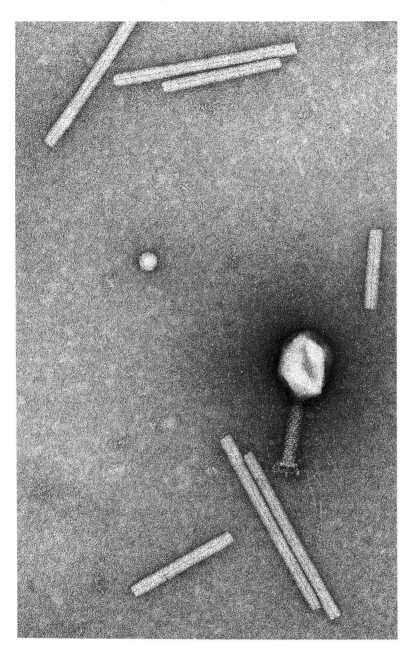

Frontispiece. Electron micrograph showing particles of bacteriophage T4, bacteriophage φX174 and tobacco mosaic virus (courtesy Dr F. Eiserling).

BASIC MICROBIOLOGY SERIES

Introduction to Modern Virology

N. J. DIMMOCK BSc, PhD

Professor of Virology
Department of Biological Sciences
University of Warwick, Coventry

S. B. PRIMROSE BSc, PhD

Amersham International plc
Amersham, Buckinghamshire

FOURTH EDITION

b

Blackwell
Science

© 1974, 1980, 1987, 1994 by
Blackwell Science Ltd
Editorial Offices:
Osney Mead, Oxford OX2 0EL
25 John Street, London WC1N 2BL
23 Ainslie Place, Edinburgh EH3 6AJ
238 Main Street, Cambridge
    Massachusetts 02142, USA
54 University Street, Carlton
    Victoria 3053, Australia

Other Editorial Offices:
Arnette Blackwell SA
1, rue de Lille, 75007 Paris
France

Blackwell Wissenschafts-Verlag GmbH
Kurfürstendamm 57
10707 Berlin, Germany

Blackwell MZV
Feldgasse 13, A-1238 Wien
Austria

First published 1974
Reprinted 1978
Second edition 1980
Third edition 1987
Fourth edition 1994

Set by Setrite Typesetters, Hong Kong
Printed and bound in Great Britain
at the University Press, Cambridge

DISTRIBUTORS

Marston Book Services Ltd
PO Box 87
Oxford OX2 0DT
(*Orders*: Tel: 01865 791155
         Fax: 01865 791927
         Telex: 837515)

USA
Blackwell Science, Inc.
238 Main Street
Cambridge, MA 02142
(*Orders*: Tel: 800 759-6102
           617 876-7000)

Canada
Oxford University Press
70 Wynford Drive
Don Mills
Ontario M3C 1J9
(*Orders*: Tel: 416 441-2941)

Australia
Blackwell Science Pty Ltd
54 University Street
Carlton, Victoria 3053
(*Orders*: Tel: 03 347-5552)

A catalogue record for this title
is available from the British Library

ISBN 0-632-03403-3

Library of Congress
Cataloging-in-Publication Data

Dimmock, N. J.
  Introduction to modern virology/
  N. J. Dimmock, S. B. Primrose. — 4th ed.
       p.   cm.
  Includes bibliographical references
  and index.
  ISBN 0-632-03403-3
  1. Virology. I. Primrose, S. B.
  II. Title. III. Series.
  [DNLM: 1. Virus Diseases.
  2. Viruses. QW 160 D582i 1994]
  QR360.D56 1994
  616'.0194 — dc20

Contents

Preface to the fourth edition

Understanding of virology has continued to grow, aided as always by major technical advances. The fourth edition aims to balance this increase in knowledge with the needs of students who are new to the field. All sections of the book have been revised, and a new chapter added on HIV and AIDS, reflecting both the importance of the disease and the explosion of knowledge in this area. The author is, as ever, grateful to generations of students at Warwick, who continue to provide criticisms and comments, which have helped in the evolution of this latest edition. I would also like to extend my thanks to colleagues, especially Keith Leppard, for their generous advice.

N. J. Dimmock

Preface to the first edition

This book has been developed from a series of ten lectures which were presented to Honours students in microbiology at the University of Edinburgh in 1971 and 1972. In these lectures I did not attempt to cover the entire field of virology. Instead, emphasis was placed on the biochemical and genetical aspects, since I was sure that the students would find this approach much more interesting than clinical details of viral diseases. Subsequent response to the lectures showed this supposition to be correct and, consequently, a similar approach has been adopted in this book. The reader will quickly realize that only a limited range of viruses is discussed; not only does this make the text easier for students, but it follows the preference of virologists for a few model systems.

Anyone who has a knowledge of virology will appreciate that research in some areas is expanding so rapidly that it is difficult to include the latest findings. However, where possible, the appropriate sections were revised shortly before submission of the manuscript.

Acknowledgement must be made of the contributions of Professor D. C. Burke and Dr J. Bennett, who read most of the chapters and suggested numerous improvements of language, style and reasoning. In addition, several other colleagues criticized individual chapters, in particular, Dr N. J. Dimmock (Chapter 7), Dr J. Gross (Chapter 6), and Dr R. A. Weiss (Chapter 11). Finally, and most important of all, I must thank my wife for her patience during the writing of this book.

1 Towards a definition of a virus

Towards the end of the last century, the germ theory of disease was formulated and pathologists were confident that for each infectious disease there would be found a micro-organism which could be seen with the aid of a microscope, could be cultivated on a nutrient medium and could be retained by filters. There were, admittedly, a few organisms which were so fastidious that they could not be cultivated *in vitro* (i.e. in the test tube), but they did satisfy the other two criteria. However, a few years later, in 1892, Iwanowski was able to show that the causal agent of tobacco mosaic passes through a bacteria-proof filter and could not be seen or cultivated. Iwanowski remained unimpressed by his discovery, but Beijerinck, upon repeating the experiments in 1898, was convinced of the existence of a new form of infective agent which he termed *Contagium Vivum Fluidum*. In the same year Loeffler and Frosch came to the same conclusion regarding the cause of foot-and-mouth disease. Furthermore, because foot-and-mouth disease could be passed from animal to animal, with great dilution at each passage, the causative agent had to be reproducing and thus could not be a bacterial toxin. Viruses of other animals were soon discovered. Ellerman and Bang reported the cell-free transmission of chicken leukaemia in 1908 and in 1911 Rous discovered that solid tumours of chickens could be transmitted by cell-free filtrates. These were undoubtedly the first indications that viruses can cause cancer.

Although studies on bacterial viruses were later to prove the most useful for investigating the virus–host relationship, they were the last to be discovered. In 1915 Twort published an account of a glassy transformation of micrococci. He had been trying to culture the smallpox agent on agar plates but the only growth obtained was that of some contaminating micrococci. Upon prolonged incubation, some of the colonies took on a glassy appearance and, once this occurred, no bacteria could be subcultured from the affected colonies. If some of the glassy material was added to normal colonies, they too took on a similar appearance, even if the glassy material was first passed through very fine filters. Among the suggestions that Twort put forward to explain the phenomenon was the existence of a bacterial virus or the secretion by the bacteria of an enzyme which could lyse the producing cells. This idea of self-destruction by secreted enzymes was to prove a controversial topic for at least the next decade. In 1917 d'Hérelle observed a similar phenomenon in dysentery bacilli. He observed clear spots on lawns of dysentery bacilli and, although realizing that this was not an original observation, resolved to find an explanation for them. Upon noting the lysis of broth cultures of pure dysentery

1

bacilli by filtered emulsions of faeces, he immediately realized he was dealing with a bacterial virus. Since this virus was incapable of multiplying except at the expense of living bacteria, he called his virus a *bacteriophage*, or *phage* for brevity.

Thus the first definition of these new agents, the viruses, was presented entirely in negative terms: they could not be seen or cultivated and, most important of all, they were not retained by bacteria-proof filters.

THE ASSAY OF VIRUSES

The observations of d'Hérelle, however, led to the introduction of two important techniques. The first of these was the preparation of stocks of bacterial viruses by lysis of bacteria in liquid cultures. This has proved particularly valuable in modern virus research, since bacteria can be grown in defined media to which radioactive precursors can be added to 'label' selected viral components. Secondly, d'Hérelle's observations provided means of assaying these invisible agents. One method was to grow a large number of identical cultures of a susceptible bacterium and to inoculate these with dilutions of the virus-containing sample. If the sample was diluted too far, none of the cultures would lyse. However, in the intermediate range of dilutions not all of the cultures lyse, since not all receive a virus particle, and the assay is based on this. For example, d'Hérelle noted that, in 10 test cultures inoculated with a volume corresponding to 10^{-11} ml, only three were lysed. Thus three cultures received one or more viable phage particles while the remaining seven received none, and it can be concluded that the sample contained between 10^{10} and 10^{11} viable phages per millilitre. It is possible to apply statistical methods to end-point dilution assays of this sort and obtain more precise estimates. The other method suggested was the *plaque assay method*, which is the most widely used and most useful. d'Hérelle observed that the number of clear spots or *plaques* formed on a lawn of bacteria (Fig. 1.1A) was inversely proportional to the dilution of bacteriophage lysate added. Thus the *titre* of a virus-containing solution can be readily determined in terms of *plaque-forming units* (PFU) and, if each virus particle in the preparation gives rise to plaque, then the *efficiency of plating* (EOP) is unity.

Both these methods were later applied to the more difficult task of assaying plant and animal viruses. However, because of the labour, time and cost involved in providing large stocks and animals, the end-point dilution assay is avoided where possible. For the assay of plant viruses, a variation of the plaque assay, the *local lesion assay* was developed by Holmes in 1929. He observed that countable necrotic lesions were produced on leaves of the tobacco plant, particularly *Nicotiana glutinosa*, inoculated with tobacco mosaic virus and that the number of local lesions depended on the virus content of the inoculum. Unfortunately, individual plants, and even individual leaves of the same plant, produce different numbers of lesions with the same inoculum. However, the opposite halves of the same leaf give almost identical numbers of lesions and it is possible to compare the same dilutions of two virus-containing samples by inoculating them on the opposite halves of the same leaf (Fig. 1.1B).

A major advance in animal virology came in 1952, when Dulbecco devised a plaque assay for animal viruses. In this case a suspension of susceptible cells, prepared by trypsinization of a suitable tissue, is placed in Petri dishes or other culture vessels. The cells attach and grow across the surface until a monolayer of cells is formed. Once the cells are present at such a density that they come into contact with one another, growth ceases by a process called *contact inhibition*; hence the formation of a monolayer. Upon formation of the monolayer, the nutrient medium bathing the cells is removed and a suitable dilution of the virus added. After a short period of incubation to allow the virus particles to adsorb to the cells, nutrient agar is placed over the cells. After a further period of incubation ranging from 24 hours to 24 days, the cells are stained by adding neutral red or some other vital dye to the agar. Living cells take up the stain but dead cells do not and the plaques are seen as unstained circular areas in the stained cell sheet (Fig. 1.1C). Since tumour viruses are not cytopathic (i.e. do not kill cells), they cannot be assayed by this means. However, cells infected with these viruses do not show contact inhibition and so form 'colonies' of cells on the surface of the monolayer, and this can also be used as an assay method.

THE MULTIPLICATION OF VIRUSES

Although methods of assaying viruses had been developed, there was still considerable doubts as to the nature of viruses. d'Hérelle believed that the infecting phage particle multiplied within the bacterium and that its progeny were liberated upon lysis of the host cell. Convincing support for this hypothesis was provided by the one-step growth experiment of Ellis and Delbruck (1939). A phage preparation was mixed with a suspension of bacteria and, after allowing a few minutes for the phage to adsorb, the culture was diluted 50-fold to stop further adsorption. The result obtained is shown in Fig. 1.2. After a latent period of 30 minutes in which no phage increase could be detected, there was a sudden 70-fold rise in PFU. This 'burst' size of 70 seen at each step in the growth curve represents the average of many different individual bursts. Samples were withdrawn at regular intervals and assayed for phage particles.

Later Delbruck extended his studies in order to settle the controversy which surrounded the growth of phage and the dissolution of cultures. d'Hérelle believed that intracellular phage growth led to the lysis of the host cell and the liberation of virus, whereas others believed that phage-induced dissolution of bacterial cultures was merely the consequence of a stimulation of lytic enzymes endogenous to the bacteria. Yet another school of thought was that phages could pass freely in and out of bacterial cells and that lysis of bacteria was a secondary phenomenon not necessarily concerned with the growth of a phage. It was Delbruck who ended the controversy by pointing out that two phenomena were involved, lysis from within and lysis from without. The type of lysis observed was dependent on the ratio of infecting phages to bacteria (*multiplicity of infection*). When the ratio of phages to bacteria is no greater than 1 : 1 or 2 : 1, i.e. low multiplicity of infection, then the phages infect the cells, multiply and lyse the cells from within. When the multiplicity of infection is high, i.e. many more phages than bacteria, the cells lyse and there is no increase in phage titre but, rather, a decrease. Lysis is due to weakening of the cell wall when large numbers of phages are adsorbed.

Although the one-step growth curve demonstrated the nature and kinetics of the process by which bacterial viruses multiply within cultures of susceptible bacteria, it gave no indication of the events taking place inside the cell. Doermann (1952) infected cells

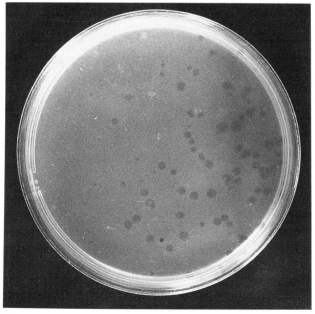

Fig. 1.1 Plaques of viruses. (A) Plaques of a bacteriophage on *Escherichia coli*.

A

of *Escherichia coli* with bacteriophage T4 in the presence of cyanide to synchronize intracellular events. By diluting the culture, no more phages can adsorb, and this also effectively removes the cyanide. Aliquots of cells were then lysed at intervals. The number of T4 pfu present inside the cell is shown in Fig. 1.3. Whereas the number of *infectious centres* (i.e. free phage + infected cells) remains constant until lysis, the number of bacteriophages inside the cell does not. Immediately after infection, there is an *eclipse* period in which no phage can be detected, followed by an increase in pfu until lysis occurs. It should also be noted that the kinetics of appearance of intracellular phage particles is *linear*, not exponential. This suggests that the particles are produced by assembly from component parts rather than by binary fission.

B

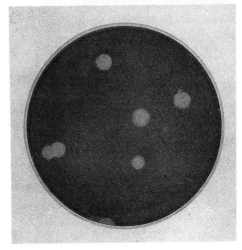

Fig. 1.1 (*Continued*) (B) Local lesions on a leaf of *Nicotiana* caused by tobacco mosaic virus (courtesy of National Vegetable Research Station). (C) Plaques of influenza virus on chick embryo fibroblasts.

C

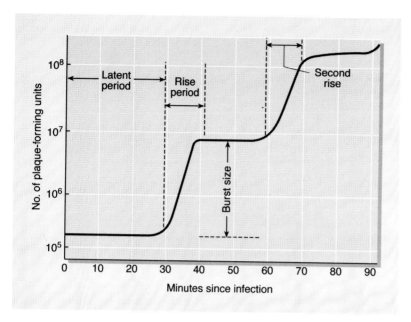

Fig. 1.2 The multiplication of bacteriophages following infection of susceptible bacteria.

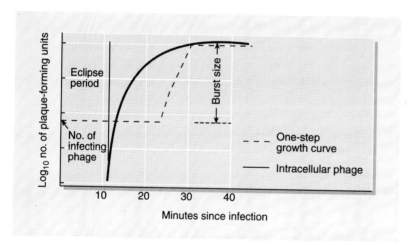

Fig. 1.3 Intracellular development of bacteriophages following infection of susceptible bacteria. Note that samples taken early (before 12 minutes) contain less infectious phage than the original inoculum.

VIRUSES CAN BE DEFINED IN CHEMICAL TERMS

The first virus was purified in 1933 by Schlessinger, using differential centrifugation. This technique, which is still widely used, consists of a cycle of low- and high-speed centrifugation. The low-speed centrifugation pellets the host cell debris and the high-speed centrifugation pellets the virus. Chemical analysis of the purified bacteriophage showed that it consisted of approximately equal proportions of protein and deoxyribonucleic acid (DNA). A few years later, in 1935, Stanley isolated tobacco mosaic virus in para-crystalline form, and this crystallization of a biological material thought to be alive raised many philosophical questions about the nature of life. In 1937 Bawden and Pirie extensively purified tobacco mosaic virus and showed it to be nucleoprotein containing ribonucleic acid (RNA).

The importance of viral nucleic acid

The purification of viruses and the demonstration that they consisted of only protein and nucleic acid came at a time when the nature of the genetic material was not known. In 1949 Markham and Smith found that the spherical particles in preparations of turnip yellow mosaic virus were of two types, only one of which contained nucleic acid. Significantly, only the particles containing nucleic acid were infective. A few years later, in 1952, Hershey and Chase demonstrated the independent functions of viral protein and nucleic acid. Bacteriophage T2 was grown in *E. coli* in the presence of ^{35}S (as sulphate) to label the protein moiety, or ^{32}P (as phosphate) to label the nucleic acid. Purified, labelled phages were allowed to adsorb to sensitive host cells and then subjected to the shearing forces of a Waring blender (Fig. 1.4).

Treatment of the cells in this way removes any phage components attached to the outside of the cell but does not affect viability. When the cells were removed from the medium, it was observed that 75% of the ^{35}S (i.e. phage protein) had been removed from the cells by blending but only 15% of the ^{32}P (i.e. phage DNA). Thus, after infection, the bulk of the phage protein appears to have no further function and consequently it must be the DNA that is the carrier of viral heredity. The release of the phage DNA from its protein envelope upon infection also accounts for the existence of the eclipse period during the early stages of intracellular virus development, since the DNA on its own cannot normally infect a cell.

In another classic experiment, Fraenkel-Conrat and Singer (1957) were able to confirm by a different means the hereditary role of viral RNA. Their experiment was based on the earlier discovery that particles of tobacco mosaic virus can be dissociated into their protein and RNA components and then reassembled to give particles which are morphologically mature and fully infective. When two different strains (differing in the symptoms produced in the host plant) were each disassociated and the RNA of one reassociated with the protein of the other, and vice versa, the type of virus which was propagated when the resulting 'hybrid' particles were used to infect host plants was always that from which the RNA was derived (Fig. 1.5).

The ultimate proof that viral nucleic acid is the genetic material comes from numerous observations that under special circumstances purified viral nucleic acid is capable of initiating infection, albeit with a reduced efficiency. For example, in 1956 Gierer and Schramm and Fraenkel-Conrat independently showed that the

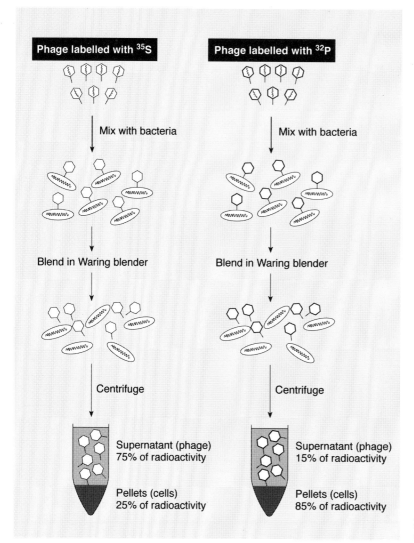

Fig. 1.4 The Hershey-Chase experiment.

purified RNA of tobacco mosaic virus can be infectious, provided precautions are taken to protect it from inactivation by ribonuclease. In fact, the causative agent of potato spindle tuber disease completely lacks a protein component and consists solely of RNA. Because these agents have no protein coat, they cannot be called viruses and are referred to as *viroids*.

The synthesis of macromolecules in infected cells

Knowing that it is the nucleic acid that is the carrier of genetic information, and that with bacteriophages only the nucleic acid enters the cell, it is pertinent to determine the events occurring inside the cell. The discovery in 1953 by Wyatt and Cohen that

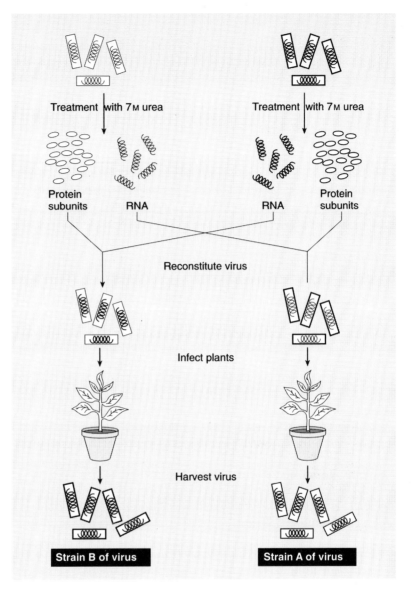

Fig. 1.5 The experiment of Fraenkel-Conrat and Singer which proved that RNA is the genetic material of tobacco mosaic virus.

the DNA of the T-even bacteriophages T2, T4 and T6 (see p. 65) contains hydroxymethylcytosine (HMC) instead of cytosine made it possible in the same year for Hershey, Dixon and Chase to examine infected bacteria for the presence of phage-specific DNA at various stages of intracellular growth. DNA was extracted from T2-infected *E. coli* at different times after the onset of phage growth and analysed for its content of HMC. Analyses of this type provided an estimate of the number of phage equivalents of HMC-containing DNA present at any time, based on the total nucleic acid and relative HMC content of the intact T2 phage particle. The results showed that, with T2, synthesis of phage DNA commences about 6 minutes after infection and then rises sharply, so that by the time the first infective particles begin to appear 6 minutes later there are 50–80 phage equivalents of HMC. Thereafter, the numbers of phage equivalents of DNA and of infective particles increase linearly and at the same rate up until lysis, even if lysis is delayed beyond the normal burst time. The observed kinetics of phage and DNA synthesis is important because it provides evidence not only for assembly of the virus from individual components but for regulation of development.

Hershey and his co-workers also studied the synthesis of phage protein, which can be distinguished from bacterial protein by its interaction with specific antibodies. During infection of *E. coli* by T2 phage, protein can be detected about 9 minutes after the onset of the latent period, i.e. after DNA synthesis begins, and by the time infectious particles begin to appear; a few minutes later there are approximately 30–40 phage equivalents inside the cell. Whereas the synthesis of viral protein starts about 9 minutes after the onset of the latent period, it was shown by means of pulse–chase experiments that the uptake of ^{35}S into intracellular protein is constant from the beginning. A small quantity (pulse) of ^{35}S (as sulphate) was added to the medium at different times after infection and was followed shortly by a vast excess of unlabelled sulphate (chase) to stop any further incorporation of label. When the pulse was made from the ninth minute onward, the label could be chased into material serologically identifiable as phage coat protein. However, if the pulse was made early in infection, it could be chased into protein but, although this was non-bacterial, it did not react with antibodies to phage structural proteins. The nature of this early protein will be discussed in Chapter 10.

Being the genetic material, the nucleotide sequence in the viral nucleic acid has to be translated into proteins. The pioneering work of McQuillen, Britten and Roberts had shown that protein synthesis takes place on the ribosomes rather than on DNA, and studies of phage-infected cells were to solve many of the other mysteries of protein synthesis. Volkin and Astrachan examined the RNA from infected and uninfected cells and compared the base ratios with the DNA from the infecting bacteriophage and from uninfected cells. From the results shown in Table 1.1, it is clear that after infection only phage-specific RNA is synthesized, and with this information Brenner, Jacob and Meselson were able to show in 1959 that this viral-specific RNA associates with pre-existing host cell ribosomes.

VIRUSES CAN BE MANIPULATED GENETICALLY

One of the easiest ways to understand the steps involved in a

Table 1.1 Base ratios of DNA and RNA from uninfected and infected cells.

Material	Ratio of $\dfrac{\text{adenine + thymine/uracil}}{\text{guanine + cytosine}}$
DNA from uninfected cell	1.0
RNA from uninfected cell	0.85
DNA from phage	1.8
RNA from infected cell	1.7

particular reaction is to isolate mutants which are unable to carry out that reaction. Like all other organisms, viruses sport mutants in the course of their growth, and these mutations can affect the type of plaque formed, the range of hosts which the virus can infect the physico-chemical properties of the virus. There is one obvious restriction, however, and this is that many mutations will obviously be lethal to the virus and remain undetected. This problem was very neatly overcome in 1963 by Epstein and Edgar and their collaborators with the discovery of *conditional lethal mutants*. One class of mutants, the *temperature-sensitive mutants* were able to grow at some low temperature, the *permissive* temperature, but not at some higher, *restrictive* temperature at which normal virus could grow. The other class of conditional lethal mutants was the *amber* mutant. This converts a codon into a triplet which terminates protein synthesis. This only grows on a *permissive* host cell, which has a transfer RNA (tRNA) that can insert an amino acid at the amber mutation. Since the first discovery of conditional lethal mutants, various other types have been described. These are *ochre* mutants, which are not unlike the amber mutants and *cold-sensitive* mutants, which are the reverse of temperature-sensitive mutants.

The drawback to conditional lethal mutants is that mutation is random, but the advent of *recombinant DNA technology* has facilitated controlled mutagenesis. With this new technology, a piece of DNA can be excised from its surroundings and modified to order by the judicious use of *restriction endonucleases* (p. 113). The key feature of these techniques is the ability of certain restriction endonucleases to produce staggered cuts at well-defined sites on DNA molecules (see Fig. 7.16 for examples). The way in which these enzymes can be used is shown in Fig. 1.6. DNAs from two different sources are cleaved with the same endonuclease, e.g. *EcoRI* (purified from *E. coli* strain RI) to produce fragments with complementary single-stranded tails. By incubating mixtures under annealing conditions and joining with DNA ligase, it is possible to produce mixed dimers. The dimer can then be inserted back into the parental DNA in the same way and the expression of the modified gene studied. (Additional reading will be necessary for the student to comprehend the scope of this area of molecular biology.)

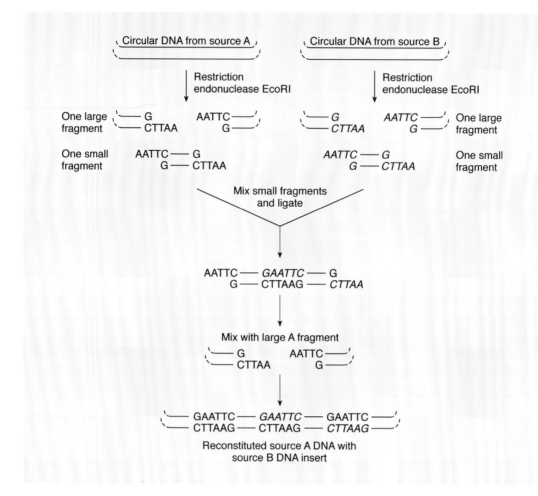

Fig. 1.6 Controlled mutagenesis by recombinant DNA technology. Note that there would be other combinations of recombinant molecules and that selection procedures would be needed to obtain the desired molecule. A, adenine; C, cytosine; G, guanine; T, thymine.

Although it is relatively easy to introduce fragments of DNA into bacteria by the process of *transformation*, such fragments will fail to replicate and will be lost from the cell. To circumvent this problem, the DNA fragment is inserted into a *vector* (or *cloning vehicle*), which is simply a DNA molecule capable of autonomous replication. Commonly used vectors are bacterial plasmids and bacteriophages such as λ. Bacteriophage λ has two advantages as a vector. Firstly, the hybrid DNA can be prepared in great quantities since each bacterial cell can produce several hundred λ DNA copies and the hybrid DNA can be purified easily as a component of bacteriophage particles. By using λ mutants that make the host lysis defective, it is possible to increase the phage yield up to 10 times and in addition the phages remain in the bacterial cell until artificially lysed. Secondly, the detailed knowledge of the λ genome is particularly useful, for it is possible to build in mutations which increase the transcription of the foreign DNA insert.

The most specific of all genetic manipulations is provided by *site-directed mutagenesis*. Here, an oligonucleotide complementary to the region that you wish to mutate is synthesized chemically but with a single mismatched base. The DNA to be mutated (it could be all or part of a viral genome) is inserted in a single-stranded form into the single-stranded DNA phage M13. Then, by annealing with the oligonucleotide, a point mutation can be made exactly where desired (Fig. 1.7). Finally, the mutated DNA is cut out of the phage with a restriction endonuclease and the effect of the point mutation on the function of the gene can then be assessed.

THE PROPERTIES OF VIRUSES

Assuming that the features of virus growth just described for particular viruses are true of all viruses, we are now in a position to compare and contrast the properties of viruses with those of their host cells. Whereas their host cells contain both types of nucleic acid, viruses only contain one type, as analyses of purified viruses have shown. However, just like their host cells, viruses have their genetic information encoded in their nucleic acid. Another difference is that the virus is reproduced solely from its genetic material, whereas the host cell is reproduced from the integrated sum of its components. Thus, the virus never arises directly from a pre-existing virus, whereas the cell always arises directly from a pre-existing cell. The experiments of Hershey and his collaborators showed quite clearly that the components of a virus are synthesized independently and then assembled into mature virus particles. In contrast, growth of the host cell consists of an increase in the amount of all its constituent parts, during which the individuality of the cell is continuously maintained. Finally, viruses are incapable of synthesizing ribosomes but, instead, depend on pre-existing host cell ribosomes for synthesis of viral proteins. These features clearly separate viruses from all other organisms, even chlamydia, which for many years were considered to be intermediate between bacteria and viruses.

PLAN OF THE REMAINDER OF THE BOOK

In the preceding discussion, we have described key experiments which led to our understanding of the nature of viruses. In addition, we have introduced some experimental techniques used by virologists. However, animal virologists often work with animal

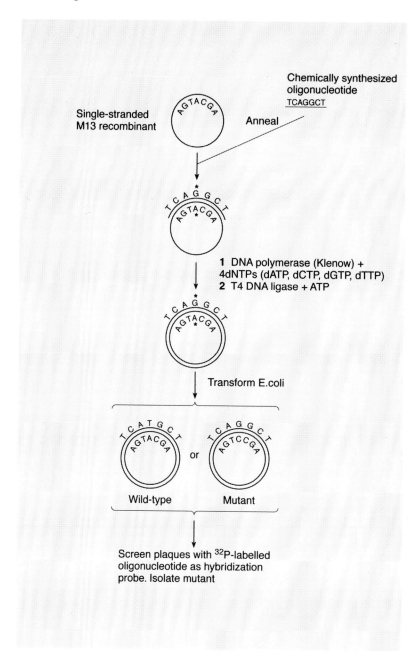

Fig. 1.7 Site-directed mutagenesis. Asterisks indicate mismatched bases. Note that, instead of a point mutation, an oligonucleotide carrying an insertion or a deletion could be used in the same way. dATP, (deoxy) adenosine 5'-triphosphate; dCTP, cytosine 5'-triphosphate, dGTP, guanosine 5'-triphosphate, dTTP, thymidine 5'-triphosphate. ATP, adenosine 5'-triphosphate.

cells in culture, and the special techniques which they use are outlined in Chapter 2.

It is clear that viruses consist of nucleic acid, the repository of the genetic information, surrounded by a protective protein coat. In Chapter 3, consideration will be given to the limited number of ways in which such protein shells can be constructed, and Chapter 4 will detail the numerous structural peculiarities exhibited by viral nucleic acids. Next attention is focused on mechanisms viruses have evolved to enable them to infect susceptible cells. Chapter 5, which examines this question, is unique in that it is the only chapter in which plant, animal and bacterial viruses are discussed separately. The properties of these three classes of virus are so similar, except for the penetration step, that in our opinion they do not warrant separate discussions.

In considering the replication of viral nucleic acid, we must bear in mind the numerous structural peculiarities of such molecules. The Baltimore classification, which facilitates such considerations, is presented in Chapter 6. The replication of viral DNA and RNA occupies the next three chapters. Not all proteins synthesized in the infected cell are structural components of the virus, and the molecular events controlling their synthesis are discussed in Chapter 10. Finally, this analysis of the process of infection is concluded in Chapter 11, when the formation of mature virus particles from their different components is described.

The second half of the book is devoted to the biological inter-actions between viruses and their hosts, as distinct from the molecular aspects of replication. Chapter 12 is devoted to lysogeny, the process whereby some bacteriophages are maintained in the cell in a state of 'suspended animation'. As well as being of intrinsic interest, lysogeny provides insights towards an understanding of the interaction between viruses and animals. These are outlined in the succeeding six chapters. Chapter 13 is a brief account of the interaction between viruses and cultured eukaryotic cells, whereas Chapter 15 is devoted to the interaction between animal viruses and their animal hosts. From the point of view of the virus, one significant difference between cells in culture and whole animals is that the latter possess an adaptive defence system, the immune system, which modulates infection (Chapter 14). The way immunity can be stimulated by vaccines is covered in Chapter 16. As a result of the widespread fear of cancer, tumour viruses are currently being subjected to intensive study, and Chapter 17 summarizes the latest information in this area. Like all living organisms, viruses continually evolve, and the effects of this evolution on the patterns of disease in man are discussed in Chapter 18. The acquired immune deficiency syndrome (AIDS) pandemic is the best known and is the subject of Chapter 19.

In any branch of science there are fashionable topics and virology is no exception. Chapter 20 contains an account of these trends and attempts to predict those topics that will be the focus of attention in the next few years. Finally, the last chapter contains a brief description of all the major groups of viruses.

FURTHER READING

Fields, B. N & Knipe, D. M. (eds) (1990) *Virology* (2nd edn), Vols 1 and

2. New York: Raven Press. (Chapters on all viruses; good but a load to carry and expensive.)

Fraenkel-Conrat, H. & Wagner, R. R. (1978) Viruses of fungi, algae and invertebrates. In: *Comprehensive virology*, Vol. 12 (3 chapters). New York: Plenum Press.

Goodsell, D. S. (1991). Inside a living cell. *Trends in Biochemical Sciences*, **16**, 203–207.

Hull, R., Brown, F. & Payne, C. (1989) *Virology: dictionary of animal, bacterial and plant viruses*. London: Macmillan.

Lewin, B. (1990) *Genes* (4th edn) New York: John Wiley & Sons.

McAllister, P. E. (1979) Fish viruses and viral infections. In: *Comprehensive virology*, Vol. 14, pp. 401–470. Fraenkel-Conrat, H. & Wagner, R. R. (eds). New York: Plenum Press.

Matthews, R. E. F. (1992) *Fundamentals of plant virology*. San Diego, CA: Academic Press.

Old, R. W. & Primrose, S. B. (1994) *Principles of genetic manipulation* (5th edn). Oxford: Blackwell Scientific Publications.

Waterson, A. P. & Wilkinson, L. (1978) *An introduction to the history of virology*. Cambridge: Cambridge University Press.

Webster, R. G. & Granoff, A. (eds) (1994) *Encyclopedia of virology*, Vols 1–3. New York: Academic Press.

Wilson, T. M. A. & Davies, J. W. (1992) *Genetic engineering of plant viruses*. Boca Raton, FL: CRC Press.

Also check Chapter 21 for references specific to each family of viruses.

2 How to handle animal viruses

Viruses are too small to be seen except by electron microscopy (EM) and this requires concentrations in excess of 10^{11} particles, or even higher if a virus has no distinctive morphology. Therefore viruses are usually detected by other *indirect* methods.

These fall into three categories: (i) *multiplication* in a suitable culture system and detection of the virus by the effects it causes; (ii) *serology*, which makes use of the interaction between a virus and antibody directed specifically against it; and (iii) detection of viral *nucleic acid*.

SELECTION OF CULTURE SYSTEM

The culture system always consists of living cells, and the choice is outlined in Table 2.1. Which is used depends on the aims of the experiment. These may be divided into isolation of viruses, biochemistry of multiplication, structural studies and study of natural infections.

The investigation of any new virus starts with ways to cultivate it. There are still many which are uncultivable, particularly those occurring in the gut, but these can be present in such numbers that they were actually discovered by EM. Often a virus is suspected of causing a disease. By definition, disease can only be studied in an animal, preferably the natural host, although for humans this may be ruled out by ethical considerations. Alternatively, organ cultures and cells can be used. Logically, these should be from the natural host and obtained from those sites where the virus multiplies in the whole animal. However, it may prove that cells from unrelated animals are susceptible, e.g. human influenza viruses were first cultivated by inoculating a ferret intranasally and grow best in embryonated chicken eggs! Frequently, viruses grow poorly on initial isolation but adapt, due to selection of mutants, on being passed from culture to culture.

Biochemical studies of virus infections require a cell system in which nearly every cell is infected. To achieve this, large numbers of infectious particles, and hence a system which will produce them, are required. Often, cells which are suitable for production of virus are different from those used for the study of virus multiplication. There is little logic in choosing a cell system, only pragmatism. Cells differ greatly and different properties make one cell the choice for a particular study and unsuitable for another. The ability to control the cell's environment is desirable, especially for labelling with radio-isotopes, since a chemically defined medium must be prepared that lacks the non-radioactive isotope. Otherwise,

Table 2.1 How to choose a culture system for animal viruses.

Culture system	Advantages	Disavantages
Animal	Natural infection	Cost of upkeep is expensive. Large variation between individuals even if inbred. Therefore large numbers needed
Organ, e.g. pieces of brain, gut, trachea	Natural infection Fewer animals needed Less variation since one animal gives many organ cultures	Many cell types present Unnatural since isolated from homeostatic processes and the immune response
Cell	Can be cloned, therefore variation between individuals is minimal Best for biochemical studies as the environment can be controlled exactly and quickly Can be immortal	Very unnatural since cells dedifferentiate when cultured

the specific activity of the radio-isotope would be reduced to an unusable level.

The investigation of natural infections can only be done in the natural host. However, these are frequently unsuitable and the nearest approximation is usually to use purpose-bred animals which, although usually not the natural host species, have a similar range of defence mechanisms and can be maintained in the laboratory. The mouse has been extensively studied and inbred strains reduce genetic variability. Although the use of animals for studying virus diseases has been criticized by organizations concerned with animal rights, the student of virology will be aware, after reading Chapters 14–19, that there is as yet no alternative for studying the complex interactions of viruses with the responses of the host. Analysis of the processes involved would be immensely easier if there were a test-tube system, but it is unlikely that any will appear in the foreseeable future, except for very specialized purposes.

Organ cultures

Organ cultures have the advantage of maintaining the differentiated state of the cell. However, there are technical difficulties in their large-scale use, and as a result they have not been widely employed.

We shall consider here only organ cultures from the trachea. Appropriately, these were first used by an ear, nose and throat surgeon, Bertil Hoorn in 1960, who was interested in respiratory viruses. Figure 2.1 shows the procedure used to prepare the cultures.

Ciliated cells lining the trachea continue to beat in co-ordinated waves while the tissue remains healthy. Virus multiplication causes the synchrony to be lost and eventually the ciliated cells to detach (Fig. 2.2). Virus is also released into fluids surrounding the tissue and can be measured if appropriate assays are available.

Cell cultures

Cells in culture are kept in an isotonic solution, consisting of a mixture of salts in their normal physiological proportions supplemented with serum (usually 5−10%), and in such a growth medium cells rapidly adhere to the surface of suitable glass or plastic vessels. Serum is a complex mixture of proteins and other

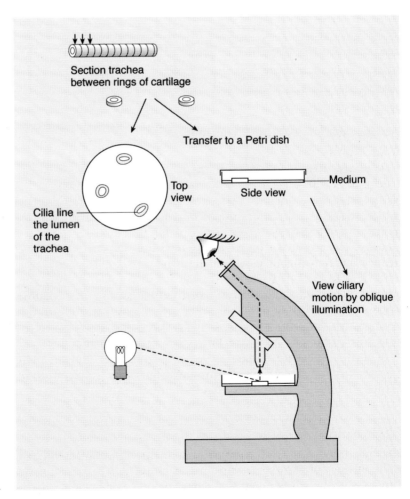

Fig. 2.1 Preparation of tracheal organ cultures.

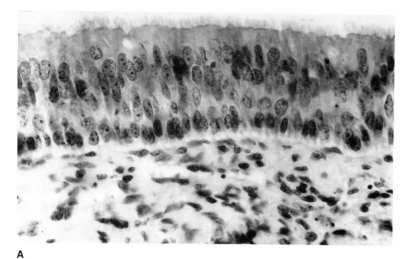

A

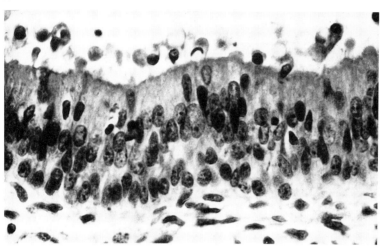

B

Fig. 2.2 Sections through tracheal organ cultures (A) uninfected, and (B) infected with a rhinovirus for 36 hours. Note the disorganization of the ciliated cells (uppermost layer) after infection. (Courtesy of B. Hoorn.)

compounds without which mitosis does not occur, but recently synthetic substitutes have become available. All components used in cell culture have to be sterile and handled under aseptic conditions to prevent the invasion of micro-organisms. Antibiotics have been invaluable in establishing cells in culture, and routine cell culture dates from the 1950s when they first appeared on the market. Figure 2.3 shows the principles of cell culture.

Cultured cells are either diploid or heteroploid (having more than the diploid number of chromosomes but not a simple multiple of it). Diploid cells will undergo a finite number of divisions, from around 10 to 100, whereas the heteroploid cells will divide for ever. The latter are known as *continuous cell lines* and they originate from naturally occurring tumours or from some spontaneous event which alters the control of division of a diploid cell. Diploid cell lines are most easily obtained from embryos by reducing kidneys or the whole body to a suspension of single cells.

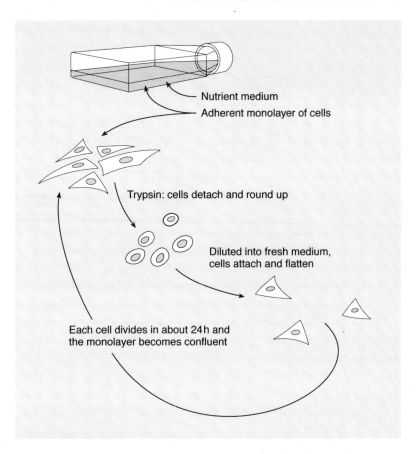

Fig. 2.3 Cell culture.

Modern methods of cell culture

The methodology described above is suited for research and clinical or diagnostic laboratories but is difficult to scale up for commercial purposes, such as vaccine manufacture. There are now various solutions to the problem, all aimed at increasing cell density. One of the earliest was to grow cells in suspension, and this has been refined, using hybridoma cells for monoclonal antibody (mab) production. However, many cells only grow when anchored to a solid surface, so the technology has sought to increase the surface area available by, for example, providing spiral inserts to fit into conventional culture bottles (Fig. 2.4C). One of the latest is to grow cells on 'microcarriers', tiny particles (about 200 μm diameter) on which cells attach and divide. The surface area afforded by 1 kg of microcarriers is about $2.5\,m^2$ and the space taken up (a prime consideration in commercial practice) is very economical. This method combines the ease of handling cell suspensions with a solid matrix for the cell to grow on (Fig. 2.4A and B).

A

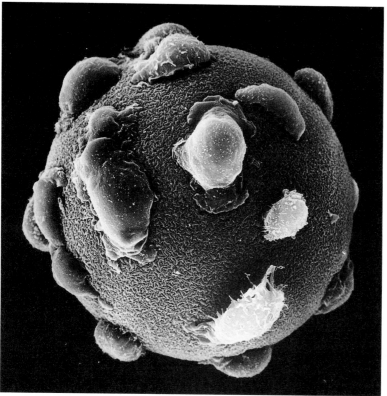

B

Fig. 2.4 (A and B) Cells growing on microcarriers. (A) Scanning electron micrograph of pig kidney cells (courtesy of G. Charlier). (B) Removal of cells from a microcarrier bead by incubation with trypsin. Each bead is about 200 μm in diameter. The microcarriers shown are Cytodex, manufactured by Pharmacia Ltd, and reproduced by permission.

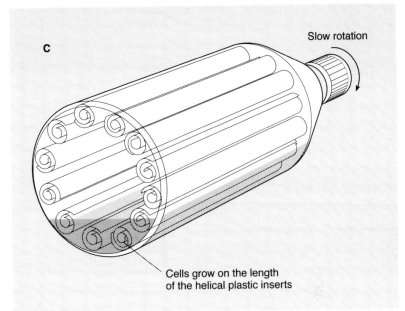

Fig. 2.4 (C) One way to increase cell density is by increasing the surface area to which cells can attach; view from the end of a bottle lined with spiral plastic coils. The bottle is rotated slowly (at about 5 revs/hour) so that a small volume of culture medium can be used.

C

Slow rotation

Cells grow on the length of the helical plastic inserts

Haemagglutination

Certain viruses attach to molecules on the surface of red blood cells (rbcs) and at a certain virus : cell ratio the rbcs are linked together by virus and the cells are agglutinated. This has nothing to do with infectivity, and inactivated virus can agglutinate efficiently, providing its surface properties are unimpaired. A quantitative test can be devised by making dilutions of virus in a suitable tray and then adding a standard amount of rbcs to each well (Fig. 2.5). The amount of virus present is estimated as the dilution at which the virus causes 50% agglutination. This test has the advantage of speed, for it takes just 30 minutes, compared with 3 days for a plaque assay. However, it is insensitive; for example at least 10^6 plaque-forming units (PFU) of influenza virus are required to cause detectable agglutination.

IDENTIFICATION OF VIRUSES

Serology, or the use of antibody

Antibodies are proteins produced by cells of the immune system of higher vertebrates in response to foreign materials (antigens) which those cells encounter. The antibodies are secreted into the body fluids and are usually obtained from the fluid part of the blood (antiserum) which remains after clotting has removed cells and clotting proteins. This is then known as an antiserum.

The principle of identifying infectious virus with antibody is

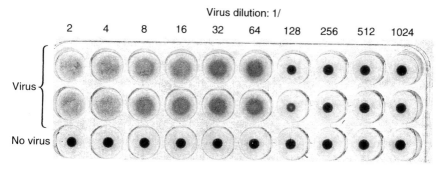

Fig. 2.5 Haemagglutination titration. Here an influenza virus is serially diluted (from left to right) in depressions in a plastic plate. Red blood cells 0.5% v/v are then added and mixed with each dilution of virus. Where there is little virus, cells settle to a button (from 1/128) indistinguishable from rbcs to which no virus was added (row 3). Where sufficient virus is present (up to 1/64), cells agglutinate and settle in a diffuse pattern. (Photograph by A. S. Carver.)

shown in Fig. 2.6. Any method of detecting or measuring virus can be used, such as inhibition of haemagglutination (Fig. 2.7) or immune EM, where, after reaction with specific antibody, viruses can be seen in aggregates. Alternatively, antibody can be employed to detect viral antigens inside the infected cell. When the cell is alive, antibodies cannot penetrate the plasma membrane and will therefore react only with antigens exposed on the surface of the cell. This permeability barrier is destroyed by 'fixing' the cell in acetone or methanol, which permeabilizes the plasma membrane and enables antibody to attach to antigens in the cytoplasm and nucleus. Antibodies are 'tagged' before use with a marker substance and hence can be detected *in situ*. Tags such as fluorescent dyes can be seen by ultraviolet (UV) microscopy (Fig. 2.8), enzymes (peroxidase, phosphatase) which leave a coloured deposit on reaction with substrate can be seen by light microscopy, radioactive substances can be detected by deposition of silver grains from a photographic emulsion, and electron-dense molecules (e.g. ferritin, an iron-containing protein, or gold particles) are visualized by EM.

Enzyme-linked antibody is now commonly used in a quantitative assay called the enzyme-linked immunosorbent assay (ELISA), as shown in Fig. 2.9. In the example, the assay is being used to measure the amount of antibody present in an unknown sample, but it can also be used to measure antigen. The coloured product is proportional to the amount of specific antibody bound and can be measured spectrophotometrically. Such assays can easily be automated to deal with large numbers of samples.

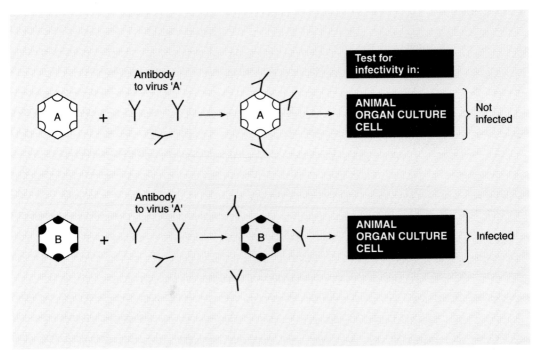

Fig. 2.6 A neutralization test. Virus 'A' loses its infectivity after combining with its homologous antibody (it is neutralized), whereas 'B' does not. The complete test requires the reciprocal reactions.

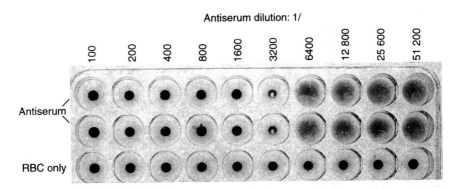

Fig. 2.7 In the haemagglutination–inhibition test above, antiserum is diluted from left to right. Four haemagglutination units (HAU) of an influenza virus are added to each well. The antibody–virus reaction goes to completion in 1 hour at 20°C. Red blood cells are then added. In this test, haemagglutination is inhibited up to an antiserum dilution of 1/3200. (Photograph by A. S. Carver.)

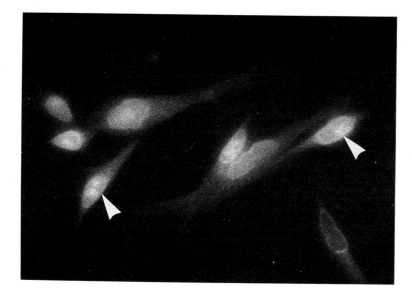

Fig. 2.8 Fluorescent antibody staining (arrowed) of an antigen present mainly in the nucleus of influenza virus-infected cells.

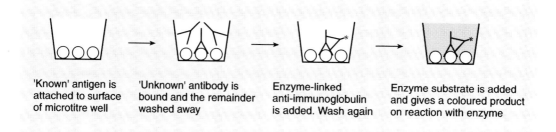

| 'Known' antigen is attached to surface of microtitre well | 'Unknown' antibody is bound and the remainder washed away | Enzyme-linked anti-immunoglobulin is added. Wash again | Enzyme substrate is added and gives a coloured product on reaction with enzyme |

Fig. 2.9 Detection of antibodies by ELISA. Serial dilutions of the antiserum are used to make the test into a quantitative assay.

Identification of antigenic polypeptides after separation by electrophoresis: Western blotting

When antibodies were first used as a probe for macromolecules separated by gel electrophoresis it seemed logical to follow the earlier precedents of Southern and Northern blotting (p. 34) by calling the technique 'Western' blotting. For this procedure virus proteins are totally denatured and separated according to size by gel electrophoresis, usually in polyacrylamide in the presence of the detergent sodium dodecyl sulphate (Fig. 2.10). The separated polypeptides are then transferred horizontally by electrophoresis on to cellulose nitrate and their presence is located by incubation with antibody. Bound antibody is detected by reaction with radio-labelled anti-antibody or enzyme-labelled anti-antibody, which leaves a coloured deposit upon reaction with substrate. Alternatively, bound antibody is detected with labelled staphylococcal protein A, which has a high affinity for the Fc portion of immuno-globulin G (IgG). A limitation of Western blotting is that it will only detect 'continuous' antigenic determinants (p. 243), which remain after the protein is denatured. 'Conformational' continuous or discontinuous determinants are lost on denaturation.

Understanding antibody–antigen reactions

Neutralization. The antibody–antigen reaction is so specific that it is unaffected by the presence of other proteins. Hence antibodies need not be extracted from crude serum, and impure virus preparations can be used. An antibody molecule recognizes and combines with part of the antigen, called an *epitope*. A protein epitope is a planar surface of about 16 amino acids which interacts with a complementary surface that forms the binding site (or *paratope*) of the reactive antibody. There are many epitopes in an antigenic site and several such antigenic sites are presented by proteins forming the coat of a virus particle. In Chapters 14 and 15 the role of antibody in the recovery from virus diseases and prevention of reinfection will be covered, but it is appropriate here to discuss how antibody renders virus non-infectious (or neutralized).

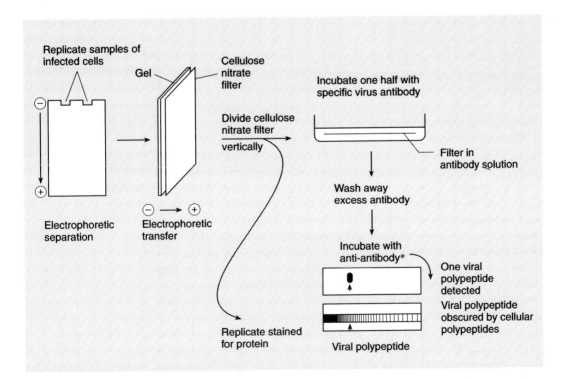

Fig. 2.10 Western blotting—using antibody to detect one viral antigen amongst many others. '*' denotes a radiolabel, enzyme label or gold label. Alternatively labelled staphyloccocal protein A, which binds specifically to the commonest forms of IgG, can be used instead of anti-antibody (see text).

Earlier it was assumed that neutralizing antibody prevented virus from attaching to receptors on the cell surface. However, when this assumption was tested with rhino-, influenza and polioviruses, only neutralized rhinovirus was prevented from attaching to monolayers of cells. Attachment of the other two viruses was unaffected. Rather surprisingly, no antibody is made to the attachment sites of any of these viruses. These sites are contained in depressions on the surface of the virus, where they are hidden from the immune system. Rhinovirus-specific neutralizing antibody attaches to and bridges amino acids on either side of the rhinovirus attachment site and blocks it indirectly (see Fig. 3.12). Antibody attached to virions can be visualized by EM as a fuzzy outer layer. However, it can be diluted to an amount which is no longer detected by EM but is still sufficient to thoroughly neutralize infectivity. Thus so few molecules of antibody per particle are unlikely to interfere with attachment. Anyhow, interference with attachment of influenza virus would be particularly inefficient, as there are 500–1000 attachment proteins (the haemagglutinin (HA)) per virion and an IgG molecule is slightly smaller than an HA spike. Also, these conclusions agree with the biochemical evidence below. *In vivo*, speed of neutralization is all-important, since antibodies cannot cross cell membranes and a virion cannot be neutralized after it has entered a cell.

So how do they lose infectivity? Recent studies have demonstrated that IgG-neutralized poliovirus is able to attach to cells and gain entry but is unable to uncoat, while neutralized influenza virus uncoats but is unable to transcribe its genome. Current thinking is that there are as many mechanisms of neutralization as there are post-attachment processes which a virus has to undergo before its genome can be expressed. Any mechanism of neutralization must be defined by all the relevant parameters, which include the neutralization protein, its epitope, the antibody paratope, the isotype of antibody, the ratio of antibody : virus and the cell receptor.

The reader will no doubt be puzzled by the discrepancy between the failure of antibody to prevent the attachment of neutralized influenza virus to tissue culture cells and the ability of the same antibody to prevent attachment to rbcs in the haemagglutination–inhibition test (Fig. 2.7). The explanation lies in the fact that different cells have different receptor molecules, and those of rbcs protrude only a short way from the cell surface, so that they are sterically prevented from attaching to viruses (Fig. 2.11).

Kinetics of neutralization. With the advent of monoclonal antibodies (mabs), it became possible to tackle the problem of how many molecules of antibody per virion were needed to cause neutralization. This can be approached through a study of the kinetics of neutralization. Kinetics of inactivation can be pseudo-first-order, where there is a straight-line relationship between the

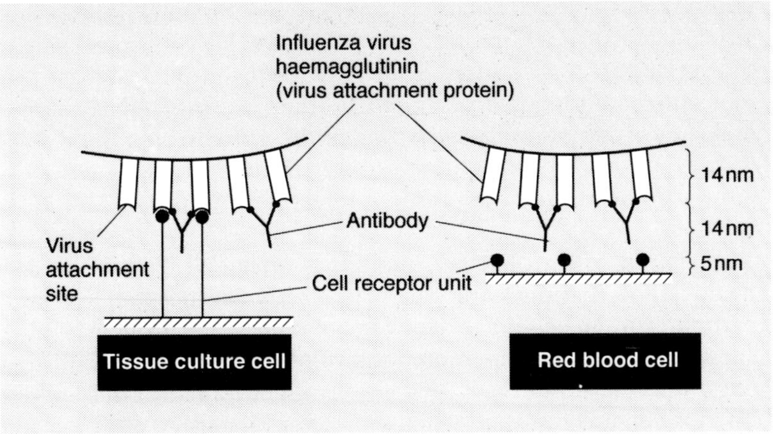

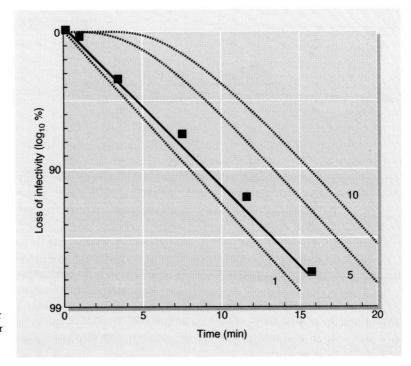

Fig. 2.11 Hypothetical scheme based on the height of the cell receptors above the cell surface to explain inhibition by antibody of attachment of influenza virus to red blood cells but not to tissue culture cells.

Fig. 2.12 Kinetics of neutralization of a virus by a monoclonal antibody. Here the observed rate of neutralization (■–■) is compared with that calculated when 1, 5 or 10 'hits' are required for neutralization (dotted lines).

$\log_{10}$ of the surviving fraction plotted against time, or second-order, where the plot has a shoulder before there is any loss of infectivity (Fig. 2.12). First-order kinetics is usually interpreted as meaning that the infectivity of one virus particle is inactivated by the binding of one molecule of antibody, while second-order kinetics requires that two or more molecules act co-operatively to bring about neutralization. Most neutralization reactions (e.g. Fig. 2.12) follow first-order kinetics. This has implications for the mechanism of neutralization since: (i) the number of attachment sites per virion ranges from 60 (picornaviruses) to around 3000 (influenza A virus); and (ii) an IgG molecule occupies a relatively small part of the surface area of a virion. Thus it is inconceivable that one IgG molecule could block attachment. Rather, we think that the IgG molecule either prevents uncoating of the virion from taking place or triggers the uncoating process prematurely; in either case, the virus particle is unable to initiate infection.

The concept of critical neutralization sites. While kinetic data demonstrate that one molecule of antibody can neutralize a virus particle, the biochemical data for poliovirus and influenza virus show that, in fact, many more molecules have attached. If, indeed, one molecule of antibody per virion is neutralizing, the Poisson distribution predicts that, when there is an average of one molecule of antibody per virion, at $1/e$ there will be 37% of virus with no attached antibody—and hence not neutralized. Conversely, 63% of virus will have bound antibody and will be neutralized. However, assay of monoclonal antibody attached to the HA of influenza virus shows that at $1/e$ there are in fact about 70 molecules of antibody per virion. This paradox can be resolved by postulating that only one of the 70 antibody molecules neutralizes and is said to have bound to a *critical site*; the other 69 antibody molecules are bound to *non-critical* sites, which, by definition, are non-neutralizing. How do critical and non-critical sites differ? Since a mab was used, all epitopes are by definition identical with respect to binding antibody; the HA bearing the critical site is postulated to differ by its association with another component of the virion (Fig. 2.13). The non-critical HA spike is not so associated. These aspects of neutralization are still being elucidated.

Non-neutralizing antibodies. Not all antibodies which attach to virus particles neutralize infectivity. There are antigenic sites (*non-neutralization sites*) on neutralization proteins which bind antibody but do not mediate neutralization. This emphasizes just how specific the neutralization reaction is. There is a further dimension to this, since these antibodies can sometimes sterically prevent the binding of neutralizing antibody and allow the virus to evade the immune response.

There are conditions under which virions can bind neutralizing antibody and not be neutralized. This occurs when the concentration of antibody is too low, or the affinity of antibody is so low that it dissociates, or because virions are non-specifically aggregated with neutralized particles and this protects them from contact with antibody. Other virions not neutralized by a mab are 'antibody-escape' mutants. These are mutants, occurring naturally at a frequency of about 1 in 100 000, which are not neutralized when mixed with a particular mab. They usually have a point mutation which results in the substitution of a crucial single amino acid within the epitope. None of the progeny of an antibody-escape

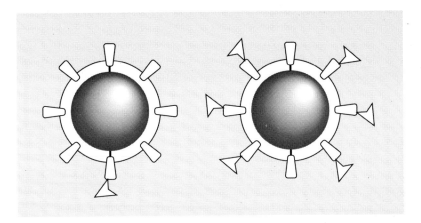

Fig. 2.13 The concept of critical and non-critical neutralization sites. All the surface spikes are identical in structure but two are critical by virtue of their connection with internal virion structures. Binding of an antibody molecule to either of these (left panel) results in neutralization by an unknown mechanism. In the right panel antibody has bound to all the non-critical sites and the particle is still infectious (see text).

Fig. 2.14 Restriction endonuclease mapping of viral DNA. In this example adenovirus DNA is extracted, digested with *Sma* I (an enzyme from *Serratia marcescens*) and the resulting fragments separated by electrophoresis in a thin slab of agarose. DNA is visualized by staining with ethidium bromide, with which it intercalates and fluoresces under UV light. Track 1 contains the known standard virus and track 2 and 3 clinical isolates. Virus in track 3 is clearly identical to the standard, while that in track 2 is different and will need to be identified by comparison with other standards.

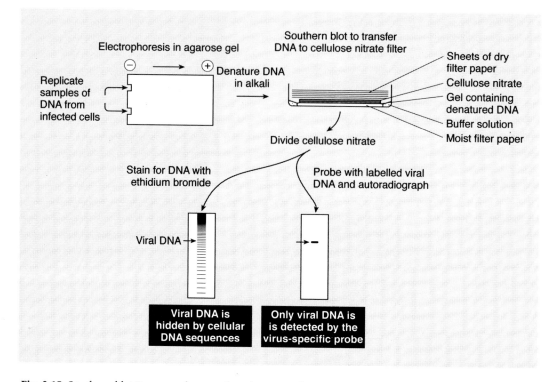

Fig. 2.15 Southern blotting procedure used to detect viral DNA amongst an excess of cellular DNA fragments. The 'blot' works by drawing buffer through the gel (top right) from wet filter paper to dry filter paper; DNA is carried to the cellulose nitrate filter, where it binds irreversibly and can be detected by labelled molecular probes.

mutant is neutralized either. Sequencing of the gene encoding the neutralization protein locates the mutation and in this way, with sufficient mabs, all the antigenic sites of a virus particle can be mapped.

Neutralization is not the only way in which viruses can be inactivated by antibody; other factors can be involved, such as the complement system or cells which have receptors for the Fc part of the antibody molecule (see Chapter 14). Under these circumstances, even non-neutralizing antibody can inactivate infectivity.

Identification by restriction endonuclease mapping

Neutralization tests are indeed simple, but can be slow, as their usefulness depends on the time taken by the virus to kill a detectable number of cells; this can take from several days to several weeks. Such a situation is far from ideal for diagnostic virology and a recent development which aims to overcome the problem is to compare the patterns of deoxyribonucleic acid (DNA) fragments

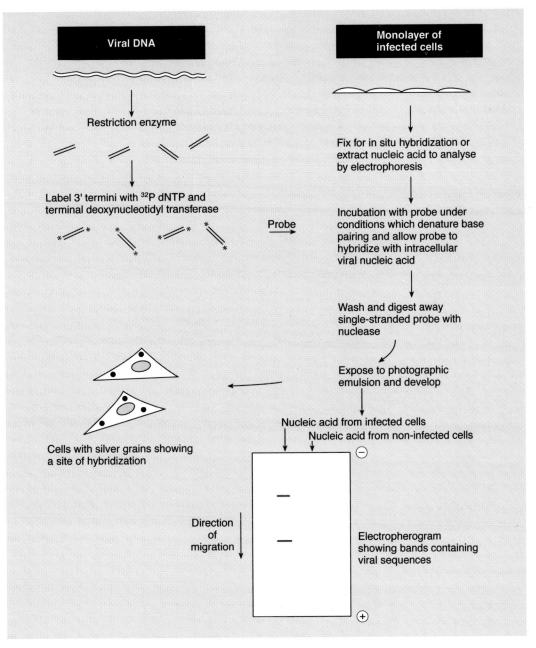

Fig. 2.16 Detection of viral nucleic acid sequences using molecular probes. dNTP represents dATP + dCTP + dGTP + dTTP.

produced by restriction endonucleases. The pattern is determined by the nucleotide sequence and is unique to a particular serotype. What happens now is that a newly isolated virus is assigned to its family by the old technologies — by reference to its cytopathic effects, its morphology by EM, or by immunofluorescent staining with antibody against a group-specific antigen. Then virus is purified and, using the new technology, its DNA is extracted and digested with one or more restriction enzymes. The resulting fragments of DNA are separated according to size and identified by comparison with known standards (Fig. 2.14).

This methodology is rapid, is no more expensive than the neutralization test and can be performed on the virus produced by a single culture tube. One important aspect is that it permits the study of strains of virus which can still not be cultivated in the laboratory.

Identification of viral nucleic acid sequences after separation by electrophoresis: Southern and Northern blotting

Until recently, one of the major problems in cell biology was the difficulty of identifying a particular nucleic acid sequence amongst a vast excess of other sequences. For example, although DNA can be isolated, using appropriate deproteinization procedures, from infected cells and analysed by gel electrophoresis, any staining technique will visualize all nucleic acid molecules, and the viral DNA will be obscured by the far larger amount of cellular material present. The problem was overcome by Ed Southern in Edinburgh, who denatured and separated nucleic acids *in situ*, blotted them on to a cellulose nitrate filter and hybridized them to a labelled nucleic acid sequence specific for the virus. Such a probe reacts only with the nucleic acid to which it is complementary and no other (Fig. 2.15). The technique bears Southern's name and a sub-

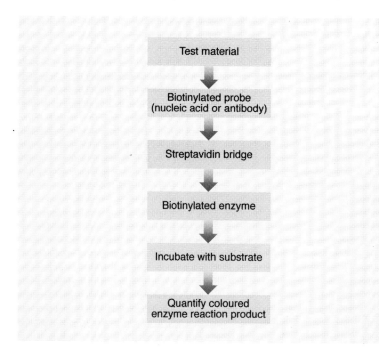

Fig. 2.17 Flow chart for use in a biotin–streptavidin system. Streptavidin is a protein obtained from *Streptomyces avidinii* which has four high-affinity binding sites for the vitamin biotin.

sequent procedure, which detects ribonucleic acid (RNA) in a similar manner with labelled DNA probes, was whimsically called Northern blotting. In this way the presence of viral genomes can be detected even when it is not possible to cultivate the virus or when the virus is latent and not actively multiplying.

Molecular probes and non-hazardous labelling

Short nucleic acid sequences (30–200 base pairs), prepared by *de novo* synthesis or isolated by recombinant DNA technology, are used to locate viral genomes in cells or tissues (Fig. 2.16). They can even show the location of virus in longitudinal sections of whole animals.

Again what was once the province of the research laboratory is rapidly adopted and adapted to diagnostic virology, but the usual radioactive probes are unsuitable as they have a short shelf-life and contribute to the radiation load of a routine laboratory. As an alternative, nucleic acid probes (and antibodies) can be biotinylated before reacting with the test substance. After washing away unbound probe, the test material is reacted with streptavidin, which binds with high affinity to biotin. Next biotinylated polymers of intestinal alkaline phosphatase are added, which bind to free sites on streptavidin (Fig. 2.17). Incubation then with 5-bromo-4-chloro-3-indolyl phosphate and nitro blue tetrazolium deposits a blue coloration at the site of the original reaction. Nucleic acid sequences in the range 1–10 pg can be detected.

FURTHER READING

Bachmann, P. (ed.) (1983) New developments in diagnostic virology. *Current Topics in Microbiology and Immunology*, **104** (many chapters on all aspects of modern diagnosis).

Dimmock, N. J. (1987) Multiple mechanisms of neutralization of animal viruses. *Trends in Biochemical Sciences*, **12**, 70–75.

Dimmock, N. J. (1993) Neutralization of animal viruses. *Current Topics in Microbiology and Immunology*, **183**, 1–149.

Fife, K. H., Ashley, R., Shields, A. F., Salter, D., Meyers, J. D. & Corey, L. (1985) Comparison of neutralization and DNA restriction enzyme methods for typing clinical isolates of human adenovirus. *Journal of Clinical Microbiology*, **22**, 95–100.

Flewett, T. H. (1980) Safety in the virology laboratory. *Recent Advances in Clinical Virology*, **2**, 169–187.

Grist, N. R., Bell, E. J., Follett, E. A. C. & Urquhart, G. E. D. (1979) *Diagnostic methods in clinical virology* (3rd edn). Oxford: Blackwell Scientific Publications.

Hames, B. D. & Rickwood, D. (eds) (1981) *Gel electrophoresis of proteins: a practical approach*. Oxford: IRL Press.

Hudson, L. & Hay, F. C. (1989) *Practical immunology* (3rd edn). Oxford: Blackwell Scientific Publications.

Johnstone, A. & Thorpe, R. (1982) *Immunochemistry in practice.* Oxford: Blackwell Scientific Publications.

Mahy, B. W. J. (ed.) (1985) *Virology, a practical approach.* Oxford: IRL Press.

Mühlbach, H.-P. (1982) Plant cell cultures and protoplasts in plant virus research. *Current Topics in Microbiology and Immunology,* **99**, 81–129.

Old, R. W. & Primrose, S. B. (1994) *Principles of genetic manipulation* (5th edn). Oxford: Blackwell Scientific Publications.

Richmann, D. D., Cleveland, P. H., Redfield, D. C., Oxman, M. N. & Wahl, G. M. (1984) Rapid viral diagnosis. *Journal of Infectious Diseases,* **149**, 298–310.

Rotbart, H. A. (1991) Nucleic acid detection systems for enteroviruses. *Clinical Microbiology Reviews,* **4**, 156–168.

Sander, E. & Mertes, G. (1984) Use of protoplasts and separate cells in plant virus research. *Advances in Virus Research,* **29**, 215–262.

Also check Chapter 21 for references specific to each family of viruses.

3 The structure of viruses

In the previous chapter we described the experiments of Hershey and Chase and of Fraenkel-Conrat and Singer, from which it was apparent that the ability of a virus to reproduce resides solely in its nucleic acid. Analysis of purified viruses shows that they contain 50–90% protein and, since nucleic acids in solution are susceptible to shearing and degradation, we can assign a protective role to the protein component. It also permits attachment to the host cell. At first sight it would appear that there is an enormous variety of ways in which the protein could be arranged round the nucleic acid in order to protect it. However, only a limited number of designs are observed and we will now discuss briefly the limiting factors.

VIRUSES ARE CONSTRUCTED FROM SUBUNITS

Before considering the architecture of viruses, it is worth remembering that, although proteins may have a regular secondary structure in the form of an α a helix, the tertiary structure of the protein is not symmetrical. This, of course, is a consequence of hydrogen bonding, disulphide bridges and the intrusion of proline in the secondary structure. Although we might naïvely think that the nucleic acid could be enveloped by a single, large protein molecule, this cannot be so since proteins are irregular in shape, as already stated, whereas most virus particles are regular in shape, at least when examined by electron microscopy (Fig. 3.1). However, this can also be deduced solely from considerations of the coding potential of nucleic acid molecules. A coding triplet has an M_r of approximately 1000 but specifies a single amino acid with an average M_r of about 100. Thus a nucleic acid can at best only specify one-tenth of its weight of protein. Since viruses frequently contain greater than 50% protein by weight, it should be apparent that more than one identical protein of smaller molecular weight must be present.

Obviously, less genetic material is required if the single protein molecule specified is to be used as a repeated subunit, but it is not essential that the coat be constructed from identical subunits, provided the combined molecular weights of the different subunits are sufficiently small in relation to the nucleic acid molecule which they protect. There is a further advantage in constructing a virus from subunits, and that is greater genetic stability, since reducing the size of the structural units lessens the chance of a disadvantageous mutation occurring in the gene which specifies

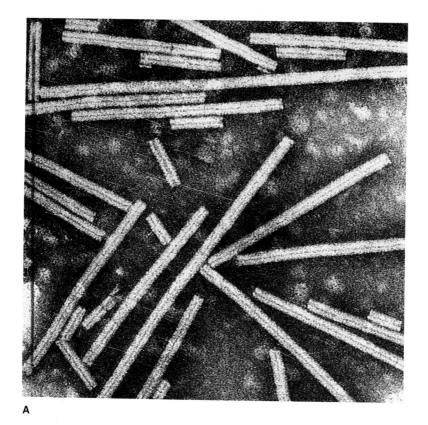

A

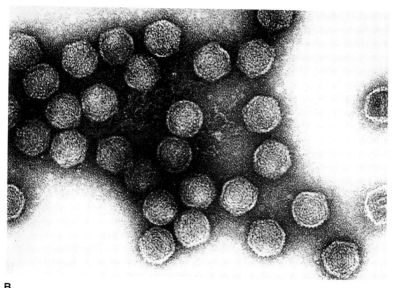

B

Fig. 3.1 Electron micrographs of viruses showing their regular shape.
(A) Tobacco mosaic virus, (B) bacteriophage Si 1.

it. If, during assembly, a rejection mechanism operates such that faulty subunits are not included in the virus particle, then an error-free structure can be constructed with the minimum of wastage.

Suspensions of pure virus can be maintained in the laboratory for long periods of time and consequently must be stable structures. The necessary physical condition for the stability of any structure is that it be in a state of minimum free energy, so we can assume that the maximum number of bonds are formed between the subunits. Since the subunits themselves are non-symmetrical, for the maximum number of bonds to be formed they must be arranged symmetrically, and there are a limited number of ways this can be done.

THE STRUCTURE OF FILAMENTOUS VIRUSES AND NUCLEOPROTEINS

One of the simplest ways of symmetrically arranging non-symmetrical components is to place them round the circumference of a circle to form discs (Fig. 3.2). This gives us a two-dimensional structure. If we stack a large number of discs on top of one another, we get a 'stacked-disc' structure. Thus we can generate a symmetrical three-dimensional structure from a non-symmetrical component such as protein and still leave room for nucleic acid. Examination of published electron micrographs of viruses reveals that some of them have a tubular structure. One such virus is tobacco mosaic virus (TMV) (Fig. 3.1). However, close examination of TMV reveals that the subunits are not arranged cylindrically, i.e. in rings, but helically. There is an obvious explanation for this. Since the nucleic acid is helical in shape, it could not be equivalently bonded in a stacked-disc structure. However, by arranging the subunits helically, the maximum number of bonds can still be formed and each subunit equivalently bonded, except,

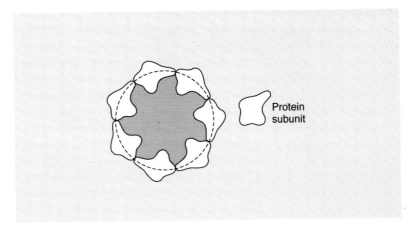

Fig. 3.2 Arrangement of identical asymmetrical components around the circumference of a circle to yield a symmetrical arrangement.

Protein subunit

of course, for those at either end. All filamentous viruses so far examined are helical rather than cylindrical and the insertion of the nucleic acid may be the factor governing this arrangement. Most nucleoprotein structures inside enveloped viruses (p. 53) are constructed in the same way.

THE STRUCTURE OF SPHERICAL VIRUSES

A second way of constructing a symmetrical particle would be to arrange the subunits around the vertices or faces of an object with cubic symmetry, e.g. tetrahedron, cube, octahedron, dodecahedron (constructed from 12 regular pentagons) or icosahedron (constructed from 20 equilateral triangles). Figure 3.3 shows possible arrangements for objects with triangular and square faces. Multiplying the number of subunits per face by the number of faces gives the minimum number of subunits which can be arranged around such an object. For example, for a tetrahedron it is 12 subunits, for a cube or octahedron it is 24 subunits and for a dodecahedron or icosahedron it is 60 subunits. Although it may not be immediately apparent, these represent the few ways in which an asymmetrical object (such as a protein molecule) can be placed symmetrically on the surface of a sphere. (The reader may care to check by using a ball and sticking on bits of paper of the shape shown in Fig. 3.3!) Examination of electron micrographs reveals that many viruses are spherical in outline but actually have icosahedral symmetry rather than octahedral, tetrahedral or cuboidal symmetry. There are two possible reasons for the selection of icosahedral symmetry over the others. First, since it requires a greater number of subunits to

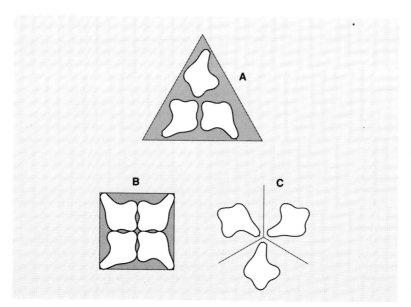

Fig. 3.3 Symmetrical arrangement of identical asymmetrical subunits by placing them on the faces of objects with cubic symmetry. (A) Asymmetrical subunits located at vertices of each triangular facet. (B) Asymmetrical subunits placed at vertices of each square facet. (C) Arrangement of asymmetrical subunits placed at each corner of a cube with face represented in (B).

provide a sphere of the same volume, the size of the subunits can be smaller, thus economizing on genetic information. Secondly, there appear to be physical restraints which prevent the tight packing of subunits required by tetrahedral and octahedral symmetry.

Symmetry of an icosahedron

An icosahedron is made up of 20 triangular faces, five at the top, five at the bottom and 10 around the middle. Each triangle is symmetrical and it can be inserted in any orientation (Fig. 3.4A). An icosahedron has three axes of symmetry: fivefold, threefold and twofold (Fig. 3.4B).

The construction of more complex icosahedral viruses

The situation is more complex than outlined above, since many viruses possess more than 60 subunits. If $60n$ subunits are put on the surface of a sphere, one solution is to arrange them in n sets of 60 units, but the members of one set would not be equivalently related to those in another set. For example, consider the arrangement of the subunits in Fig. 3.5. If all the subunits, represented by open and closed circles, are identical, then those represented by closed circles are related equivalently to those represented by open circles. However, open circle units do not have the same spatial arrangement of neighbours as closed circle units and so cannot be equivalently related. Of course, if the structure were built out of n different subunits there would be no conceptual difficulty and, indeed, no problem. However, accepting the restriction that we must build the structure out of identical subunits, how can we regularly arrange more than 60 asymmetrical subunits? The solution to the problem was inspired by the geodesic domes constructed by Buckminster Fuller (Fig. 3.6). Fuller's dome designs involve the subdivision of the surface of a sphere into triangular facets, which are arranged with icosahedral symmetry. The device of triangulating the sphere represents the optimum design for a closed shell built of regularly bonded identical subunits. No other subdivision of a closed surface can give a comparable degree of equivalence. Thus, this is a minimum-energy structure and hence a further reason for the preponderance of icosahedral viruses.

The triangulation of spheres

It is possible to enumerate all the ways in which this subdivision can be carried out, but, before doing so, let us consider one simple example. If we start with an icosahedron and arrange the subunits around the vertices, there will be 12 groups of five subunits (Fig. 3.7A). Now we can subdivide each triangular face into four smaller and identical equilateral triangles and incorporate 240 sub-

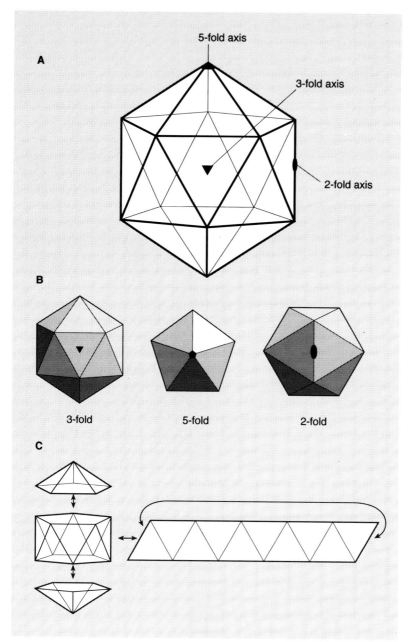

A

5-fold axis

3-fold axis

2-fold axis

B

3-fold   5-fold   2-fold

C

Fig. 3.4 Properties of a regular icosahedron (A). Each triangular face is equilateral and has the same orientation whichever way it is inserted. Axes of symmetry intersect in the middle of the icosahedron. There are 12 vertices, which have fivefold symmetry, meaning that rotation of the icosahedron by one-fifth of a revolution achieves a position such that it is indistinguishable from its starting orientation; each of the 20 faces has a threefold axis of symmetry and each of the 30 edges has a twofold axis of symmetry: see (B). The icosahedron is built up of five triangles at the top, five at the bottom and a strip of 10 around the middle (C). From Branden & Tooze (1991).

units at the vertices of those smaller triangles (Fig. 3.7B). At the vertices of the original icosahedron there will be rings of five subunits, called *pentamers* (solid circles). However, at all the other vertices there will be rings of six subunits, called *hexamers* (open circles). Since some of the subunits are arranged as pentamers and others as hexamers, it should be apparent that they cannot be equivalently related; hence the use of the term *quasi-equivalence*, but this still represents the minimum-energy shape.

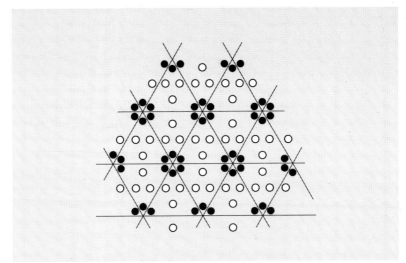

Fig. 3.5 Spatial arrangement of two identical sets of subunits. Note that any member of the set represented by closed circles does not have the same neighbours as a member of the other set represented by open circles.

Fig. 3.6 An example of a geodesic dome — the United States Pavilion at Expo '67 in Montreal (courtesy of the US Information Service).

The ways in which each triangular face of the icosahedron can be subdivided into smaller, identical equilateral triangles are governed by the laws of solid geometry. These can be calculated from the expression $T = Pf^2$, where T, the *triangulation number* = the number of smaller, identical equilateral triangles, f may = 1, 2, 3, 4, etc. and P is given by the expression $h^2 + hk + k^2$, h and k being any pair of integers without common factors.

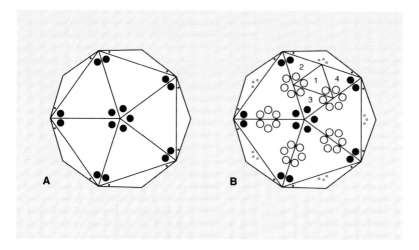

Fig. 3.7 Arrangement of 60*n* identical subunits on the surface of an icosahedron. (A) *n* = 1 and the 60 subunits are distributed such that there is one subunit at the vertices of each triangular facet. Note that each subunit has the same arrangement of neighbours and so all the subunits are equivalently related. (B) *n* = 4. Each triangular facet is divided into four smaller, but identical, equilateral triangles and a subunit is again located at each vertex. In total, there are 240 subunits. Note that, in contrast to the arrangements shown in Fig. 3.5, each subunit, whether represented by an open or closed circle, has the identical arrangement of neighbours. However, since some subunits are arranged in pentamers and others in hexamers, the members of each set are only 'quasi-equivalently' related.

For viruses so far examined, the values of P which have been found are 1 ($h = 1$, $k = 0$), 3 ($h = 1$, $k = 1$) and 7 ($h = 1$, $k = 2$). Representative values of T are shown in Table 3.1. Once the number of triangular subdivisions is known, the total number of subunits can easily be determined since it is equal to 60T.

The smallest virus: satellite tobacco necrosis virus (STNV)

No independently replicating virus is known which consists of only 60 protein subunits, but satellite viruses do (Table 3.1). These encode one coat protein but depend upon coinfection with a helper virus to provide missing replicative functions. The M_r of the single-stranded ribonucleic acid (RNA) of STNV is 0.3×10^6 (about 1000 nucleotides). Presumable the volume of a 60-subunit structure is too small to accommodate the genome of an independent virus. The virion is only 18 nm in diameter.

Determination of the structure of a virus depends on being able to grow crystals of purified virus. We do not understand how to make crystals and many viruses will not form crystals at all. Large stable crystals are bombarded with X-rays, which are diffracted by atoms within the virion, and the image captured on film. Knowing the amino acid sequence makes it possible to determine the three-

Table 3.1. Values of capsid parameters in a number of icosahedral viruses. The value of T was obtained from examination of electron micrographs, thus enabling the values of P and f to be calculated.

P	f	T $(= Pf^2)$	No of subunits $(60T)$	Example
1	1	1	60	Satellite tobacco necrosis virus
3	1	3	180	Tomato bushy stunt virus, picornaviruses
1	2	4	240	Inner shell of reovirus, Sindbis virus
1	3	9	540	Outer shell of reovirus
1	4	16	960	Herpesviruses
1	5	25	1500	Adenoviruses

dimensional crystal structure. High-powered computers are used to make the calculations necessary for this process.

The morphological units seen by electron microscopy are called capsomers and *the number of these need not be the same as the number of protein subunits*. The numbers of morphological units seen will depend on the size and physical packing of the subunits and on the resolution of electron micrographs.

Molecular basis for quasi-equivalent packing of chemically identical polypeptides

Some plant viruses achieve the $T = 3$, 180-subunit structure (Fig. 3.8) while encoding only a single virion polypeptide. They compensate for the physical asymmetry of quasi-equivalence by each polypeptide adopting one of three subtly different conformations. The virion polypeptide of tomato bushy stunt virus is folded into three domains P, S and R: P and S are external and hinged to each other, while R is inside the virion and has a disordered structure. An arm (a) connects S to R (Fig. 3.9).

Each triangular face is made of three identical polypeptides, but these are in different conformations to accommodate the quasi-equivalent packing. For example, the C subunit has the S and P domains orientated differently from the A and B subunits, while the arm (a) is ordered in C and disordered in A and B. The S domains form the viral shell with tight interactions, while the P domains (total = 180) interact across the twofold axes of symmetry to form 90 dimeric protrusions. This virion is 33 nm in diameter and can accommodate a single-stranded RNA genome about fourfold larger than that of the satellite viruses. Thus a larger particle can be achieved without any more genetic cost.

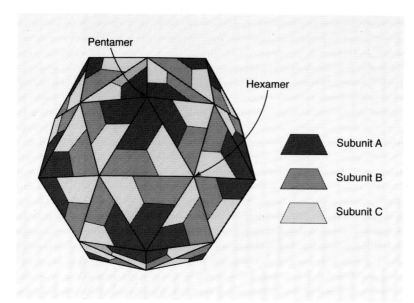

Fig. 3.8 Schematic diagram of a $T = 3$, 180-subunit virus. Each triangle is composed of three subunits, A, B and C, which are asymmetric by virtue of their relationship to other subunits (pentamers or hexamers). From Branden & Tooze (1991).

Icosahedra constructed of four different polypeptides — the picornaviruses

Picornaviruses are made of 60 copies of each of the four polypeptides VP1, VP2, VP3 and VP4 (see p. 153). VP4 is entirely internal. The 180 subunits of these $T = 3$ viruses are arranged exactly like those of tomato bushy stunt virus (Fig. 3.8), but by employing different polypeptide chains rather than conformational variants of one polypeptide. The basic triangle consists of a trimer of VP1, VP2 and VP3. These are assembled into pentamers containing 15 polypeptides, with five molecules of VP1 forming a central vertex; these pentamers are the building-blocks in the cell from which the virion is assembled. Use of three polypeptides gives a chemically more diverse structure and is probably an adaptation to cope with the immune system of animal hosts.

Common structure of plant and animal virion proteins: the antiparallel β barrel

All the virus proteins considered so far, STNV, tomato bushy stunt virus and the VP1, VP2 and VP3 of picornaviruses, have the same antiparallel β barrel structure, of a type sometimes called a 'jelly-roll'. Its formation from a linear polypeptide can be visualized in three stages: firstly, as a hairpin structure, where β strands are hydrogen-bonded to each other: 1 with 8, 2 with 7, 3 with 6 and 4 with 5 (Fig. 3.10A); secondly, these pairs are arranged side by side, so that further hydrogen bonds can be formed by newly adjacent β strands (e.g. 7 with 4) (Fig. 3.10B); thirdly, the pairs wrap around an imaginary barrel (Fig. 3.10C). The eight β strands are arranged in

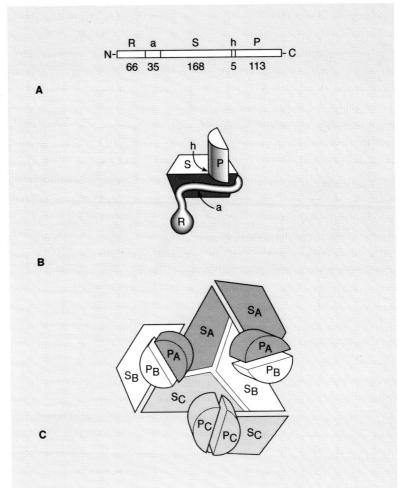

Fig. 3.9 (A) The linear arrangement of domains in the single virion polypeptide of tomato bushy stunt virus. (B) Conformation of the polypeptide. The S domain forms the shell of the virion, while P points outwards and R is internal. (C) This shows a triangular face, composed of subunits A, B and C, and the interaction of the P domains to form dimeric projections. From Branden & Tooze (1991).

two sheets, each composed of four strands: strands 1, 8, 3 and 6 form one sheet and strands 2, 7, 4 and 5 form the second sheet. The dimensions are such that each protein forms a wedge, and these are the structures assembled into virus particles (Fig. 3.11).

The attachment site of picornaviruses

Together, crystallographic, biochemical and immunological data have identified a depression within the β barrel of VP1, which is thought to be the attachment site of picornaviruses. There are 60 attachment sites per virion. Apart from its intrinsic interest, the structure of the attachment site is important as the prime target for antiviral chemicals which can stop attachment of virus to the host cell. The arrangement of the β strands of VP1 is such that an annulus is formed around each fivefold axis of symmetry

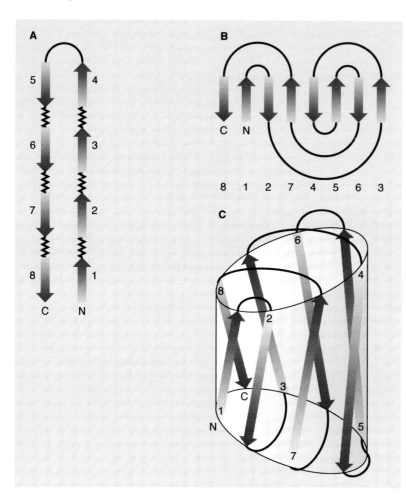

Fig. 3.10 (A) Initial arrangement of the eight β strands before they adopt a 'jelly-roll' β barrel conformation. Note the antiparallel pairs. β strands are separated by loop regions of variable length. (B) Two-dimensional arrangement of the four pairs of β strands. (C) The three-dimensional jelly-roll β barrel. From Branden & Tooze (1991).

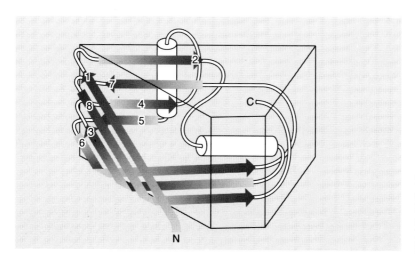

Fig. 3.11 Wedge-shaped form of the jelly-roll β barrel adopted by subunits of viruses. From Branden & Tooze (1991).

(Fig. 3.12). In the rhino (common cold) viruses, this is particularly deep and is called a 'canyon'. The canyon lies within the structure of the β barrel. Amino acids within the canyon are invariant, as expected from their requirement if they have to interact with the cell receptor, while amino acids on the rim of the canyon are variable. Only the latter interact with antibody. It is thought that the floor of the canyon has evolved so that it physically cannot interact with antibody. This is necessary to avoid immunological pressure to accumulate mutations in order to escape from reaction with antibody, since these would at the same time render the attachment site non-functional and hence be lethal for the virus.

Viruses with 180 + 1 subunits and no jelly-roll β barrel

The leviviruses are 24-nm icosahedral RNA bacteriophages, which include MS2, R17 and QB. They encode two coat proteins. There are 180 subunits of one of these arranged with $T = 3$, but only a single copy of the second 'A' protein in each particle. This is the attachment protein. It is not known how the single subunit is incorporated into the particle. The main coat protein does not form a jelly-roll β barrel like the others described above, but instead has five antiparallel β strands arranged like the vertical elements of battlements. Two subunits interact to form a sheet consisting of 10 antiparallel β strands.

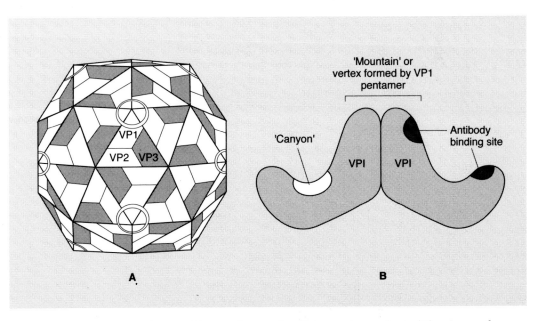

Fig. 3.12 The attachment site of picornaviruses shown (A) as an annulus around the fivefold axis of symmetry formed by VP1 (from Smith *et al.*, 1993), and (B) as a vertical section through a VP1 pentamer (from Branden & Tooze, 1991).

More complex animal virus particles

Careful examination of electron micrographs of adenoviruses ($T = 25$) shows that the 1500 subunits are arranged into 240 hexamers and 12 pentamers and that there is a fibre projecting from each vertex of the virus (Fig. 3.13). It can be shown serologically that the fibres, the pentamers and the hexamers are constructed from different proteins. We are thus faced with the problem of arranging not one, but three different proteins in a regular fashion, while adhering to the design principles outlined above. This can be achieved by arranging the pentamers and the fibres at the vertices of the icosahedron and the hexamers on the faces of the icosahedron (Fig. 3.13). Bacteriophage φX174, which has much shorter spikes than adenoviruses, is probably constructed in a similar fashion.

A different structural arrangement is found in another class of spherical viruses, the reoviruses. Here the capsid is constructed from nine different proteins, but these are arranged in two layers, both of which have icosahedral symmetry. Three of the proteins are arranged in the outer capsid and the remaining six proteins in the inner capsid. However, one of the latter protrudes to the exterior at each vertex to form part of a small spike (Fig. 3.14).

ENVELOPED VIRUSES

Many of the larger animal viruses and a few plant and bacterial viruses are enveloped by a 40 Å-thick lipid bilayer. This envelope, which is derived from host cell membranes, can be disrupted by treatment with ether or detergents and this destroys the infectivity of the virus. These are sometimes referred to as ether-sensitive viruses. Enveloped viruses are formed by budding from cell membranes, but most contain no cell proteins. How cell proteins are excluded and why retroviruses, the exception, do not exclude cell proteins from their virions are not understood.

Sindbis virus

Sindbis is a togavirus and consists of an icosahedral nucleocapsid, consisting of a single 'core' protein, surrounded by an envelope from which viral spike proteins protrude. Sindbis virus core has $T = 3$ and 180 subunits, exactly like tomato bushy stunt virus described above. Surprisingly, the envelope also has icosahedral symmetry, but to everyone's surprise this has $T = 4$ and 240 subunits. The apparent discrepancy was resolved when it was found that the two structures are complementary, the internal ends of the spike proteins fitting exactly into holes between the subunits of the nucleocapsid (Fig. 3.15). So far, this is the only enveloped virus which is known to have a geometrically symmetrical envelope.

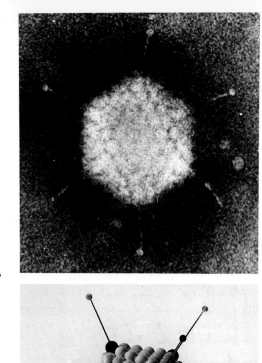

A

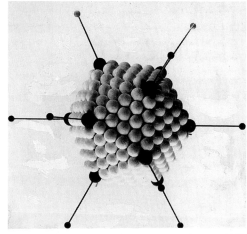

B

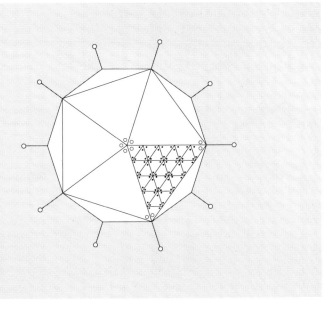

C

Fig. 3.13 The structure of adenoviruses. (A) Electron micrograph of adenovirus (courtesy of Dr N. G. Wrigley). (B) Model of adenovirus to show arrangements of the capsomers (courtesy of Dr N. G. Wrigley). (C) Schematic diagram to show the arrangements of the subunits on one face of the icosahedron. Note the subdivision of the face into 25 smaller equilateral triangles.

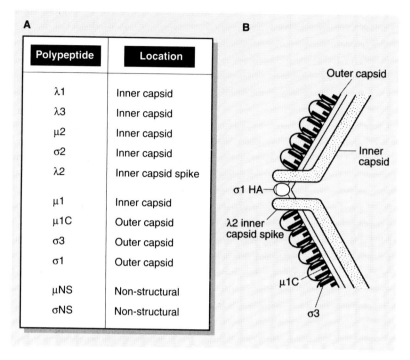

Fig. 3.14 The double capsid structure of reovirus showing (A) the virus-encoded polypeptides and (B) their location in the virion (from Fields & Green, 1982).

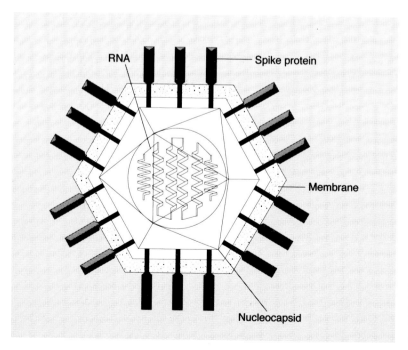

Fig. 3.15 Sindbis virus: the only known icosahedral enveloped virus. The core is $T = 3$, but the envelope is $T = 4$. See text for the explanation.

The influenza viruses

One of the best studied groups of enveloped viruses is the influenza viruses. These viruses are roughly spherical and are normally described as pleomorphic (Fig. 3.16). In electron micrographs (Fig. 3.16), a large number of surface protrusions, or spikes, projecting about 135 Å from the viral envelope can be observed. These spikes, which have an overall length of 175 Å, are transmembrane glycoproteins like those of the cell. The spike layer consists of only virus-specified glycoproteins and has the haemagglutinating and neuraminidase activities (Chapter 5). These reside in morphologically different spikes. Neuraminidase spikes are arranged non-randomly in clusters (Fig. 3.16D). The structure of the influenza virus haemagglutinin is given in Fig. 3.17.

Internal to the envelope is a core composed of the M1 protein (matrix protein). Finally, inside the M1 core are eight ribonucleoproteins; these are flexible rods of RNA and protein (constructed as described on p. 39) and arranged in a twisted hairpin structure. The influenza virus particle thus comprises four major virion proteins, which are assembled by a variety of strategies.

The rhabdoviruses

The viruses of this large group are either bullet-shaped or bacilliform and often display very regular cross-structures, which represent the internal helical nucleoprotein structure. Animal rhabdoviruses are bullet-shaped. Like the influenza viruses, rhabdoviruses are covered with a layer of regularly arranged spikes and have an underlying membrane-associated protein (matrix protein). To account for the unusually rigid structure of the rhabdoviruses, compared with the influenza viruses, it has been suggested that there could be a direct geometrical relationship between the protein subunits. Otherwise the influenza and rhabdoviruses have fundamentally a very similar structure (Fig. 3.16).

Enveloped icosahedral viruses

Whereas the influenza and rhabdoviruses have a helical nucleocapsid, some other enveloped bacterial and animal viruses have an icosahedral nucleocapsid. The various permutations are summarized in Table 3.2.

VIRUSES WITH HEAD–TAIL MORPHOLOGY

The only viruses built on the head–tail architectural principle (Fig. 3.18) are bacteriophages and it may be that this is connected with the way in which these bacterial viruses infect susceptible cells (Chapter 5). The number of different bacteriophages with a

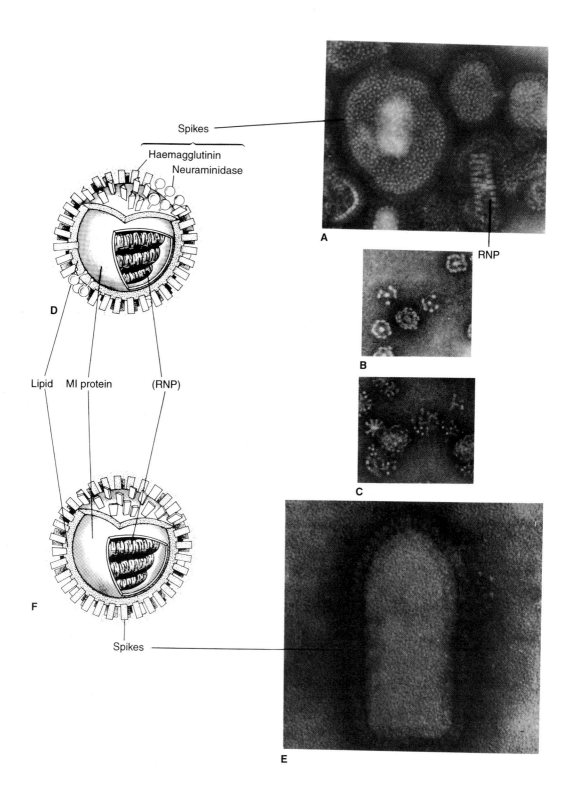

Spikes

Haemagglutinin
Neuraminidase

D

Lipid   MI protein   (RNP)

F

Spikes

A

RNP

B

C

E

Table 3.2 Distribution of the various types of virus architecture amongst viruses of bacteria, plants and animals.

Virus architecture	Distribution amongst viruses of		
	Bacteria	Plants	Animals
Helical	Relatively rare. All examples so far are male-specific	Very common	Not known
Icosahedral	Relatively rare. Best examples are the φX174-related phages and the RNA-containing male-specific phages of *Escherichia coli*	Common	Common
		Rare. Tomato spotted wilt virus	
Enveloped	Three families known	Rare; tospo- and plant rhabdoviruses	Common
Icosahedral + helical			
Head + tail	Most commonly found	Not known	Not known

* All have a 'spherical' core underlying the lipid bilayer but only that of Sindbis virus p. 50 is known to be truly icosahedral.

head—tail type of architecture is very large and they can be sub-divided into those with short tails, those with long non-contractile tails and those with complex contractile tails (see pp. 358–9). A number of other structures, such as base plates, collars, etc., may also be present (Fig. 3.19). Despite their more complex structure, the design principles involved are identical to those outlined earlier for the viruses of simpler architecture. Heads usually possess octahedral or icosahedral symmetry, whereas tails usually have

Fig. 3.16 Comparative structure of an orthomyxovirus (influenza) and a rhabdovirus (vesicular stomatitis virus). Although the electron microscopic morphology is different, they are constructed in the same way. (A) Electron micrograph of influenza virus showing the internal ribonucleoprotein (RNP) and the surface spikes. (B) Aggregates of purified neuraminidase. (C) Aggregates of purified haemagglutinin. Note the triangular shape of the spikes when viewed 'end-on'. (D) Schematic representation of the structure of influenza virus. (E) Electron micrograph of vesicular stomatitis virus. (F) Schematic representation of the structure of vesicular stomatitis virus. (Electron micrographs courtesy of N. G. Wrigley and C. J. Smale.)

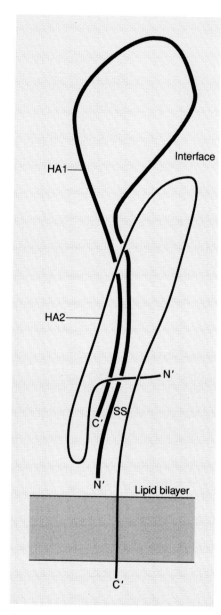

A

Fig. 3.17 The influenza virus
haemagglutinin (HA). This is a
homotrimer but only a monomer
is shown for simplicity. The HA
is synthesized as a single
polypeptide which is
proteolytically cleaved into the
membrane-bound HA2 and the
distal HA1. (A) An outline
structure showing that HA1 and
HA2 are both hairpin structures.

helical symmetry. All other structures, such as base plates, when
present, also possess a defined symmetry.

OCCURRENCE OF DIFFERENT VIRUS MORPHOLOGIES

The different virus morphologies discussed above do not occur
with equal frequency among bacterial, plant and animal viruses.

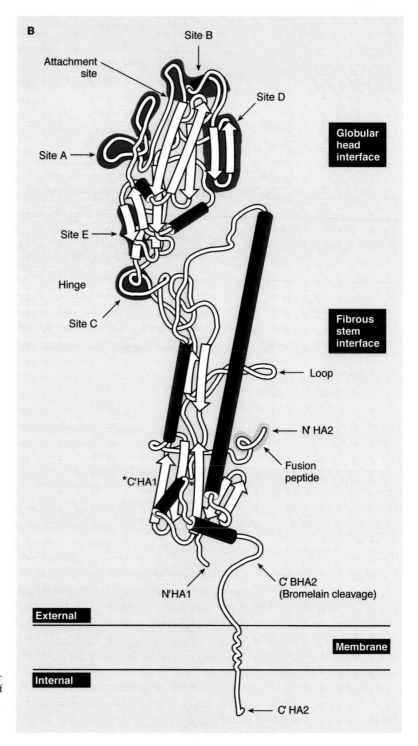

B

Site B

Attachment site

Site D

Globular head interface

Site A

Site E

Hinge

Site C

Fibrous stem interface

Loop

N′ HA2

Fusion peptide

*C′HA1

C′ BHA2 (Bromelain cleavage)

N′HA1

External

Membrane

Internal

C′ HA2

Fig. 3.17 (*Continued*) (B) The crystal structure. The globular head of HA1 bears all the neutralization sites (A–E; shaded) and is made of a distorted jell-roll β barrel like most of the icosahedral viruses. From Wiley *et al.* (1981).

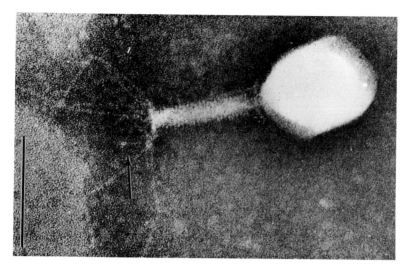

Fig. 3.18 Electron micrograph of bacteriophage T2 (courtesy of Dr L. Simon). Six long tail fibres are evident. Tail pins cannot be seen but a short fibre (indicated by the arrow) can be seen. The bar is 1000 Å.

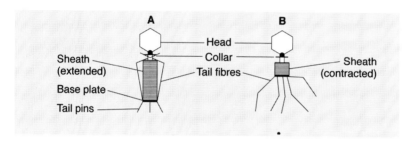

Fig. 3.19 Schematic representation of the structure of some tailed bacteriophages.

Information on the frequency with which they occur is summarized in Table 3.2.

THE PRINCIPLE OF SELF-ASSEMBLY

If viruses are constructed according to principles of efficient design, they should be able to assemble themselves without any external organizer, due to the formation of a large number of weak bonds when they are brought into the right configuration by random thermal movements. That this is indeed the case has been best shown by TMV. Self-assembly is economical in that it requires no specific genetic information, and it may also have the advantage of providing a built-in rejection mechanism for any faulty subunits which may be produced during replication. However, with some bacteriophages, such as T4, assembly of virion proteins is assisted by 'scaffolding' proteins, which are removed as the particle takes its final form (Chapter 11).

FURTHER READING

Branden, C. & Tooze, J. (1991) *Introduction to protein structure*. New York: Garland Publishing.

Butler, P. J. G. (1984) The current picture of the structure and assembly of tobacco mosaic virus. *Journal of General Virology*, **65**, 253–279.

Caspar, D. L. D. & Klug, A. (1962) Physical principles in the construction of regular viruses. *Cold Spring Harbor Symposia on Quantitative Biology*, **27**, 1–24.

Eiserling, F. A. (1979) Bacteriophage structure. In: *Comprehensive virology*, Vol. 13, pp. 543–580. Fraenkel-Conrat, H. & Wagner, R. R. (eds). New York: Plenum Press.

Fields, B. N. & Green, M. I. (1982) Genetic and molecular mechanisms of viral pathogenesis: implications for prevention and treatment. *Nature (London)*, **300**, 19–23.

Finch, T. J. & Holmes, K. C. (1967) Structural studies of viruses. In: *Methods in virology*, Vol. 3, pp. 352–474. Maramorosch, K. & Koprowski, H. (eds). London: Academic Press.

Harrison, S. C. (1984) Structure of viruses. In: 'The Microbe 1984'. *Society for General Microbiology Symposium*, **36**, Pt 1, 29–73.

Hogle, J. M. (ed.) (1990) Virus structure. *Seminars in Virology*, **1**, no. 6.

Klug, A. (1983) Architectural design of spherical viruses. *Nature (London)*, **303**, 378–379.

McKenna, R. *et al.* (1992) Atomic structure of single-stranded DNA bacteriophage φX174 and its functional implications. *Nature (London)*, **355**, 137–143.

Smith, T. J., Olson, N. H., Cheng, R. H., Liu, H., Chase, E. S., Lee, W. M., Leippe, D. M., Mosser, A. G., Rueckert, R. R. & Baker, T. S. (1993) Structure of human rhinovirus complexed with Fab fragments from a neutralizing antibody. *Journal of Virology*, **67**, 1148–1158.

Wiley, D. C., Wilson, I. A. & Skehel, J. J. (1981) Structural identification of the antibody-binding sites of Hong Kong influenza haemagglutinin and their involvement in antigenic variation. *Nature (London)*, **289**, 373–378.

Also check Chapter 21 for references specific to each family of viruses.

4 Viral nucleic acids

The nucleic acid of a virus contains both the specific information and the operational potential such that upon entering a susceptible cell it can subvert the biosynthetic machinery of that cell and redirect it towards the production of virus particles. Viral nucleic acids display a remarkable array of structural and compositional varieties and, since any peculiarities of the nucleic acid must have a bearing on the process of replication, we shall discuss some of them in detail.

THE PHYSICAL CHEMISTRY OF THE NUCLEIC ACIDS

Nucleic acids contain the nucleosides: adenosine, guanosine, cytidine and either uridine or thymidine. In double-stranded molecules, base pairing occurs between guanosine and cytidine and between adenosine and either uridine or thymidine. It is possible to denature these double-stranded molecules, i.e. separate the two strands, by heating or by alkali treatment. When heat is applied to nucleic acid solutions, the temperature at which 50% denaturation occurs (Tm) is most easily measured by following the ultraviolet (UV) absorbance of the sample as the temperature is increased, a sudden increase in absorbance indicating strand separation (denaturation or melting) (Fig. 4.1). The temperature at which melting occurs is dependent upon ionic strength and on the guanine plus cytosine (% GC) content of the nucleic acid. If denatured deoxyribonucleic acid (DNA) is incubated at a temperature 25°C below Tm, the two strands will slowly anneal, but it is usually impossible to get complete annealing, since a certain amount of mispairing occurs. If ribonucleic acid (RNA) complementary to either strand of DNA is added to the annealing mixture, it is possible to obtain RNA–DNA hybrids.

When a small amount of macromolecule (DNA, RNA, etc.) in a concentrated solution of a heavy-metal salt, such as CsCl, is centrifuged until equilibrium is reached, the opposing forces of sedimentation and diffusion produce a stable concentration gradient with a continuing increase in density along the direction of the centrifugal force. The macromolecules are driven by the centrifugal field into the region where the solution density is equal to their own buoyant density. At equilibrium, a single species of macromolecule is distributed over a narrow band. If several different density species of macromolecule are present, each will form a band at the position where the density of the CsCl equals the buoyant density of the species (Fig. 4.2). Components banding at

60

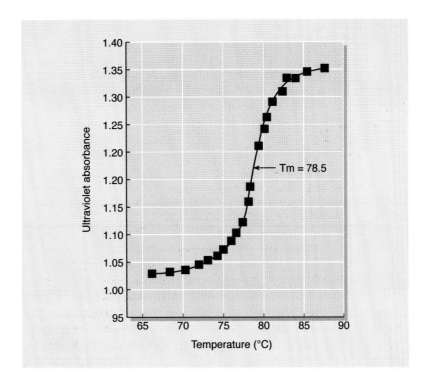

Fig. 4.1 Increase in ultraviolet absorbance of DNA with increase in temperature.

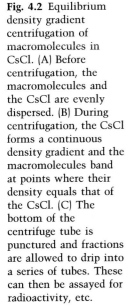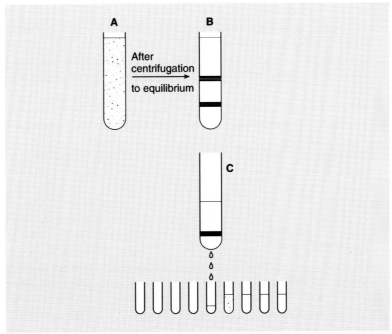

Fig. 4.2 Equilibrium density gradient centrifugation of macromolecules in CsCl. (A) Before centrifugation, the macromolecules and the CsCl are evenly dispersed. (B) During centrifugation, the CsCl forms a continuous density gradient and the macromolecules band at points where their density equals that of the CsCl. (C) The bottom of the centrifuge tube is punctured and fractions are allowed to drip into a series of tubes. These can then be assayed for radioactivity, etc.

different places in the gradient by virtue of their differences in density may be isolated by puncturing the bottom of the centrifuge tube and collecting fractions of one or several drops in size.

When the five possible classes of nucleic acid (double- and single-stranded DNA, double- and single-stranded RNA, and DNA–RNA hybrids) are centrifuged in CsCl gradients, they are all found to have characteristic buoyant densities.

It is possible to introduce a density label into DNA by growing the DNA source (phage, cell, etc.) in a medium containing either heavy isotopes (^{15}N, ^{18}O, ^{2}H) or the thymine analogue 5-bromouracil (Fig. 4.3). DNA which is given such a label will have a higher buoyant density than normal DNA, and this can be very useful when it is desirable to determine the fate of a particular molecule.

The buoyant density of dehydrated DNA and RNA is greater than 2 g/ml, but in the presence of heavy-metal ions DNA is extensively hydrated, leading to lower observed buoyant densities. However, the degree of hydration differs with the different salts and hence the buoyant density also varies. In CsCl, the buoyant density of RNA is approximately 1.9, and densities as high as this are only produced close to the saturation point of aqueous CsCl solutions. Consequently, most RNA is centrifuged to equilibrium in Cs_2SO_4, in which its buoyant density is much lower. However, CsCl is preferred for DNA, since density is not a linear function of base composition in Cs_2SO_4.

Analysis on sucrose gradients
(velocity or rate-zonal centrifugation)

Whereas the separation of nucleic acids in CsCl gradients is a function of base composition, their separation in sucrose gradients is a function of size and shape. Unlike CsCl gradients, sucrose gradients are not self-generating in centrifugal fields and so must be preformed. There are several ways of doing this, but the commonest method is the use of a gradient-maker (Fig. 4.4). This consists of two reservoirs, joined by a capillary tube which is fitted with a valve. In addition, one of the reservoirs has an outlet tube

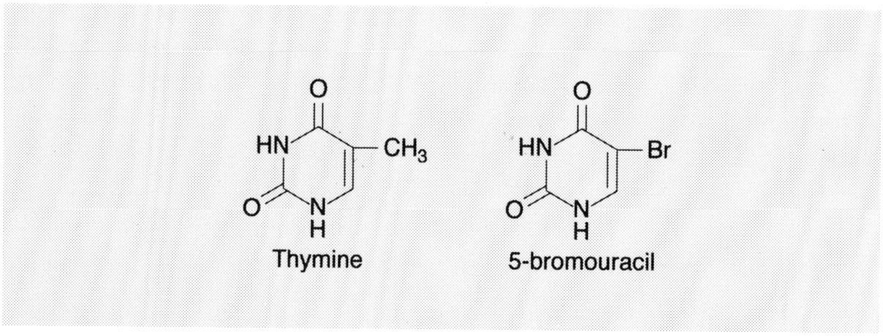

Fig. 4.3 Structure of the pyrimidine base, thymine, and its analogue, 5-bromouracil.

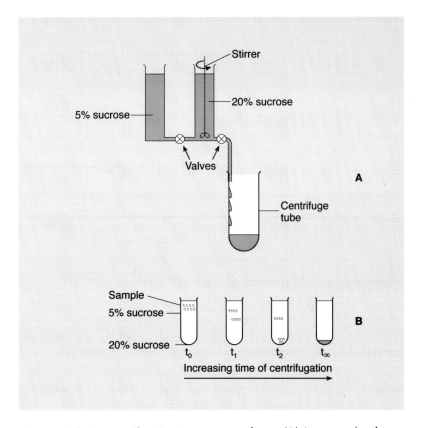

Fig. 4.4 Velocity centrifugation in sucrose gradients. (A) Apparatus for the preparation of sucrose gradients (see text for details). (B) Separation of two components in a sucrose gradient. At time t_0 the mixture of two components is layered on top of the gradient. After centrifugation for time t_1 the two components have separated and by time t_2 the faster of the components has pelleted. Note that if centrifugation is continued both components will be pelleted and separation will not be achieved. (Compare with Fig. 4.2.)

and contains a stirring device. If a 5–20% sucrose gradient is desired, 20% sucrose is put into the mixing reservoir and 5% sucrose in the other. Stirring is begun and the exit line opened. As 20% sucrose leaves the mixing reservoir, the valve separating the two reservoirs is opened to allow 5% sucrose to flow in and gradually dilute the 20% sucrose. A small volume of macro-molecule-containing solution is then gently layered on top of the gradient. Upon application of a centrifugal field, the macromolecules sediment through the solution at a rate dependent on their size and shape. The rate of sedimentation is usually expressed as the sedimentation coefficient, S:

$$S = \frac{dx/dt}{\omega^2 x}$$

where x is the distance from the centre of rotation, ω the angular velocity in radians per second and t the time in seconds. Nucleic acids have sedimentation coefficients in the range between 1×10^{-13} and 100×10^{-13} seconds. A sedimentation coefficient of 1×10^{-13} seconds is called a Svedberg and is abbreviated S. Thus a sedimentation coefficient of 42×10^{-13} seconds would be denoted 42S.

There is no theoretical reason why macromolecules cannot be separated by sedimentation through a column of water instead of sucrose. Practically, however, good separations are hard to achieve in this way because the slightest disturbance of the centrifuge tube will cause remixing: the presence of a gradient of sucrose minimizes such disturbances. In addition, the increasing concentration of sucrose counteracts the increasing centrifugal force imposed on nucleic acids as they move further from the centre of rotation; in this way the rate of sedimentation is kept constant.

Electrophoresis of nucleic acids

For the separation of different RNA molecules, the use of sucrose gradients has largely been superseded by zone electrophoresis in gels of polymerized acrylamide (polyacrylamide). In this method the centrifugal field is replaced by an electrical field, and separation is again based on size and shape. Polyacrylamide gels are hydrated and porous but are mechanically rigid and so function in an analogous manner to sucrose, by preventing convectional and vibrational disturbances. DNA molecules, particularly those smaller than $10^7 M_r$, can also be separated by electrophoresis (Fig. 4.5), but agarose is the material of choice.

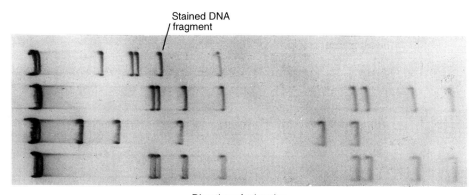

Stained DNA
fragment

Direction of migration

Fig. 4.5 Separation of DNA molecules of different sizes by electrophoresis in an agarose gel. DNA from bacteriophage λ was cleaved with different restriction endonucleases (see p. 113) and after electrophoresis the DNA was stained by immersing the gel in ethidium bromide. (Photograph courtesy of Professor K. Murray.)

Armed with this brief introduction to nucleic acids, we are now in a position to examine some of the structural peculiarities of those isolated from viruses.

TYPES OF NUCLEIC ACIDS FOUND IN VIRUSES

There are four possible kinds of viral nucleic acid: single-stranded DNA, single-stranded RNA, double-stranded DNA and double-stranded RNA. Their distribution among the different types of virus is shown in Table 4.1. Nucleic acid is classified by determining its base composition, nuclease sensitivity, buoyant density, etc. Single-stranded nucleic acids are detected by the absence of a sharp melting profile upon heating and the non-equivalence of the molar proportions of adenine (A) and thymine (T) (or uracil (U)) or guanine (G) and cytosine (C).

The four types of nucleic acid may exist as linear molecules or form circles, as described later in this chapter. However, these forms are not automatically interchangeable and many viral genomes exist only in the linear form. Circular and linear molecules can be distinguished by their sensitivity to exonucleases which require a free 5′ or 3′ terminus and by their sedimentation and electrophoretic behaviour (see Fig. 4.10).

Unusual bases

The DNA from certain bacteriophages contains unusual bases. Examples are the replacement of thymine by uracil* (bacteriophage PBS1) or hydroxymethyluracil (bacteriophage SP8) or the replacement of cytosine by hydroxymethylcytosine (bacteriophages T2, T4, T6). In the T-even phage series, the hydroxymethylcytosine may be further substituted with glucose or gentiobiose. Since these bases do not appear in the DNA from uninfected host cells, the information specifying their synthesis must be carried by the virus. Like the host cell DNA, most viral DNA molecules are also partially methylated, and subtle differences in methylation may help enzymes to distinguish between host and viral DNA. Since the relationship between base composition and buoyant density and Tm no longer holds if the DNA is modified in any way, the presence of such bases can be determined by measuring the latter two parameters. If these two methods yield significantly different values for the base composition, it is a good indication of the presence of modified bases (Table 4.2).

---

* Note that replacement of thymine with uracil does not convert DNA into RNA. The type of nucleic acid, i.e. DNA or RNA, is specified by the sugar moiety.

Table 4.1 Distribution of the four basic types of nucleic acid among viruses of bacteria, plants and animals.

Type of nucleic acid	Distribution		
	Bacterial viruses	Plant viruses	Animal viruses
Single-stranded DNA	Not very common. Found in Microviridae and Inoviridae	Rare. Only found in geminiviruses	Only found in Parvoviridae
Double-stranded DNA	Most common	Rare. Only found in caulimoviruses	Common
Single-stranded RNA	Only in Leviviridae	Most common type found	Common
Double-stranded RNA	Rare. Only found in Cystoviridae	Only found in Reoviridae and viruses of fungi	Common

Table 4.2 Effect of unusual bases on melting temperature and buoyant density of DNA.

Phage	%GC as determined from			Unusual base
	Tm	Buoyant density	Chemical composition	
SP8	17.5	84	43	5-hydroxymethyluracil replaces thymine
PBS1	17.5	63	28.2	Uracil replaces thymine
T4	35.9	41.3	34	Hydroxymethylcytosine replaces cytosine. Also glucosylated

Single-stranded ('sticky') ends

The DNA from bacteriophage λ sediments as a single component in sucrose gradients. Since macromolecules are separated on the basis of size and shape in sucrose gradients, it can be concluded that the DNA from λ is homogeneous. When the DNA is heated to temperatures below its *Tm* and slowly cooled, two new components appear, one sedimenting 1.13 times faster and the other 1.41 times faster than native λ DNA. Since the formation of both species increases as the temperature and salt concentration are raised, and since both disappear after melting and quick cooling, it is likely that hydrogen bonding is involved. There is a stretch of single-stranded DNA at either end of the molecule and the single-stranded segments are complementary. The two new components observed thus consist of hydrogen-bonded circles and dimers (Fig. 4.6) and,

when such DNA preparations are examined in the electron microscope, circles can be seen. Furthermore, if the DNA is treated with
DNA polymerase, which has a preference for single-stranded DNA,
the product cannot form circles, since the single-stranded ends are
removed. Incubation with exonuclease restores the ability to form
circles by regenerating the single-stranded ends.

Terminal redundancy or terminal repetition

Genetic studies with bacteriophage T4 suggested that its DNA is
terminally redundant; that is, the first few genes at one end of the
chromosome also appear at the other end (Fig. 4.7). This was
shown biophysically by treating the DNA with exonuclease to
produce single-stranded segments followed by incubation under
annealing conditions. If the DNA is indeed terminally redundant,
then 'sticky' ends should have been produced by the exonuclease
treatment, which should enable the molecules to circularize during
the incubation period. Such circles were indeed seen by electron
microscopy.

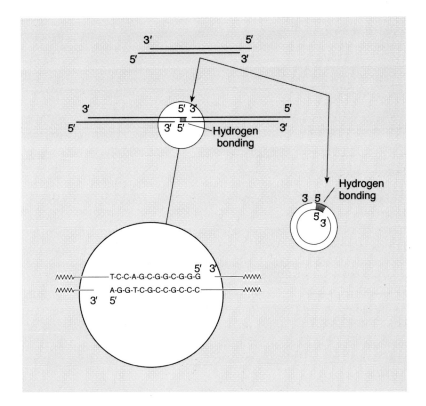

Fig. 4.6 Formation of
dimers and circles of
DNA after incubation
under conditions
favouring annealing.
The base composition
of the 'sticky ends' is
also shown.

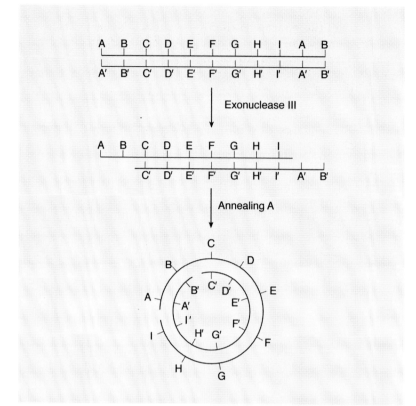

Fig. 4.7 Demonstration of terminal redundancy. (A) A terminally redundant molecule is depicted by two paralled lines and the different genes by letters of the alphabet. Following exonuclease treatment, 'sticky ends' are produced, which can circularize on incubation under conditions favouring annealing.

Inverted repeats

Treatment of adenovirus DNA with exonuclease does not allow the formation of double-stranded DNA circles. This means that the single-stranded termini exposed by nuclease digestion cannot base-pair and that the ends of adenovirus DNA are not terminally repetitious in the conventional manner. However, when a solution of adenovirus DNA is denatured with alkali and then neutralized, both of the DNA strands are capable of forming circles. Since these circles are always of unit length, they must be formed by interaction between the 3′ and 5′ termini of the same strand, i.e. there must be an inverted terminal repetition. Two kinds of inverted repeat can be envisaged (Fig. 4.8), which give rise to different kinds of single-stranded circles. In model A, self-annealed strands would be closed by an 'in-line' duplex segment, whereas strand closure in model B produces a duplex projection or 'panhandle'. In the case of adenovirus DNA, self-annealing produces 'in-line' circles, whereas self-annealing of adeno-associated virus DNA produces 'panhandles'.

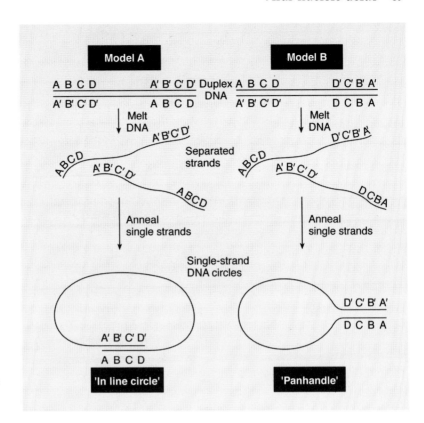

Fig. 4.8 Detection of the two kinds of inverted repeat by the formation of single-stranded circles.

Circular permutation

Once again, genetic studies with T4 suggested an unusual feature of its nucleic acid. This virus has a linear DNA molecule and yet its genetic map is circular. An answer to this paradox is to propose that the genes in bacteriophage T4 are circularly permuted. If we place all the genes of the bacteriophage round the circumference of a circle and break the circle at any point the result is a collection of linear molecules in which the gene order is unchanged but which have different genes at the ends of the molecules (Fig. 4.9). If such a collection of molecules is denatured and then annealed, there is a high probability of complementary strands from different molecules coming together. This will result in the formation of single-stranded ends which are complementary, i.e. 'sticky' ends, which should enable the molecules to circularize. Indeed, when such experiments are done, it is possible to observe circular molecules, using the electron microscope.

Circular DNA

Not only are some viral DNA molecules capable of circularizing but some are actually circular when extracted from the virus. Such

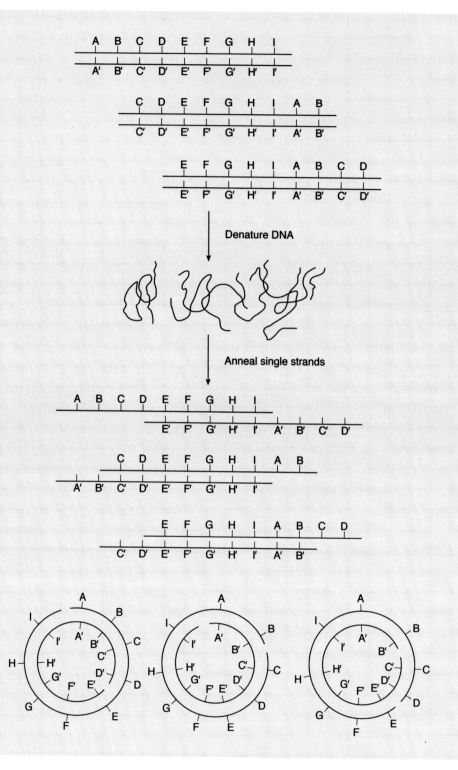

Fig. 4.9 Demonstration of circular permutation. A collection of circularly permuted molecules is depicted by parallel lines, with the different genes indicated by letters of the alphabet. By denaturing the DNA and then annealing the single strands so produced, it is possible to generate molecules with 'sticky ends', which are capable of circularizing.

molecules are resistant to exonuclease III (which can only act on molecules that have a free end) but become sensitive after a brief endonuclease treatment. Circular DNA molecules are found with both single-stranded DNA viruses (plant geminiviruses, bacteriophage φX174) and double-stranded DNA viruses (polyoma, simian virus 40). The double-stranded circular molecules can be of two types: a covalently closed double-stranded structure and a double-stranded circular structure in which there is a break in one strand (Fig. 4.10). There is usually a second difference between these two structures in that the closed circular form I isolated from viruses such as SV40 or from φX174-infected cells is supercoiled due to a deficiency of turns in the double helix (Fig. 4.10).

Because the shapes of the two molecules are so different, they are readily separable in sucrose gradients or by electrophoresis in

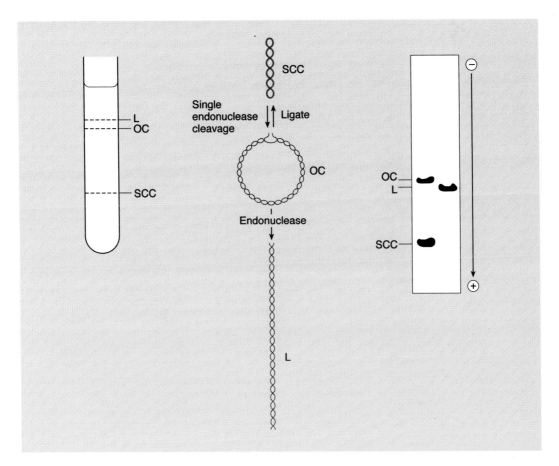

Fig. 4.10 Structure of double-stranded closed circular DNA and its conversion to the linear form. Separation by velocity centrifugation on a sucrose gradient is shown on the left and by electrophoresis in 0.4% agarose gel on the right. SCC, supercoiled closed circle; OC, open circle; L, linear form.

agarose gel (Fig. 4.10). The supercoiled form has the most compact configuration and hence sediments and electrophoreses most rapidly.

It is also possible to separate these circular molecules from linear molecules of the same buoyant density by centrifugation in density gradients containing the drug ethidium bromide. Ethidium bromide is capable of intercalating between the stacked bases of the double helix and, as the amount bound increases, the helix untwists until the open form of the circular molecule is produced. Further intercalation introduces excess turns in the double helix, resulting in supercoiling in the opposite sense (Fig. 4.11). Because of the 'opposite' supercoiling, there is a limit to the amount of drug that can be bound by a circular molecule, but this restraint does not operate with linear molecules. As a consequence of the circular and linear forms binding different amounts of drug, their buoyant densities are different and hence they can be separated in CsCl gradients.

Circularization through protein linkers

Examination in the electron microscope of partially disrupted adenoviruses reveals circular DNA, sometimes in the form of supercoils, whereas DNA extracted from adenoviruses by treatment with proteolytic enzymes, detergent and phenol consists of linear duplex molecules. Adenovirus DNA is not permuted and lacks

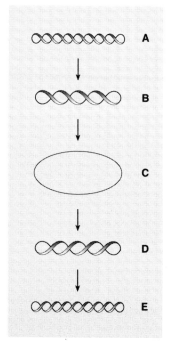

Fig. 4.11 Effect of intercalation of ethidium bromide on supercoiling of DNA. As the amount of intercalated drug increases, the double helix (shown here as a single line) untwists, with the result that the supercoiling in A decreases until the open form of the circular molecule is produced (C). Further intercalation introduces excess turns in the double helix resulting in supercoiling in the opposite sense (note the direction of coiling at B and D).

sticky ends and terminal repetitions, structural peculiarities found in other viral DNA molecules and which enable them to form circles. So far, the only structural peculiarity detected in adenovirus DNA is the presence of an inverted terminal repetition (see Fig. 4.8), but this does not permit circularization of duplexes. The explanation for the circular DNA molecules found in partially disrupted virions, but not in extracted DNA, is the presence of a protein linked covalently to the 5' end of each DNA strand which circularizes the genome by binding non-covalently to the 3' terminus. This protein linker is destroyed by the proteolytic enzymes and detergent used in extracting the DNA. When the virus is disrupted by guanidinium chloride and deproteinized with chloroform and isoamyl alcohol, the protein linker can be kept intact. DNA isolated in this way sediments in sucrose gradients faster than the linear duplexes obtained by conventional extraction methods (Fig. 4.12A), and electron microscopy reveals that up to 90% of the DNA is in the form of circles and oligomers. Treatment of these circles and oligomers with detergent or proteolytic enzymes converts them to linear duplex monomers (Fig. 4.12B).

Circular RNA of viroids

The RNAs of the viroids responsible for potato spindle tuber (PSTV) and citrus exocortis (CEV) diseases exhibit hyperchromicity upon heating, indicating the presence of base-pairing. When examined in the electron microscope, native RNA from PSTV and CEV can

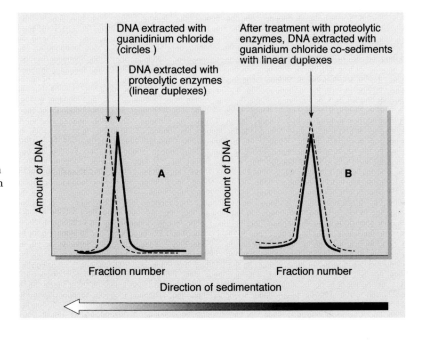

Fig. 4.12 Sucrose gradient sedimentation of DNA extracted from adenovirus by two different methods. (A) Separation of circles and oligomers from linear duplexes. (B) Conversion of circles and oligomers to linear duplexes by treatment with proteolytic enzymes.

be seen as rod-shaped or dumb-bell-shaped molecules. Denatured RNA from PSTV and CEV, on the other hand, exhibits a variety of structures, ranging from rod-shaped molecules (possibly due to renaturation during preparation for electron microscopy) to single-stranded circles. All attempts to label the 5' and 3' ends, using the techniques outlined for reovirus (p. 76), have been unsuccessful, suggesting that there are no free ends. These and more recent sequence data suggest that RNAs from PSTV and CEV exist as single-stranded covalently closed circular molecules with considerable intramolecular base-pairing (Fig. 4.13). It is worth noting that, with only 359 nucleotides, PSTV is the smallest self-replicating pathogen known. Its replication is a mystery and there are no viroid proteins, as evidenced by the absence of any initiation AUG codon or protein synthesis in an *in vitro* translation system. Viroids exist and are transmitted as coat-free nucleic acid and appear to be replicated by a cellular enzyme which normally recognizes a DNA template (DNA-dependent RNA polymerase II). They are thought to produce disease by interfering with the processes which control expression of the host genome.

Terminal caps

Messenger RNAs are synthesized, translated and possibly also degraded in a 5'−3' direction. Consequently, their 5' termini are of some interest, particularly with regard to their ability to regulate genetic expression (see Chapter 10). In prokaryotes, and viruses of prokaryotes, the 5' end of many messenger RNAs (mRNAs) is a triphosphorylated purine corresponding to the residue that initiated transcription. In contrast, most eukaryotic cellular and viral mRNAs, as well as native nucleic acid from some RNA viruses of eukaryotes, have been found to be modified at the 5' end. The modification consists of a 'cap' that protects the RNA at its 5' terminus from attack by phosphatases and other nucleases and promotes mRNA function at the level of initiation of translation (see also p. 131).

The general structural features of the 5' cap are shown in Fig. 4.14. The terminal 7-methylguanine and the penultimate nucleotide are joined by their 5' hydroxyl groups through a triphosphate bridge. This 5'−5' linkage is inverted, relative to the normal 3'−5' phosphodiester bonds in the remainder of the polynucleotide chain.

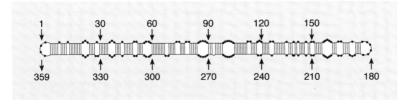

Fig. 4.13 Diagram of the circular single-stranded 359-nucleotide RNA of potato spindle tuber viroid, showing the maximized base-paired structure.

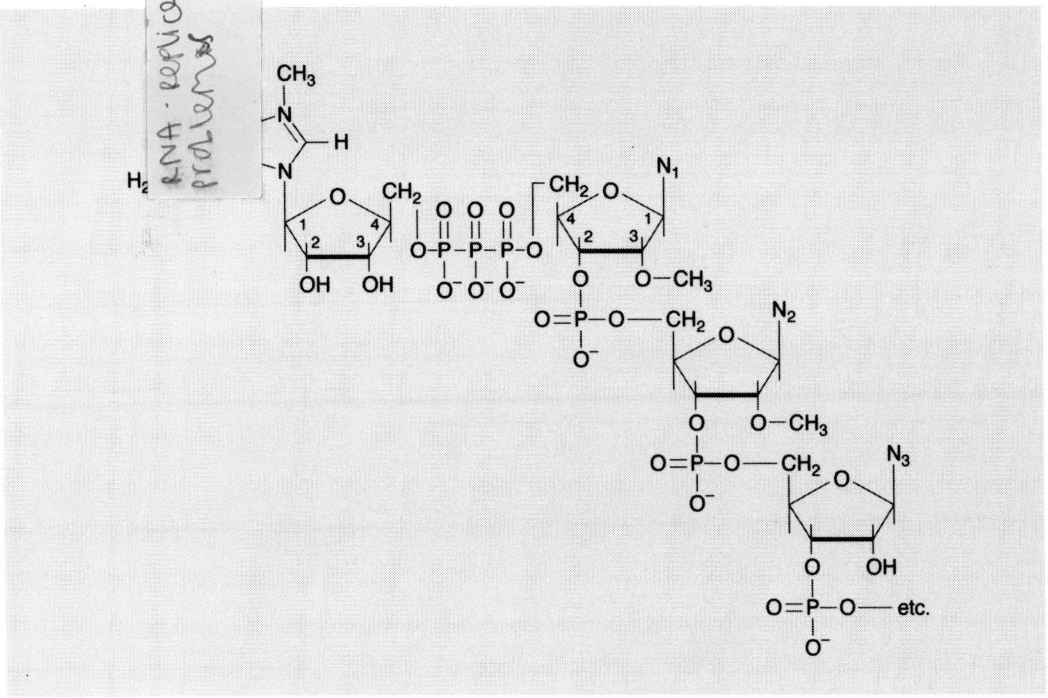

CH₃
N
H
N
H₂

(handwritten in margin: RNA - replication problems)

Fig. 4.14 Structure of a capped RNA molecule. Note especially the 5'−5' phosphodiester linkage. N_1, N_2 and N_3 can be any of the four nucleic acid bases. The 2'-O-methyl group on the ribose of N_1 nucleotide is always present (cap 1 and 2 structures) and the methyl group on N_2 is present in cap 2 structures only.

Segmented genomes

Chemical analysis of the reovirus genome showed it to have an M_r greater than 10^7, but the observed sedimentation rate of viral RNA was lower than that expected for a molecule of this size. Further studies showed that, regardless of the amount of care exercised in extracting the viral RNA, a trimodal distribution of sizes was always found by velocity sedimentation of the RNA on sucrose gradients. When subjected to electrophoresis, each size class could be further separated, resulting in the recognition of 10 different molecules of RNA. These 10 fragments showed no base sequence homology in hybridization tests and so could not have arisen by random fragmentation of the genome.

The sum total of the M_r of the 10 fragments corresponds to the size of the viral genome as estimated by chemical means, suggesting that the intact virus contains one copy of each fragment. Consequently, does the genome exist in the intact particle as a single molecule, which is susceptible to breakage at fixed points during extraction, or as a collection of 10 different fragments? The reovirus genome consists of double-stranded RNA and we can propose four models for its structure (Fig. 4.15). In trying to decide between

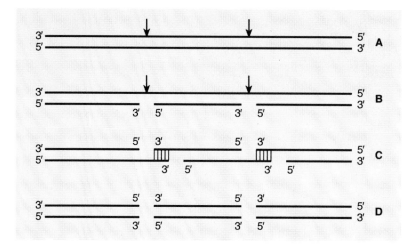

Fig. 4.15 Schematic models for the double-stranded RNA genome of reovirus. The arrows represent hypothetical weak points in the viral genome.

these models, a technique has been developed for specifically label-ling the 3' ends of the RNA, regardless of whether the RNA has been extracted or is still in the intact virion. The technique involves oxidizing the RNA with periodate, resulting in the formation of a dialdehyde at the free 2',3'-OH positions. This dialdehyde is then reduced with tritiated borohydride, which introduces four tritium atoms per 3' terminal (Fig. 4.16). In the actual experiments, reovirus was labelled with ^{32}P to provide an RNA marker and purified and the virus preparation divided into two. RNA was extracted from one part, oxidized with periodate and reduced with tritiated boro-hydride. The other part of the virus was oxidized and the RNA was extracted and then analysed by gel electrophoresis. Results showed

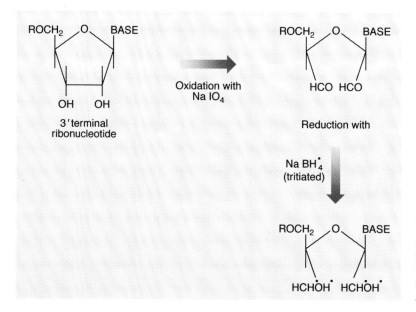

Fig. 4.16 Tritium labelling of 3' terminal ribonucleotides. The terminal ribonucleotide is oxidized to a dialdehyde and this is then reduced with tritiated borohydride of high specific activity. Tritiated atoms are indicated with asterisks.

that all 10 fragments were labelled, even when the RNA had not been extracted prior to oxidation, and so we can eliminate model A immediately. According to model B, one strand is continuous and the other is nicked in nine places. Oxidation of this structure *within the virion* will modify only the left-hand segment at both 3' terminals, while the other nine segments will be modified at one 3' terminus only. This means that during the reductive step, nine of the 10 subunits will incorporate half as much tritium as when oxidation is done on extracted double-stranded RNA. In the actual experiments, quantitative analysis of the tritium distribution showed that both samples of double-stranded RNA incorporated tritium to about the same extent in all classes, and we can thereby eliminate model B.

If model C were correct, then one would expect to find short single-stranded tails on the RNA fragments. When RNA labelled with ^{32}P is extracted from the virion and treated with nucleases which attack single-stranded RNA at either the 3' or 5' end, no release of ^{32}P was observed. We can thus rule out model C, suggesting that model D represents the structure of the reovirus genome.

Other RNA viruses which have segmented genomes include phage φ6, which also has double-stranded RNA, the single-stranded RNA animal ortho-, bunya- and arenaviruses and several of the plant virus families (pp. 350–7). However, there is an important difference, as in animal viruses all segments are present within one particle but in plant viruses each particle usually contains only one segment. Plant viruses have up to four RNAs. For example, the four RNAs of brome mosaic virus are contained in three particles, all of which are essential for infectivity. Frequently, the smallest RNA of multicomponent plant viruses duplicates information contained in one of the other segments and is co-encapsidated (Fig. 4.17). The significance of this is not known. In brome mosaic virus, segment 4 is an exact copy of part of segment 3. Details of this and other multicomponent viruses are given in Chapter 21.

HERPESVIRUS DNA HAS A NUMBER OF STRUCTURAL PECULIARITIES

DNA from herpes simplex virus type 1 (HSV-1) is terminally redundant, since incubation of exonuclease-treated DNA under annealing conditions results in the formation of circles (Fig. 4.7). However, HSV-1 DNA is considerably more complex than this. When the isolated *single* strands are self-annealed, two kinds of molecules of particular interest can be seen by electron microscopy. The first of these, designated type A, are those in which one terminus is annealed to an internal region to form a structure consisting of a small single-stranded loop, a short double-stranded region and a long single-stranded tail (Fig. 4.18). Molecules designated type B

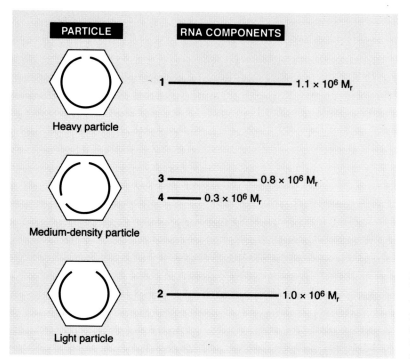

Fig. 4.17 Relationships between the RNA and virus particles of brome mosaic virus. The RNA components are numbered from 1 to 4 on the basis of size.

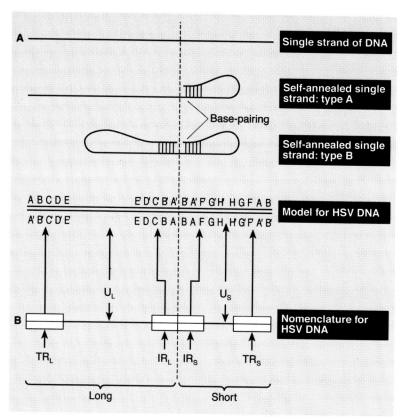

Fig. 4.18 (A) Detection of inverted repeats in herpes simplex virus (HSV) type 1 DNA and the proposed structure of the DNA. Note that the sequence AB appears at both ends of the molecule (terminal redundancy). (B) Conventional nomenclature to describe the HSV genome; L = long, S = short, U = unique sequence, TR = terminally redundant sequence, IR = internally repeated sequence. The genome is not drawn to scale; the short sequence should represent about 7.5%.

are those in which both termini are annealed to internal inverted sequences to yield a structure consisting of a long and a short single-stranded loop bridged by a double-stranded region. These results indicate that the two terminal sequences ABCDE and HGFAB are also found internally in the form of adjacent inverted repeats (Fig. 4.18) and that the unique long DNA segment, called U_L, separating ABCDE from E'D'C'B'A' is longer than the unique short segment, called U_S, which separates B'A'F'G'H' from HGFAB. Although the two terminal sequences are not identical, the sequence AB must be present in both, since the molecule is known to be terminally redundant.

OVERLAPPING GENES

In recent years the genomes of a number of viruses have been completely sequenced and the list of those sequenced grows monthly. The first complete sequence of a viral RNA was that of bacteriophage MS2, which was achieved by W. Fiers group in 1976. In 1977 Sanger and his collaborators published the complete sequence of the genome of the single-stranded DNA bacteriophage φX174. The elegant techniques used in sequencing genomes are too complex to be considered here, but the interested reader should consult Old and Primrose, 1994.

Once the complete sequence of a genome is known, it is possible to scrutinize it for information on its overall genetic organization. For example, it is possible to locate the start signals for each gene, since the initial amino acid in all protein chains is methionine, which is specified by a single codon, ATG. Not all methionine residues are at the start of a protein, so not all ATG sequences will be signals for initiating a protein. Rather, the ATG codon must be preceded by a sequence which tells the ribosome to begin translation. Although the exact sequence of these ribosome recognition signals varies, they are similar enough to be easily recognizable. Once the start of a gene is located, it is possible to locate its terminus by moving along the sequence until one of the three termination codons, TGA, TAA or TAG, in the correct reading frame is encountered.

Detailed studies of a number of bacterial operons suggested that coding regions of DNA were clearly separated by non-coding regions. This idea was strengthened when sequencing of the complete MS2 genome showed that the three genes were separated by two non-coding regions, one 26 nucleotides long, the other 36 nucleotides long. Examination of the complete sequence of the φX174 genome suggests that there is no absolute requirement for non-coding spacer regions between genes. Firstly, in three instances the termination codon of one region overlaps the initiation codon of the next gene, leaving no space for an untranslated region (Fig. 4.19). Thus the end of gene A overlaps the start of gene C, the end of gene C

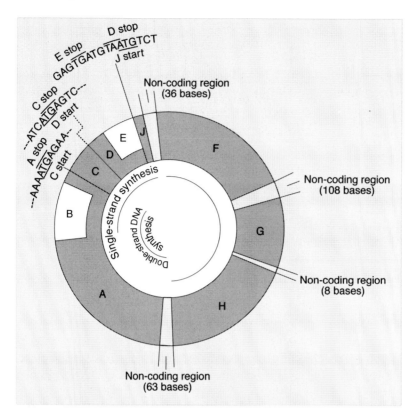

Fig. 4.19 Arrangement of the nine genes of bacteriophage φX174 showing the DNA sequences in the vicinity of the overlaps between genes A, C, D and J. Note that genes B and E are entirely located within genes A and D respectively. Note also that genes with similar function are located adjacent to each other.

overlaps the start of gene D and the end of gene D overlaps the start of gene J. Secondly, and much more surprising, two small genes are entirely contained within two larger genes (Fig. 4.19). Thus gene B is contained within gene A and gene E within gene D. In both cases, however, the smaller genes are encoded in different reading frames from the larger genes (Fig. 4.20).

How were overlapping genes discovered? Conventional genetic mapping of φX174 had located gene E between genes D and J, but correlation of the amino acid sequence of D and J proteins with the nucleotide sequence clearly indicated that genes D and J were contiguous. This apparent paradox was resolved by DNA sequence analysis of an amber (termination) mutant of gene E. In the mutant DNA, a guanine residue was changed into an adenine residue and this change was in a region of the DNA that codes for the D protein. In a particular reading frame, the effect of this change was to convert a tryptophan codon, TGG, into the termination codon TAG. The reason that the D gene product is unaffected in the E gene mutant is simple. The reading frame for the D gene is different (Fig. 4.20) and the change from guanine to adenine converts the leucine codon CTG into CTA, which also codes for leucine. Overlapping genes are now known to occur widely in both pro- and eukaryotic viruses and represent a very economic use of coding potential.

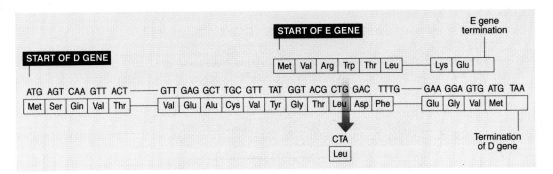

Fig. 4.20 DNA sequence at the start and finish of the overlapping genes D and E showing how the two genes are translated in different reading frames. The shaded guanine residue is the base that is converted to an adenine residue in the E gene amber mutant described in the text.

THE IMPORTANCE OF STRUCTURAL PECULIARITIES

The study of the structural peculiarities of viral nucleic acids is important for several reasons. Firstly, these structural peculiarities, particularly the presence of unusual bases, may help the virus to subvert the cell's biosynthetic machinery and redirect it to the production of new virus. Secondly, any models which may be proposed for the replication of these nucleic acids must take into account their structural peculiarities. In this context, it is worth pointing out that many viral nucleic acids are either circular or capable of circularizing and thus resemble the chromosomes of *E. coli*, *Streptomyces* and *Mycoplasma*. We shall return to this theme in Chapter 7 when we discuss the 'rolling circle' model for DNA replication.

FURTHER READING

Clements, J. B., Cortini, R. & Wilkie, N. M. (1976) Analysis of herpesvirus DNA sub-structure by means of restriction endonucleases. *Journal of General Virology*, **30**, 243–256.

Fraenkel-Conrat, H. (1969) *The chemistry and biology of viruses*. London and New York: Academic Press. (Approximately half of this book is devoted to the chemistry of nucleic acids. Although somewhat dated, it is still very readable.)

McGeoch, D. J. (1984) The nature of virus genetic material. In: 'The Microbe 1984', *Society for General Microbiology Symposium*, **36**, Pt 1, 75–107.

Old, R. W. & Primrose, S. B. (1994) *Principles of gene manipulation* (5th edn). Oxford: Blackwell Scientific Publications.

Sanger, F., Air, G. M. *et al.* (1977) Nucleotide sequence of bacteriophage φX174 DNA. *Nature (London)*, **265**, 687–695.

Sanger, H. L. (1984) Minimal infectious agents: the viroids. In: 'The Microbe 1984', *Society for General Microbiology Symposium*, **36**, Pt 1, 281–334.

Sheldrick, P. & Berthelot, N. (1974) Inverted repetitions in the chromosome of herpes simplex virus. *Cold Spring Harbor Symposium on Quantitative Biology*, **39**, 667–679.

Symons, R. H. (ed.) (1990) Viroids and related pathogenic RNAs. *Seminars in Virology*, **1**, no. 2.

Also check Chapter 21 for references specific to each family of viruses.

5 The process of infection:
I. Attachment and penetration

The process of infection begins with the coming together of a virus particle and a susceptible host cell, but this union comes about by different means with each of the three types of virus, namely, plant viruses, animal viruses and bacteriophages. The initial interaction of phages with bacteria in liquid culture occurs by simple diffusion, since particles the size of bacteria and viruses are in constant Brownian motion when suspended in liquid. Diffusion of animal viruses is probably also the force influencing their union with animal cells in tissue culture, since their attachment is independent of temperature except in so far as this affects Brownian motion. In most cases plants become infected with viruses following mechanical damage to the plant, very often as a result of the wind, or by the activities of virus-carrying insects. Consequently, the way in which the union of virus and cell occurs is not so important in plant systems. However, when viruses are transmitted in plants as a result of grafting, diffusion through the vascular system is most likely to be responsible.

Terminology relating to attachment can be confusing. Consequently we use the terms *virus attachment protein* (of the virus) and *cell receptor unit* (of the cell) to describe the interacting components. In some situations it is useful to refer to *cell receptor sites*, which may be multivalent and consist of several cell receptor units.

ATTACHMENT

Experiments on attachment of phages to bacteria, like that detailed on p. 7, really only measure phage particles that are *irreversibly* bound to bacteria. Phage particles can also be *reversibly* attached and this can be detected by adsorbing phages to cells in suspension and centrifuging the complexes into a pellet. When the pellet is resuspended in fresh medium and the phage–bacterium complexes are again centrifuged down, many infective phages are left in the supernatant. These infective phages, initially reversibly bound to the cells, were released from the cells when the fresh medium was added.

Attachment of bacteriophages to the cell wall

Most bacteriophages attach to the cell wall, but there are other cell receptor units on the pili, flagella or capsule of the host. Most of the tailed bacteriophages attach to the cell wall and do so by the tip of their tail. With those phages whose head is much larger than the tail, diffusion will tend to cause the tail to oscillate more than the head, making it more likely that the tail will collide first with the cell. The chemical nature of the cell receptor unit on the cell wall has been elucidated for some *Salmonella* phages and has proved most instructive. Figure 5.1 shows the structure of the

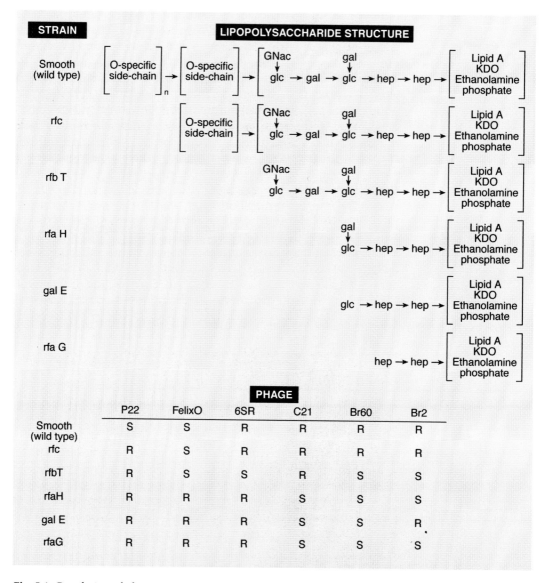

Fig. 5.1 Correlation of phage sensitivity with lipopolysaccharide structure of the cell wall. The upper part of the diagram shows the structure of the lipopolysaccharide in different mutants of *Salmonella typhimurium*.
Abbreviations: glc, glucose; gal, galactose; G Nac, N-acetylglucosamine; hep, heptose; KDO, 2-keto-3-deoxyoctonic acid. The table in the lower part shows the sensitivity of the different mutants to several bacteriophages.
Abbreviations: S, sensitive; R, resistant.

O antigen of wild-type *Salmonella typhimurium* and certain cell wall mutants derived from it. When the mutants are tested for sensitivity to different phages, it becomes apparent that each has a characteristic sensitivity pattern. For example, the presence of the O-specific side-chain makes a cell resistant to phages 6SR, C21, Br60 and Br2 but sensitive to P22 and Felix O. The absence of the O-specific side-chain, as in *rfb* T mutants, now makes the cell sensitive to 6SR, Br60 and Br2 and resistant to P22 and Felix O. Thus, if the lipopolysaccharide is the cell receptor unit for a particular phage, it is possible to determine precisely the composition and structure of the receptor by testing the different mutants for sensitivity to that phage. Experiments of this type are not only of interest to virologists but also to cell wall chemists. For example, suppose we wished to isolate an *rfc* mutant of *Salmonella*. In the absence of a selective procedure, this would be a tedious task. However, if a bacterial lawn showing plaques of P22 is incubated for several days, small colonies arise within the plaques which represent the growth of phage-resistant mutants. It is clear that these resistants could belong to any one of the mutant classes shown in Fig. 5.1. However, by simultaneously selecting for resistance to 6SR, C21, Br60 and Br2, we can eliminate all but the *rfc* class, which should be sensitive to Felix O.

The best-documented example of phage attachment is that of phage T2 and T4. As will be recalled from Chapter 3, these viruses have a complex structure, including a tail, base plate, pins and tail fibres. The initial attachment of these phages to the receptors on the bacterial surface is made by the distal ends of the long tail fibres (Fig. 5.2). The long tail fibres which make the first attachment, bend at their centre and their distal tips contact the cell wall only some distance from the midpoint of the phage particle. After attachment, the phage particle is apparently brought closer to the cell surface. When the base plate of the phage is about 100 Å from the cell wall, contact is made between the short pins extending from the base plate and the cell wall, but there is no evidence that the base plate itself is attached to the cell wall.

Fig. 5.2 Schematic illustration of the major steps in the attachment of bacteriophage T4 to the cell wall of *E. coli*. (A) Unattached phage showing tail fibres and tail pins (cf. Fig. 3.18). (B) Attachment of the long tail fibres. (C) The phage particle has moved closer to the cell wall and the tail pins are in contact with the wall.

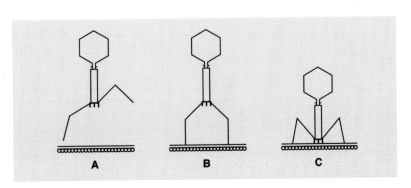

Other attachment sites for bacteriophages

Not all tailed phages attach to the cell wall. Some, such as phage χ(chi) and PBS1, attach to flagella. The tip of the tail of these phages has a kinky fibre on it, which wraps around the filament of the flagellum and the phage then 'slides' down the filament till it reaches the base of the flagellum (Fig. 5.3). Other tailed phages attach to the capsule of the cell.

The other important cell receptor units are on the sex pili. Bacteria which harbour the sex factor (F) or certain colicins or drug resistance factors produce pili, and two classes of phage have been shown to attach to these pili. The filamentous single-stranded deoxyribonucleic acid (DNA) phages attach to the tips of the pili, while the spherical ribonucleic acid (RNA) phages attach along the sides of the pili (Fig. 5.4). These phages are particularly useful to microbial geneticists because they offer a ready means of establishing whether cells harbour pili of these types.

The attachment of animal viruses

A cell cannot be infected unless it expresses the molecule which serves as a receptor for that particular virus on its outer surface. These are usually proteins, but carbohydrates and, very occasionally, lipids are also used (Table 5.1). Receptors are molecules essential

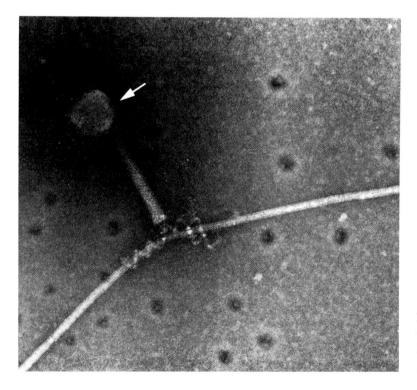

Fig. 5.3 Attachment of a single bacteriophage χ to the filament of a bacteria flagellum (courtesy of Dr J. Adler).

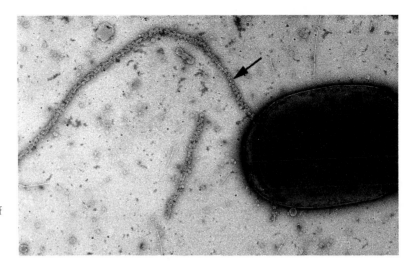

Fig. 5.4 Attachment of many spherical RNA phages to the sex pili of *Escherichia coli* (courtesy of Dr C. C. Brinton).

to the normal functioning of a cell, and viruses have evolved to take advantage of these. Unequivocal identification of protein receptors has come from expressing the gene for the receptor in a cell to which a virus does not normally bind or, alternatively, by blocking attachment with a monoclonal antibody to the receptor. Usually receptors are highly specific for a particular virus. One notable exception is the sugar, *N*-acetylneuraminic acid, which often forms the terminal moiety of a carbohydrate group of a glycoprotein or glycolipid, and this is used by members of several different families of viruses. Occasionally, viruses can use different molecules as receptors (e.g. reoviruses) (Table 5.1).

PENETRATION

Penetration by bacteriophage T2

The experiments of Hershey and Chase (see p. 7) indicated that only nucleic acid entered the cell during bacteriophage T2 infection. The way in which this occurs has now been elucidated and is a complex but fascinating story, which we can only sketch briefly here. The tail of the bacteriophage is contractile and in the extended form consists of 24 rings of subunits surrounding a core. Each ring consists of five subunits of a larger size. Following attachment, the tail contracts, resulting in a merging of the small and large subunits to give 12 rings of 12 subunits. The tail core, which is not contractile, is pushed through the outer layers of the bacterium with a twisting motion, and contraction of the head results in the injection of the DNA into the cell. This process is probably aided by the action of the lysozyme which is built into the phage tail. There are 144 molecules of adenosine 5'-triphosphate (ATP) built into the sheath and the energy for contraction most probably comes from their conversion to adenosine diphosphate (ADP). The phage has

Table 5.1 Molecules used by viruses as receptors.

Molecule	Normal function	Virus
Protein		
ICAM-1	Adhesion to other cells	Rhinoviruses
Unknown; member of the immunoglobulin superfamily	Not known	Poliovirus
MHC I	Ligand for CD8 Presents peptides to T cells	Semliki Forest virus Adenovirus Human cytomegalovirus
MHC II	Ligand for CD4 Presents peptides to T cells	Lactate dehydrogenase-elevating virus
CR2	Receptor for complement component C3d	Epstein–Barr virus
IgA receptor	Binds IgA for transport across the cell	Hepatitis B virus
CD4	Ligand for MHC II	HIV-1, HIV-2, SIV
IgM	B cell receptor	Murine leukaemia virus
IgG bound to cells by Fc receptors	Binds to virus	Dengue virus and many others
β-Adrenergic receptor	Binds the hormone adrenaline	Reoviruses
Acetylcholine receptor	Binds the neurotransmitter acetylcholine	Rabies virus
Carbohydrate		
N-Acetylneuraminic acid	Gives cells their negative charge	Influenza virus A, B, C, paramyxoviruses, polyomavirus, encephalomyocarditis virus, reoviruses
Lipid		
Phosphotidylserine Phosphoinositol	Constituent of cell membrane	Vesicular stomatitis virus

ICAM, intercellular adhesion molecule; MHC, major histocompatibility complex; CR, complement receptor; IgA, immunoglobulin A; HIV, human immunodeficiency virus; SIV, simian immunodeficiency virus.

been likened to a hypodermic syringe. The various steps in penetration are shown in Fig. 5.5.

Penetration by RNA bacteriophages

Many bacteriophages do not possess contractile sheaths and so the way in which their nucleic acid enters the cell is not known. Hershey–Chase-type experiments with the filamentous DNA phages, which attach to the sex pili, suggest that both the DNA and the coat protein enter the cell and it has been postulated that after attachment the phage–pilus complex is retracted by the host

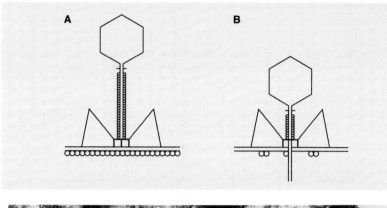

Fig. 5.5 Schematic representation of the mechanism of penetration of the phage T4 core through the bacterial cell wall. (A) The phage tail pins are in contact with the cell wall and the sheath is extended. (B) The tail sheath has contracted and the phage core has penetrated the cell wall; phage lysozyme has digested away the cell beneath the phage. (C) Electron micrograph of T4 adsorbed to an *E. coli* cell wall, as seen in thin section. The needle of one of the phages can be seen to penetrate just through the cell wall (arrow). The thin fibrils extending on the inner side of the cell wall from the distal tips of the needles are probably DNA.

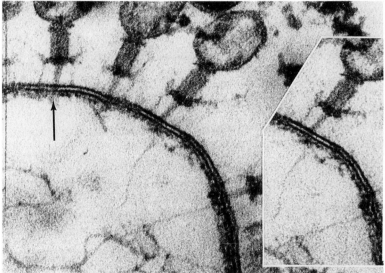

C

cell. However, when similar experiments are performed on the RNA phages, which also attach to the sex pili, the results obtained are similar to those for T2, i.e. the phage coat protein does not enter the cell. Indeed, following attachment, the RNA phages are rapidly eluted again. Examination of the eluted phage in sucrose gradients shows that 70% of them sediment as intact 78S particles, while the remaining 30% lack RNA and sediment as 42S particles. Treatment of the eluted 78S particles with ribonuclease (RNase) converts them to 42S particles, whereas phages which have not been allowed to interact with pili remain RNase-resistant. An analogous sequence of events occurs when poliovirus interacts with cells carrying the appropriate receptors, except that the RNA of the particles is still RNase-resistant.

Since the A protein of the single-stranded RNA phages (see p. 49) is involved in attachment, it is likely that the RNase sensitivity of the eluted 78S particles stems from loss of this protein. It is, in fact, possible to deduce the course taken by the A protein by locating the radioactivity of ^{3}H-histidine-labelled phage following its interaction with host cells. Such experiments are based on the fact that the coat protein of the RNA phages lacks the amino acid histidine, whereas the A protein contains approximately 4.5 residues of this amino acid per molecule. When ^{3}H-histidine-labelled phages are allowed to adsorb and elute from pili, the histidine label is completely absent from the 42S particles and only a small amount is present in the 78S particles. It can thus be concluded that the A protein is released from the phage particles following attachment.

Further experiments on the RNA phages have shown that the viral RNA and the A protein are taken up by the cell in approximately equimolar amounts, and that their kinetics of penetration is similar. This was shown by infecting cells with phage labelled with ^{32}P (RNA) and ^{3}H-histidine (A protein). After attachment of the phage has taken place, the cells are depilated by forcing them through a narrow-gauge hypodermic needle, and freed from the pili by repeated centrifugation and resuspension in fresh medium. The ^{32}P and ^{3}H radioactivity remaining with the cells was determined and, after conversion to phage equivalents, it was clear that similar molar amounts of RNA and A protein were taken up by the cells.

Penetration by animal viruses

Animal cells do not have a rigid cell wall but are bounded by a plasma membrane, which is a very mobile and active structure (Fig. 5.6A). Thus cells are constantly taking samples of their immediate environment by pinocytosis or carrying out the reverse process to export from the cell substances such as enzymes, hormones or neurotransmitters (Fig. 5.6B).

Viruses can only infect a cell if it has an appropriate receptor molecule on its outer surface, and different receptors are required by different viruses. However, the precise details of how viruses enter animal cells after attaching to a cell receptor are not clear. One of the major difficulties in this study is that the majority of virus particles do not successfully initiate multiplication. These 'non-infecting' particles often outnumber the infecting particles by 1000 : 1. Neither electron microscopy nor biochemical studies can distinguish between non-infecting particles and infecting particles, which both contain the viral genome. However, it is clear that enveloped viruses can penetrate into cells by fusing with the plasma membrane, while both enveloped and non-enveloped viruses can penetrate by receptor-mediated endocytosis. Fusion with the plasma membrane (Fig. 5.7A) results in the release of the viral genome into the cytoplasm, but, after endocytosis (Fig. 5.7B), the virus is contained in a vesicle of plasma membrane. Here uncoating is dependent upon the decrease in pH of the internal environment of the vesicle to pH 5–6, which is achieved by fusion with an intracellular vesicle called an endosome. Protons are concentrated in the endosome by means of the proton pump present in its

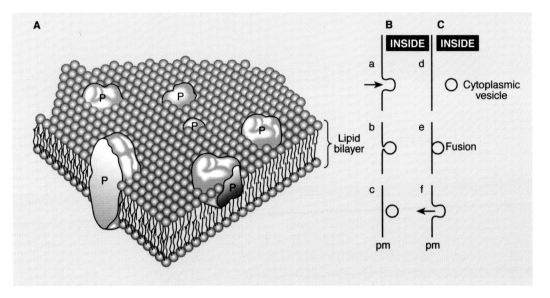

Fig. 5.6 (A) 'Lipid sea' model of plasma membrane structure devised by Singer and Nicholson. Proteins (P) may span the lipid bilayer or may not, and are free to move laterally like icebergs. (B) Pinocytosis by a plasma membrane (pm) inwards (a, b, c) and exocytosis outwards (d, e, f). ((A) redrawn from Singer, S. J. & Nicholson, G. L. (1972) *Science*, **175**, p. 723.)

membrane. The acid pH promotes conformational changes in viral proteins which permit the escape of the viral genome of enveloped viruses by fusion with the wall of the vesicle and of non-enveloped viruses by permeabilizing the particle (Fig. 5.7).

It should be pointed out that penetration and uncoating do not end with naked nucleic acid free in the infected cell. The nucleic acid may remain associated with internal structural proteins, which could include enzymes necessary for nucleic acid synthesis, or, if the genome also serves as message, will associate with cellular structures such as ribosomes. Equally, the nucleic acid will not be stretched out (as illustrators frequently draw it!) but together with its associated proteins will have a defined conformation upon which its functions depend.

One of the striking aspects of penetration of animal cells by viruses is its inefficiency. In poliovirus infections, the majority of the RNA of the infecting virus is degraded as a result of interaction with cells, presumably because only a few of the many viral receptors lead on to successful infection. Thus we can account for the very high physical particle : infectious particle ratio mentioned earlier.

An unusual situation is the uncoating of poxviruses. This takes place in two stages, as shown by the following experiment. Cells were infected with rabbit poxvirus, which was labelled in both the nucleic acid and phospholipid components, and the susceptibility

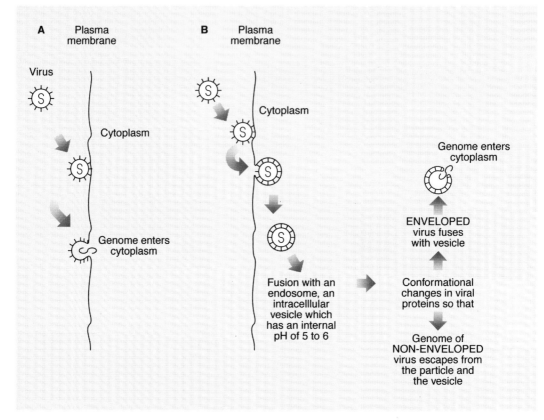

Fig. 5.7 Penetration and uncoating of viruses in animal cells.
(A) By fusion of enveloped viral and plasma membranes.
(B) By endocytosis followed by a decrease in the pH of the vesicle, which promotes conformational changes in viral proteins, leading to the release of the genome. Note the differences between enveloped and non-enveloped viruses.

of the labelled DNA to the action of deoxyribonuclease (DNase) was used as an index of uncoating. When cells were disrupted at intervals following infection, it was observed that, whereas phospholipid is released without appreciable delay, release of the DNA, in a form that was susceptible to DNase, did not occur for 1–2 hours. The lag in appearance of DNase sensitivity could be abolished by infecting cells with unlabelled virus before adding labelled virus. However, the lag could be prolonged indefinitely by the addition of puromycin and fluorophenylalanine (inhibitors of protein synthesis) or actinomycin D (inhibitor of RNA synthesis). An explanation for these results is provided by the isolation from highly purified virions of a DNA-dependent RNA polymerase activity which can transcribe viral DNA. Release of the DNA in a form which is susceptible to DNase probably depends on the synthesis of an enzyme whose mRNA is transcribed from viral

DNA early in infection by the virus-associated polymerase. Release of phospholipid, on the other hand, is mediated by a pre-existing cellular enzyme.

Penetration of plants by viruses

The infection of plants occurs by quite different means from those we have discussed for bacteria and animals. Most plants have rigid cell walls of cellulose and consequently viruses must be introduced into the host cytoplasm by some traumatic process. Thus plant viruses can be introduced by *vectors*, namely, animals that feed on plants or invading fungi, or by mechanical damage caused by the wind or passing animals, all of which allow viruses to enter the thin-walled cells of internal tissues. This is why the application of carborundum to leaves increases the number of lesions produced during a local lesion assay of virus-containing material (see Fig. 1.1B). Many plants that become infected naturally do so because virus-carrying animals feed upon them. However, this transmission is not a casual process which occurs whenever any animal chances to feed on an uninfected host plant after feeding on an infected one. Rather, the transmission of most plant viruses is a highly specific process, requiring the participation of particular animals as vectors. Although some viruses, such as tobacco mosaic virus (TMV), require no vector and can be transmitted mechanically, most have a specific association with their animal vectors, which may include leafhoppers, aphids, thrips, whiteflies, mealy bugs, mites and nematodes. It should be pointed out that most of these animals feed by piercing plant tissues with their mouthparts and not by biting them. Biting insects have limitations as vectors, since biting not only damages leaves excessively but is a method most likely to succeed only with viruses that can be spread by mechanical inoculation. Once inside the cell, uncoating of the virus probably takes place in a similar fashion to that of animal viruses (see Wilson 1985; 1992).

Arthropod-transmitted animal viruses

The infection of animals with *arboviruses* resembles in some ways the infection of plants by the feeding activities of virus-carrying insects. The word arbovirus is an abbreviation for 'arthropod-borne virus of vertebrates' and includes members of such diverse groups as the Rhabdoviridae, Bunyaviridae, Reoviridae (orbiviruses), Alphaviridae and Flaviviridae. These viruses are maintained in nature principally through biological transmission between suscep-tible vertebrate hosts by haematophagous arthropods. The virus cannot be transmitted immediately to a new host following feeding on a viraemic host, since the virus must multiply in the vector before transmission can occur. However, while virologists first learned of these viruses as a result of people becoming infected, usually humans are incidental to the natural insect–vertebrate relationship. The natural vertebrate host of these viruses, which occur mostly in the tropics, is commonly a rodent, a monkey or a bird. Infection of humans appears to be accidental and may well break the chain of transmission.

PREVENTION OF THE EARLY STAGES OF INFECTION

One of the goals of studying the cell–virus relationship is to develop methods of aborting viral infections, particularly those of humans and their domestic animals, and how better than by preventing virus from attaching and penetrating. Cellular receptors are likely to be proteins or glycoproteins, which perform functions vital to the normal metabolism of the cell and serve only incidentally the needs of the virus. Thus these cannot be attacked without endangering the cell itself. As to the viruses there are only a few atomic resolution, three-dimensional structures of virions/coat proteins currently known, so the search for antiviral drugs which inhibit attachment and penetration can only be achieved by random screening. So far, such procedures have only uncovered one drug suitable for clinical use. This is amantadine, which has been used to prevent influenza. Amantadine acts at high concentrations by elevating the internal pH of endosomes, which prevents the low pH-mediated fusion of virus and vesicle membrane referred to above, but at low concentrations it affects the virion M2 envelope protein which functions as a proton channel. This prevents proper uncoating of the infecting virion, and also upsets the pH environment of the nascent HA protein so that it converts to the low pH-conformation within the cell and becomes non-functional.

For the protection of plants, it should be apparent that the best preventive measure is a reduction in the number of appropriate vectors, but this is not necessarily an easy task. Bacteriophages can also cause problems, since they are capable of infecting certain organisms of industrial fermentations, resulting in lysis and loss of product. The best measure to adopt here is the use of resistant strains and the simple measure of reducing the concentration of divalent cations, since the latter are frequently important for phage attachment.

FURTHER READING

Bretscher, M. S. & Pearse, B. M. F. (1984) Coated pits in action. *Cell*, **38**, 3–4.

Dimmock, N. J. (1982) Initial stages of infection with animal viruses. *Journal of General Virology*, **59**, 1–22.

Fuchs, P. & Kohn, A. (1983) Changes induced in cell membranes adsorbing animal viruses, bacteriophages and colicins. *Current Topics in Microbiology and Immunology*, **102**, 57–99.

Kohn, A. (1985) Membrane effects of cytopathogenic viruses. *Progress in Medical Virology*, **31**, 109–167.

Lentz, T. L. (1990) The recognition event between virus and host cell receptor: a target for antiviral agents. *Journal of General Virology*, **71**, 751–766.

Lindberg, A. A. (1977) Bacterial surface carbohydrate and bacteriophage adsorption. In: *Surface carbohydrates of the prokaryotic cells*, pp. 289–356. Sutherland, I. W. (ed.). London and New York: Academic Press.

Marsh, M. & Pelchen-Matthews, A. (1993) Entry of animal viruses into cells. *Reviews in Medical Virology*, **3**, 173–185.

Nomoto, A. (ed.) (1992) Viral receptors and cell entry. *Seminars in Virology*, **3**, no. 2.

Randall, L. L. & Philipson, L. (1980) *Virus receptors. Part 1: Bacterial viruses*. London: Chapman and Hall.

Wilson, T. M. A. (1985) Nucleocapsid disassembly and early gene expression by positive-strand RNA viruses. *Journal of General Virology*, **66**, 1201–1207.

Wilson, T. M. A. (ed.) (1992) Early events in RNA virus infection. *Seminars in Virology*, **3**, no. 6.

Also check Chapter 21 for references specific to each family of viruses.

6 The process of infection: IIA. The Baltimore classification

Viruses exhibit diversity of morphologies, nucleic acid structure, mode of infection, regulation of development, etc. In such circumstances, it might be thought impossible to uncover any unifying concept which would simplify a discussion of the process of replication. However, Nobel laureate David Baltimore has proposed an elegant classification of all viruses, based on the mode of gene replication and expression. In this classification, messenger ribonucleic acid (mRNA) is assigned a central role, since protein synthesis takes place by the same mechanism in all cells. All viruses are then divided into groups, assignment to a group being determined by the pathway of mRNA synthesis (Fig. 6.1).

In order to maintain unity, all mRNA is designated as 'plus' RNA. Strands of viral deoxyribonucleic acid (DNA) and RNA which are complementary to the mRNA are designated as 'minus' and those that have the same sequence are termed 'plus'. Using this terminology, six groups of viruses can be distinguished:

Class I consists of all viruses that have a double-stranded DNA genome. In this class the designation of 'plus' and 'minus' is not meaningful since different mRNA species may come from either strand.

Class IIa consists of viruses that have a single-stranded DNA genome of the same sense as mRNA.

Class IIb viruses have DNA complementary to mRNA. These do not have a class to themselves as they (the autonomous parvoviruses) were discovered after Baltimore devised the classification. Before the synthesis of mRNA can proceed, the DNA must be converted to a double-stranded form.

Class III consists of viruses that have a double-stranded RNA genome. All known viruses of this type have segmented genomes but mRNA is only synthesized on one strand of each segment.

Class IV consists of viruses with a single-stranded RNA genome of the same sense as mRNA. Synthesis of a complementary strand precedes synthesis of mRNA (positive-strand RNA viruses or RNA$^+$).

Class V consists of viruses that have a single-stranded RNA$^+$ genome which is complementary in base sequence to the mRNA (negative-strand RNA viruses or RNA$^-$).

Class VI consists of viruses that have a single-stranded RNA genome and which have a DNA intermediate during replication.

Because the replication of class II viruses involves a double-stranded DNA intermediate, they will be considered along with class I viruses in Chapter 7. For similar reasons, the replication of classes III, IV and V will be discussed together in Chapter 8. Class VI RNA viruses have a DNA intermediate and will be the subject

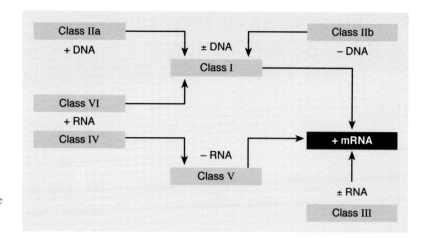

Fig. 6.1 The Baltimore classification (see text for details).

of Chapter 9, together with 'mirror-image' DNA viruses, which have an RNA intermediate in replication.

Finally, it must be emphasized here that classifications of viruses based on other criteria are equally valid and frequently far more useful. A weakness of Baltimore's scheme is that it takes no account of many important properties. Thus, classified together in class I are bacteriophage T2 and variola virus (the cause of small-pox), agents which are totally dissimilar in structure and biology. The reader should refer to other texts (below) and Chapter 21 to see how the problems of classification have been approached by 'non-molecular' virologists.

FURTHER READING

Baltimore, D. (1971) Expression of animal virus genomes. *Bacteriological Reviews*, **35**, 235–241.

Francki, R. I. B., Fauquet, C. M., Knudson, D. L. & Brown, F. (eds) (1991) *Classification and nomenclature of viruses* (Archives of Virology Supplementum 2; 5th edn). Vienna: Springer-Verlag.

7 The process of infection: IIB. The replication of viral DNA

The basic mechanism whereby a deoxyribonucleic acid (DNA) molecule is duplicated in a cell is the same, regardless of whether the DNA is of cellular or viral origin. Indeed, because viral DNA molecules can be manipulated with ease they have often been the substrate preferred by biochemists attempting to unravel the mysteries of DNA replication. At first sight, the replication of DNA would appear to be a simple process. A polymerase traverses the molecule and makes two daughter strands, using the parental strands as templates (Fig. 7.1), but there is still much to learn. This simple model for DNA replication implies that each of the daughter molecules contains one parental strand and one newly synthesized strand, i.e. that DNA replication is semi-conservative. Proof of this was obtained from the now classical Meselson–Stahl experiment.

PROBLEMS WITH POLYMERASES

Since the two strands of a DNA duplex are antiparallel, a second implication of our model for DNA replication is that one of the daughter strands is synthesized in the 3' to 5' direction and the other in the 5' to 3' direction (Fig. 7.1). However, of the many cellular and viral-specified DNA polymerases which have been purified, none has the capacity to synthesize DNA in the 3' to 5' direction. All are incapable of adding monodeoxyribonucleotides to 5' hydroxyl termini. One solution to the problem is for synthesis to proceed in the 5' to 3' direction along one parental strand, the *leading* strand, and for discontinuous 5' to 3' synthesis to occur along the other, or *lagging*, strand (Fig. 7.2). The fragments produced by discontinuous synthesis could ultimately be tied together by a ligase.

Evidence supporting discontinuous synthesis has been obtained by Okazaki. A culture of *Escherichia coli* which had been infected with phage T4 was pulsed with radioactive thymidine and the resulting labelled DNA examined by velocity sedimentation in a sucrose gradient. Immediately after the pulse, most of the label was found in DNA fragments (called 'Okazaki fragments') approximately 1000–2000 nucleotides long. By 1 minute later, the radioactivity had been chased into material which sedimented much faster, as would be expected if the fragments had been covalently linked together by ligase action.

A second problem with our simple model is the apparently

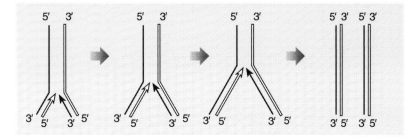

Fig. 7.1 Simple model for DNA replication. In this and subsequent diagrams, the arrow heads indicate the positions of the growing points. Note that, in the model shown, one strand is synthesized in the 5' to 3' direction and the other in the 3' to 5' direction, but no known enzyme can carry out the latter function.

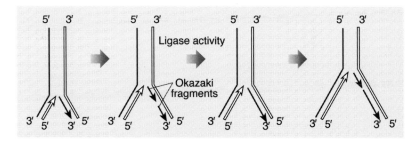

Fig. 7.2 Model for discontinuous synthesis of DNA. Both strands are synthesized in a 5' to 3' direction but only one strand is synthesized continuously. Experimental evidence (see text) suggests that both strands may be synthesized discontinuously.

perverse inability of known DNA polymerases to start DNA chains *de novo*. Such new chain starts are required at frequent intervals on the lagging side of the fork. A solution to our problem is to invoke the action of ribonucleic acid (RNA) polymerase, which does not require a primer since its basic function is to start poly-nucleotide synthesis. This enzyme could synthesize a short RNA primer on the lagging strand and this could be extended by DNA polymerase and ligated to the previously synthesized Okazaki fragment (Fig. 7.3). In support of this model is the observation that the antibiotic rifampicin, which inhibits the enzyme RNA polymerase, is known to inhibit the conversion of single-stranded (SS) M13 DNA to the double-stranded form. No such inhibition is observed in an *E. coli* mutant with a rifampicin-resistant RNA polymerase. Furthermore, many DNA polymerases do not discriminate against extending a chain with a ribonucleotide at the growth end. Clearly, the synthesis of each Okazaki fragment will require a short RNA primer and this primer has to be excised after synthesis of the fragment and the space left filled by extension of the next fragment.

Experimentally it is found that remnants of the RNA primer disappear so rapidly that one suspects that primer degradation may be tightly coupled to primer utilization. Both processes might be catalysed by the same enzyme complex, thereby maximizing the efficiency of primer removal.

WHY THE REQUIREMENT FOR AN RNA PRIMER?

The model of DNA replication shown in Fig. 7.3 is so much more complex than that shown in Fig. 7.1 that one wonders if it is possible that nature has been foolish and wasteful in the creation of DNA replication systems. This is unlikely and a rationale for the complexity is provided by the observation that the fidelity of DNA replication is such that only one mistake is made in 10^9-10^{10} base-pair replications, compared with one in 10^3-10^4 for RNA genomes. Error-free replication arises from the ability of DNA polymerases to 'proof-read' the DNA which they have just synthesized. For example, T4 and *E. coli* DNA polymerases have a very strong requirement for a Watson−Crick base-paired residue at the 3′ OH primer terminus to which nucleotides are being added. When confronted with a template primer with a terminal mismatch, these polymerases make use of their built-in $3′ \rightarrow 5′$ exonuclease activities to clip off unpaired primer residues by hydrolysis until a base-paired terminus is created. As a result, these polymerases will efficiently remove their own polymerization errors. This self-correcting feature allows DNA polymerase to select for the proper template base-pairing of each added nucleoside triphosphate in a separate backward reaction, in addition to its strong selection for base-pairing of nucleoside triphosphates during the initial polymerization. If DNA polymerase had to initiate new DNA chains in the absence of a primer, this proof-reading function would not be operative at the initiation regions. In contrast, RNA polymerases need not be self-correcting, inasmuch as relatively high error rates can be tolerated during transcription. Indeed, it seems reasonable to suggest that the use of ribonucleotides in the synthesis of the primer was an evolutionary step forward, for it automatically marks these sequences as 'bad copy' to be removed.

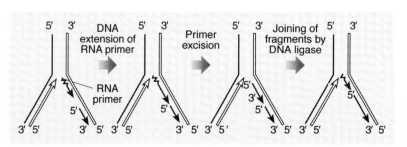

Fig. 7.3 Involvement of an RNA primer in synthesis of Okazaki fragments.

THE MECHANISM OF PRIMING

The small DNA phage, M13, has been of great use in studying the priming of DNA synthesis, as its SS circular genome has no free 3' OH termini. When the conversion of M13 SS DNA to the double-stranded replicative form (RF) was followed in crude extracts, two stages could be recognized. The initial stage involved RNA synthesis and produced primed SS molecules which could be isolated. Such primed SS molecules were converted to RF molecules in a second stage in the absence of ribonucleotide triphosphates even if rifampicin was present. Resolution and purification of the components needed for the conversion of M13 SS to RF disclosed requirements for five proteins (Fig. 7.4). RNA polymerase binds to M13 SS DNA, but only in the presence of a DNA unwinding protein, and synthesizes a short RNA primer. The RNA-primed SS DNA is extended to form RF II molecules by the *E. coli* DNA polymerase III holoenzyme. DNA polymerase I then excises the RNA primer and fills the gap thus created, and DNA ligase joins the two ends of the newly synthesized polynucleotide.

Studies on the replication of polyomavirus DNA have provided some of the most convincing evidence that nascent DNA fragments are primed by RNA *in vivo*. The replication of polyomavirus DNA, which is a circular duplex of M_r 3×10^6 and similar in size to ϕX174 RF, proceeds discontinuously with synthesis of small DNA fragments, 100–140 nucleotides long. Such fragments synthesized in mouse cells, or in nuclei isolated from them, appear to have RNA at their 5' ends, but with no sequence specificity at the covalent linkage of RNA to DNA. The RNA primer appears to be a decanucleotide with adenosine 5'-triphosphate (ATP) or guanosine triphosphate (GTP) at the 5' terminus.

PROBLEMS WITH PRIMERS

The requirement for a primer causes a further complication which we have so far ignored. If synthesis in the 5' to 3' direction is initiated with an RNA primer which is later digested away, there exists no mechanism for filling the gap left by the primer at the 5' end of the leading strand. To fill this gap would require 3' to 5' synthesis and we know that this cannot occur. When the replicating fork reaches the other end, a similar problem arises with the lagging strand (Fig. 7.5). The net result is the synthesis of two daughter molecules, each with a 3' SS tail. If such molecules were to undergo further rounds of replication the result would be smaller and smaller 3' tailed duplexes. One solution to this problem is concatemer formation.

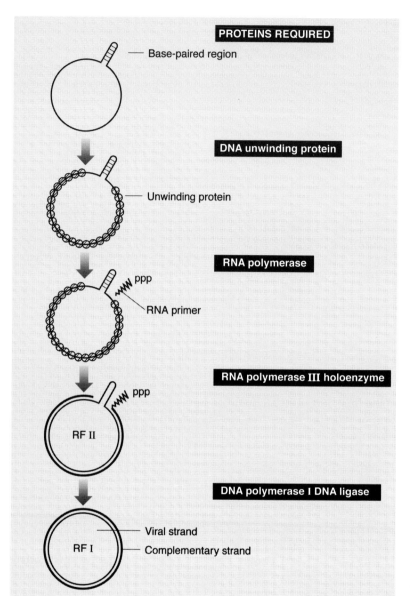

Fig. 7.4 Proteins required for the conversion of phage M13 single-stranded DNA to replicative form I.

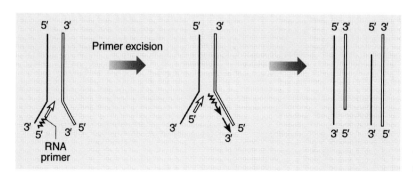

Fig. 7.5 Formation of two daughter molecules with complementary single-stranded 3′ tails.

Concatemer formation

It will be recalled from Chapter 4 that many viruses have terminally repetitious genomes and that exonuclease treatment of such genomes results in the production of 'sticky ends', enabling concatemers to be produced. Consequently, if a pair of tailed molecules are produced during replication, they too should be able to form concatemers. Concatemer formation would be independent of the length of the terminal repetition and the replication tail. If the tails were longer than the redundant region, then gaps would be produced which could only be filled with the aid of a DNA polymerase and a DNA ligase. If the tails were shorter than the redundant stretch, then concatemer formation would be the result of exonuclease and ligase action (Fig. 7.6).

Formation of unit-length molecules would occur by the reverse process to concatemer formation. Two specific enzymatic breaks would be introduced, one in each strand but at opposite ends of the repetitious sequence (Fig. 7.7). Creation of such cuts would probably not cause separation of the two halves, since there would be enough hydrogen bonds to hold them together. If the cuts were introduced such that the longer of the two chains has a 5′ terminus, then a DNA polymerase could add deoxyribonucleotides to the 3′ hydroxyl end. This would displace a 5′-ended tail. When the growing 3′ ends meet somewhere in the middle of the redundant section, there would no longer be any hydrogen bonds holding the two halves together; they would thus separate. The two halves could either reform into a hydrogen-bonded concatemer or else be converted to a mature terminally repetitious molecule by the continuing action of DNA polymerase (Fig. 7.7).

Experimental evidence in favour of the above model comes from a study of T7 replication. If T7-infected cells are pulsed with ^{3}H-thymidine, all of the label is found in phage DNA, since host DNA synthesis ceases shortly after infection. When analysed in sucrose gradients, most of this label sediments faster than DNA from purified T7. Measurement of the sedimentation rate of labelled molecules indicates that they are dimers, trimers, tetramers, etc. of T7 DNA. Furthermore, the label can be chased from such concatemers into molecules the same size as mature phage DNA.

Fig. 7.6 Dimer formation by molecules with complementary tails. (A) Molecule in which the tails are longer than the redundant region. (B) Molecules in which the tails are shorter than the redundant region.

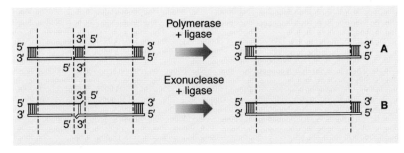

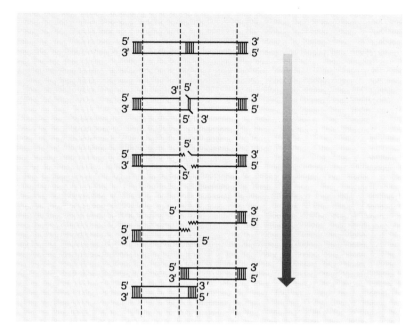

Fig. 7.7 Conversion of dimer to two unit-length molecules with terminally redundant ends. Terminally redundant regions of the molecule are represented by hatching.

The smaller concatemers have been subjected to intensive analysis and the results support the model just outlined. Denaturation of dimers and trimers and sedimentation of the resultant chains in alkali revealed unit-length molecules, as expected. Chromatographic separation of sheared monomers, dimers and trimers showed that they have SS regions spaced at intervals corresponding to the length of a mature phage DNA molecule. Furthermore, annealing experiments with exonuclease-treated phage DNA demonstrated that SS regions correspond to the ends of a mature molecule. Finally, studies with an endo-nuclease and an exonuclease specific for SS DNA showed that both tails and gaps are present in the concatemers.

The model represented above should be applicable to any virus with a terminal repetition, e.g. the T-even phages, P22 and SP50. However, some of these viruses have genomes which are circularly permuted and, during the formation of 'phage lengths' of DNA, the cutting nuclease apparently does not recognize any specific sequence. Instead, some spatial factor probably fixes the site of nuclease action. In the case of T4, empty head shells are assembled and then subsequently filled with DNA. When a headful has been packaged, a nuclease snips off the remaining unpackaged conca-temer. This model is supported by the observation that aberrant T4 particles with giant heads contain continuous lengths of DNA much longer than the usual genome.

THE REPLICATION OF CIRCULAR DNA

Both SS and double-stranded circular DNA genomes have been found in viruses. Since the replication of SS DNA always involves a double-stranded intermediate, we shall first turn our attention to the replication of double-stranded circular genomes, as exemplified by bacteriophage λ.

About 30 minutes after infection with λ, pulse-labelled DNA from infected cells can be separated into three classes by sedimentation — linear monomers, covalently closed circles and concatemers. Most of the label is found in the concatemers and can be chased into linear monomers. What type of structure is the concatemer and how might it be generated? The two simplest concatemeric structures are linear multimers and multimeric circles, and ways of generating such structures are shown in Fig. 7.8. Concatemers may also be generated by replication of a rolling circle (Fig. 7.8D), in which the circle corresponds to a unit genome and the tail is a linear multimer. These concatemers are probably produced by replication, for they are still formed when a recombination-deficient mutant of λ infects a recombination-

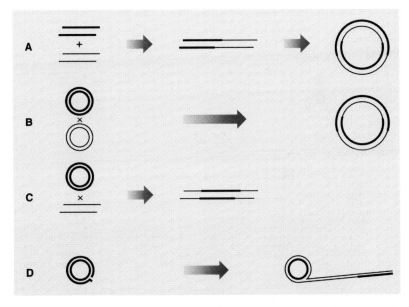

Fig. 7.8 Models for the generation of concatemers. (A) End-to-end joining of monomers. Circular concatemers can be generated by joining the two free ends of the linear form. (B) Reciprocal recombination between circular monomers generates a circular dimer. (C) Reciprocal recombination between a circular monomer and a linear monomer generates a linear dimer. (D) Replication of a rolling circle generates a structure in which the circular duplex is monomeric in length but the tail may be of any length, depending upon how long replication has proceeded.

deficient host cell, a procedure which ensures the absence of recombination functions.

Figure 7.9 summarizes the replication cycle of λ. The DNA in the virion is a linear monomer, which is converted to a closed circle after infection. During the period of early replication, λ DNA replicates in a circular form, usually bidirectionally, to generate circular progeny. Late replication is initiated by the conversion of circular DNA to the rolling circle, but the events responsible for this transition are not known. Replication of the rolling circle generates a concatemeric tail, which is cleaved during packaging into molecules of the correct length and possessing cohesive ends. The bacteriophage λ thus uses two strategies to overcome the problem of SS tails at the 5′ end of the leading DNA strand. By replicating via a circular molecule early in infection, the gap formed by removal of the RNA primer is filled by the polymerase, after circumnavigating the circle. Late in infection, the synthesis of concatemers again prevents any problems arising, by the removal of the terminal RNA primers.

Replication of single-stranded circular genomes

The genomes of some bacterial viruses consist of SS circular DNA (see p. 66), but the only well-studied example is that of bacteriophage φX174. In all cases, the infecting single strand is converted to a

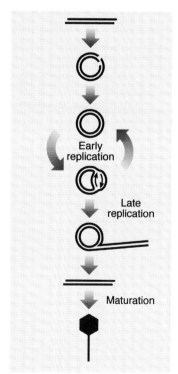

Early replication

Late replication

Maturation

Fig. 7.9 A model for the replication of bacteriophage λ DNA. Upon entering the cell, the linear duplex is converted to an open circle and then to a closed circle. In early replication, the circular template assumes a θ (theta) shape and replication proceeds bidirectionally. This generates circles, which do not act as precursors for mature linear duplexes. During late replication, there is a change in the mode of replication to a rolling circle, which generates a concatemeric linear tail, which is cleaved during packaging into virus particles.

double-stranded RF by the mechanism outlined on p. 102. This parental RF then replicates to generate progeny RF, but at present it is not clear whether this occurs via a theta structure or a rolling circle, since evidence in favour of both models has been presented. What is certain is that the leading strand is synthesized on the complementary strand of the RF, while the Okazaki fragments are synthesized on the viral strand of the RF (Fig. 7.10). This has been shown in a number of different ways, the simplest being the observation that denaturation of RF synthesized in cells in the absence of DNA ligase releases short fragments from the complementary strand, while the viral strands are of unit length.

THE REPLICATION OF LINEAR SINGLE-STRANDED DNA

A linear SS genome is found in both the autonomous and the defective parvoviruses. The autonomous parvoviruses, such as the minute virus of mice (MVM), use host cell enzymes for replication and transcription and package a minus DNA strand. Defective parvoviruses, such as the adeno-associated viruses (AAV), are entirely dependent on adenovirus coinfection for their own replication and package both plus and minus DNA strands in separate virions. The parvovirus genome contains terminal hairpins

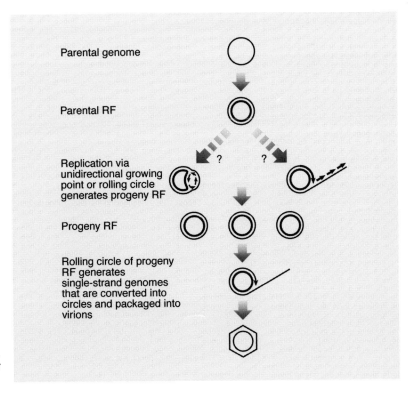

Parental genome

Parental RF

Replication via unidirectional growing point or rolling circle generates progeny RF

Progeny RF

Rolling circle of progeny RF generates single-strand genomes that are converted into circles and packaged into virions

Fig. 7.10 The replication of bacteriophage φX174. The encapsidated DNA strand is represented by the thin line.

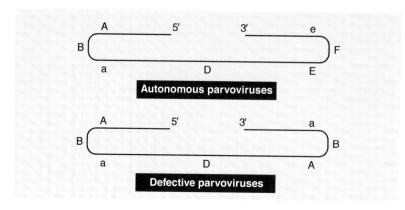

Fig. 7.11 Schematic representation of the genome structure of autonomous and defective parvoviruses. In this and the subsequent two figures, the letters A and a, etc., represent complementary sequences.

(Fig. 7.11), and this is an important feature, for it provides a 3′ OH terminus at which DNA elongation can be initiated. However, the overall genome structure is different for the defective and autonomous parvoviruses (Fig. 7.11), and this provides an explanation for the observation that viruses of one group package both strands of DNA and the others do not.

Cells infected by either type of parvovirus accumulate double-stranded forms of DNA, which appear to be intermediates in replication. A large fraction of such double-stranded molecules cannot be irreversibly denatured, suggesting that the two strands are covalently linked. Linear dimeric double-stranded DNA molecules which spontaneously renature have also been observed. The monomeric and dimeric double-stranded DNAs appear to involve plus and minus strands linked end to end: one of each in the monomer and four alternating plus and minus strands in the dimer. These observations suggested a model for autonomous parvovirus DNA replication (Fig. 7.12). The first step involves gap-fill synthesis to generate complement d of sequence D. This is followed by displacement synthesis and gap-fill synthesis, to copy the 5′ terminal hairpin sequence aBA. The structure now undergoes rearrangement to form a 'rabbit-eared' structure which recreates the hairpin originally present at the 5′ end of the parental genome. More important, it simultaneously creates a copy of this hairpin at the 3′ end of the complementary strand, which can serve as a primer. The dimer-length duplex is then completed by displacement synthesis. The resulting molecule comprises a single polynucleotide chain from which two viral genomes could be generated by endonuclease action. However, in the absence of endonuclease action, continued replication could occur to generate larger multimers (as shown in Fig. 7.12).

The dimers and multimers created by the process just described could serve as replicative intermediates from which progeny viral DNA would be excised by a similar method of displacement synthesis. An SS break would be introduced at the 5′ end of a genome within the concatemer. The 3′ OH terminus then acts as a primer for displacement synthesis of progeny DNA strands, possibly driven

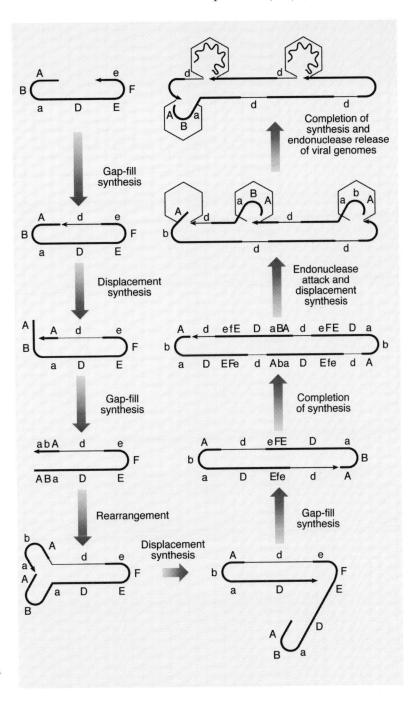

Fig. 7.12 A scheme for the replication of an autonomous parvovirus (see text for details).

by the packaging process. After the 3' terminal sequence e had been displaced, excision of the progeny genome could be completed by another site-specific endonuclease, resulting in the release of an intact virus particle and termination of the displacement synthesis. This model neatly accounts for two experimental observations.

Firstly, duplex molecules with SS tails have been observed in cells infected with MVM, an autonomous parvovirus. Secondly, no free SS DNA has been detected in such infected cells.

SYNTHESIS OF DNA CONTAINING MODIFIED BASES

You will recall from Chapter 4 that the DNA from certain bacterio-phages contains modified bases. For example, 5-hydroxymethyl-cytosine (HMC) replaces cytosine in T-even phage DNA, and some of these HMC residues are glucosylated. How is cytosine excluded from the DNA of these phages and how are the modified bases synthesized? The asking of these and other questions about the biosynthesis of the DNA of the T-even phages stimulated several laboratories to study the biochemistry of virus infection. The outcome was the identification of a whole series of new enzymes. Since these enzymes must be synthesized prior to the initiation of DNA replication, they make their appearance early in the infection process. In fact, they constitute the bulk of the early proteins identified by Hershey (p. 10).

Approximately 4 minutes after infection, an activity appears in *E. coli* infected with T2 that cleaves dCTP and dCDP to dCMP. This appears to be the major mechanism by which dCTP is prevented from being incorporated into phage DNA (Fig. 7.13). The function of this dCDP-dCTPase is the prevention of accumulation of dCTP as a DNA polymerase substrate and the maintenance of a reasonable pool of dCMP. The utilization of dCMP is via two reactions: deamination to dUMP, which is a precursor of dTMP, and hydroxy-methylation to dHMCMP. Hydroxymethylation of dCMP is a

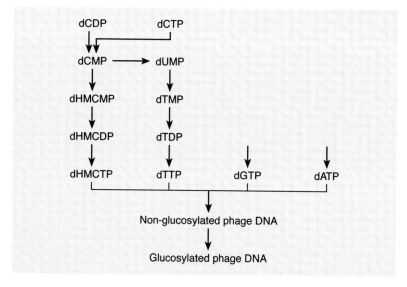

Fig. 7.13 Biosynthesis of phage T2 DNA and the exclusion of cytosine (see text).

reaction unique to cells infected with T-even phages and requires tetrahydrofolic acid. Hydroxymethylase activity is found neither in uninfected *E. coli* nor in cells infected with T5, whose DNA contains cytosine.

If an extract of a T-even phage-infected *E. coli* is assayed for DNA polymerase, very little activity is found. However, if dCTPase is inhibited by fluoride, or dHMCTP is used instead of dCTP, DNA polymerase activity can be detected. Since this enzyme is not inhibited by antiserum prepared against *E. coli* DNA polymerase and has different template requirements, it appears to be a new enzyme. Failure of the T2-induced polymerase to utilize dCTP further ensures that cytosine is absent from phage DNA. However, this polymerase does not produce T2 DNA, since the HMC residues are not glucosylated. Glucosylation is achieved through the appearance of yet another set of phage-specified enzymes, which catalyse the transfer of glucose from uridine diphosphate (UDP)-glucose to the HMC residues in the DNA.

All of the reactions described above are outlined in Fig. 7.13. Similar enzymes are known to be responsible for the synthesis of the other modified bases found in phage DNA molecules (Table 4.2).

HOST-CONTROLLED RESTRICTION AND MODIFICATION

The phenomenon of host-controlled restriction is most easily observed as the inactivation of bacteriophages following transfer from one bacterial strain to another. Thus phage λ grown on strain

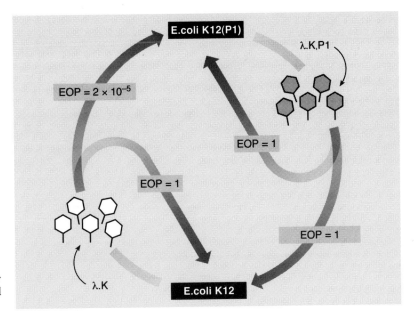

Fig. 7.14 Restriction of phage by *E. coli* strain K12 (P1). The efficiency of plating is abbreviated to EOP.

K12 *E. coli* (λ.K) forms plaques on strain K12 (P1) but with a much lower efficiency. Since λ.K enters both strains of cell equally well, multiplication of the phage must be restricted in strain K12 (P1). Those phages (λ.K, P1) that do survive restriction and are propagated on strain K12 (P1) are modified in such a way that they are no longer restricted by this host (Fig. 7.14). By labelling λ.K DNA with ^{32}P, it has been shown that the reduced efficiency of plating on strain K12 (P1) is due to the degradation of phage DNA shortly after infection. Since no degradation occurs upon infection of strain K12, restriction must act at the level of DNA. As we shall see later, modification also acts at this level.

Most of the progeny issuing from strain K12 after infection with λ.K, P1 have lost the capacity to grow on K12 (P1). Apparently, those phages which can still grow on K12 (P1) have at least one

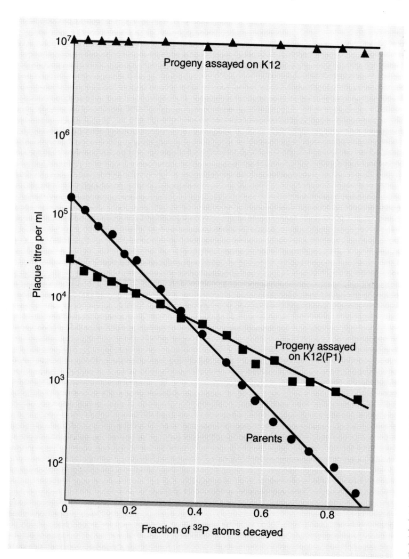

Fig. 7.15 Correlation between ability to grow on K12 (P1) and sensitivity to autoradiolysis by ^{32}P incorporated in progeny phage particles produced in strain K12 after infection with ^{32}P labelled λ.K, P1. See text for details.

conserved parental DNA strand. This was shown by growing λ.K, P1 phage in strain K12 in the presence of ^{32}P and then correlating the ability of the progeny to grow on K12 (P1) and their sensitivity to autoradiolysis by incorporated ^{32}P (Fig. 7.15). An aliquot of heavily ^{32}P-labelled λ.K, P1 was stored at 4°C and assayed from day to day. Plated on either K12 or K12 (P1), it gave the same plaque titres. Another aliquot was allowed to grow for one cycle in K12 bacteria in non-radioactive medium. Dilutions of this lysate were also stored at 4°C and assayed on K12 and K12 (P1). Since strain K12 permits the growth of all progeny, the titres obtained were independent of the time of storage, an indication that a large majority of the particles contained little or no ^{32}P. Strain K12 (P1) permits only the growth of phage with the λ.K, P1-specific modification and the assays showed that all such particles contained ^{32}P, since their activity decreased with time of storage. The slope of the curve, roughly half that observed with the parental phage stock, indicates that the λ.K, P1 progeny phage contained half as much ^{32}P as its parents. The modification that phage DNA undergoes is the enzymatic methylation of adenine.

Breakdown of unmodified DNA calls for the presence of specific nucleases in strains exhibiting restriction, and recently a number of such enzymes have been purified. All degrade unmodified DNA but not modified DNA. Restriction sites have been determined by sequence analysis. Several restriction enzymes cut at palindromic sequences (Fig. 7.16), with scissions being introduced at the same point on either strand.

Enzyme	Source	Sequence recognized
EcoRI	Escherichia coli strain RY13	5′ GAATTC 3′ CTTAAG
HindIII	Haemophilus influenzae strain Rd	5′ AAGCTT 3′ TTCGAA
SalI	Streptomyces albus strain G	5′ GTCGAC 3′ CAGCTG
TaqI	Thermus aquaticus strain YT1	5′ TCGA 3′ AGCT
HaeIII	Haemophilus aegyptius	5′ GGCC 3′ CCGG

Fig. 7.16 Palindromic DNA sequences recognized by restriction enzymes. The arrows indicate the point at which scissions occur, and the asterisks indicate the bases modified by methylation. The enzyme nomenclature derives from the first letters of the bacterial species from which the enzyme was isolated.

As yet, restriction and modification have not been described for animal virus systems. However, restriction enzymes are now used extensively in sequencing DNA and in the construction of novel DNA molecules (p. 12; Chapters 16 and 20).

DEPENDENCE AND INDEPENDENCE AMONG DNA VIRUSES

The autonomy of viruses as regards the replication of their DNA varies between wide limits and is a function of the size of the viral genome. At one end of the scale are viruses, such as the T-even phages and the poxviruses, with genomes of M_r $1-2 \times 10^8$, corresponding to $100-200$ genes. Such viruses probably require little more from their host cells than an enclosed environment, protein synthesizing machinery, a supply of amino acids and deoxyribonucleotide triphosphates and an energy source. Some may not even require this much, for we have already seen how T2-specified proteins are involved in the biosynthesis of modified bases. Other viruses, such as herpes simplex and vaccinia, specify a new thymidine kinase and several other enzymes.

At the other end of the scale are viruses, such as φX174 and minute virus of mice (MVM), whose genomes (M_r 1.7×10^6) cannot specify more than 10 proteins. Since there may be as many as four different coat proteins, not many genes are left to code for functions essential to replication. Such viruses probably rely on the host not only for nucleic acid precursors but also for polymerases, ligases, nucleases, etc.

One approach which can be taken to ascertain the contribution of the host and the virus towards viral DNA replication is to isolate host cell mutants which no longer permit viral replication. This approach has largely been confined to the study of phage—bacterium relationships and usually involves investigating the ability of a phage to grow in a known bacterial mutant. For example, most coliphages grow in mutants lacking DNA polymerase I, but P2 does not. Obviously a functional DNA polymerase I is a prerequisite for P2 replication. Again, T4 can replicate in mutants lacking DNA polymerase III, but φX174 cannot.

Many instances are known where animal viruses fail to replicate in a particular cell line, although they are capable of initiating infection. For example, human adenoviruses can infect monkey kidney cells, but fail to grow in them. Curiously enough, coinfection of the cells with simian virus type 40 (SV40) permits adenovirus replication through the formation of progeny adenovirus—SV40 hybrids. These have a variable amount of the adenovirus and SV40 genomes covalently linked.

In other cell lines, the adenoviruses themselves can act as 'helpers'. Many adenovirus preparations contain a contaminating virus which is much smaller. This virus, adeno-associated virus

(AAV), has an SS genome of M_r 1.7×10^6; hence the coding potential of this genome is identical to that of ϕX174 or MVM. We would thus expect AAV to replicate independently of the parent adenovirus, but this has not been observed. Perhaps AAV can replicate independently in some other type of cell. Alternatively, the virus may lack some essential function which only adenovirus can provide. Such dependent viruses are called 'satellite' viruses.

A satellite virus of a different type is phage P4, which is a satellite of phage P2. Unlike most satellite viruses, this phage can replicate its nucleic acid in the absence of helpers. Instead, it lacks all known genes for phage morphogenesis and its DNA is packaged in a head composed of helper phage proteins.

FURTHER READING

Challberg, M. (ed.) (1991) Viral DNA replication. *Seminars in Virology*, **2**, no. 4.

Fraenkel-Conrat, H. & Wagner, R. R. (1977) Reproduction: bacterial DNA viruses. In: *Comprehensive virology*, Vol. 7. New York: Plenum Press.

Gilbert, W. & Dressler, D. (1968) DNA replication, the rolling circle model. *Cold Spring Harbor Symposia on Quantitative Biology*, **33**, 473–484. (Many other articles in the same volume are also worth reading, even though they are now dated.)

Kornberg, A. & Baker, T. (1991) *DNA replication* (2nd edn). San Francisco: W. H. Freeman.

Lewin, B. (1977) *Gene expression — Vol. 3. Plasmids and phages*. London and New York: John Wiley and Son. (Excellent chapters dealing with the replication of selected DNA bacteriophages.)

Ogawa, T. & Okazaki, T. (1980) Discontinuous DNA replication. *Annual Review of Biochemistry*, **49**, 421–457.

Also check Chapter 21 for references specific to each family of viruses.

8 The process of infection: IIC. RNA synthesis by RNA viruses

The synthesis of ribonucleic acid (RNA) by RNA viruses involves: (i) replication, which is defined as the production of progeny virus genomes; and (ii) transcription, which is the production of RNA complementary to the genome. Since RNA genomes may be either 'plus' (e.g. class IV) or 'minus' (class V) sense, transcription is not always synonymous with the synthesis of messenger RNA (mRNA) as it is in deoxyribonucleic acid (DNA) viruses.

SYNTHESIS OF THE RNA OF CLASS IV VIRUSES

Just as there are several ways in which a single-stranded DNA molecule could be replicated, there are several ways in which this can be achieved by a single-stranded RNA^+ virus. Since most RNA viruses can replicate in the presence of inhibitors of DNA synthesis, no DNA intermediate is involved. This is not true, however, for the Retroviridae, and their replication will be discussed in Chapter 9.

When RNA is extracted from infected cells, a ribonuclease (RNase)-resistant fraction can be isolated, which is double-stranded RNA. This RNase-resistant material also bands in Cs_2SO_4 at the position expected for double-stranded RNA and exhibits a steep thermal denaturation profile, confirming its double-stranded nature. This double-stranded form, also called replicative form (RF) by analogy with the ϕX174 system, was first discovered with encephalomyocarditis (EMC) virus and has since been found in all the systems studied.

When RNA is extracted from infected cells and analysed in a sucrose gradient or, better still, by polyacrylamide gel electrophoresis, a third kind of RNA molecule is found (Fig. 8.1). Mild RNase treatment causes this new form to sediment as double-stranded RNA, while it is completely degraded with more concentrated RNase. This suggests that it consists of a double-stranded molecule with single-stranded tails. Mild treatment of the replicative intermediate (RI), as this new form of RNA is called, with RNase removes the 'tails', leaving a double-stranded core (Fig. 8.2). Concentrated RNase degrades the continuous template, resulting in completely fragmented RNA.

As yet, the roles of RF and RI in the replication of viral RNA are not clear, since these two classes of molecule appear to have different roles in bacterial and animal virus systems. We know

116

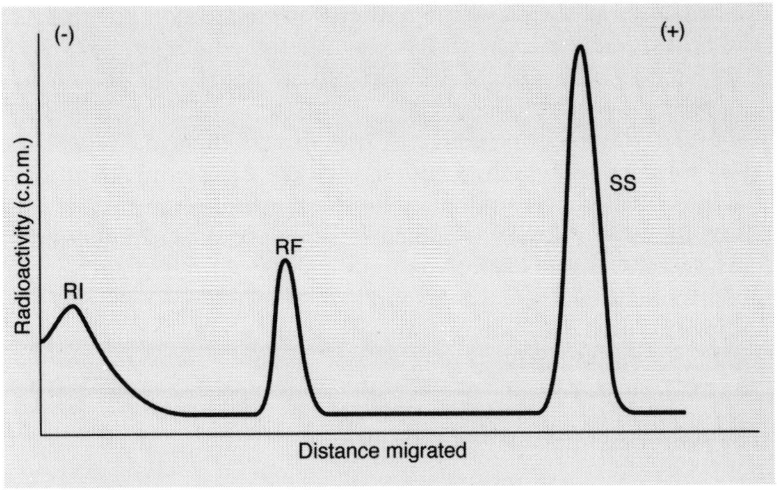

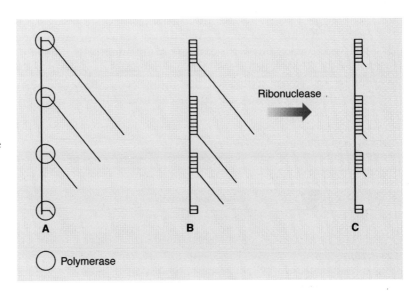

Fig. 8.1 Separation of radioactive RNA components from virus-infected cells by polyacrylamide gel electrophoresis. RI, replicative intermediate; RF, replicative form; SS, single-stranded RNA.

Fig. 8.2 Proposed structure for a molecule of replicative intermediate before deproteinization (A), after deproteinization (B) and the effect of treating deproteinized RI with RNase (C). Horizontal lines represent hydrogen bonding.

most about the replication of bacteriophage Qβ and this is the system which we shall discuss most fully. However, the evidence from other systems will be introduced where this differs significantly from that obtained with Qβ.

Both the RF duplex and the extensively hydrogen-bonded RI of Qβ and other RNA phages are formed only after deproteinization (Fig. 8.2A, B). RF probably results from the annealing of free plus and minus strands during deproteinization. Thus, during RNA synthesis, the region of hydrogen bonding is confined to the sequences covered by the replicase, and the enzyme makes and breaks hydrogen bonds as it proceeds. In other words, the RF and

RI structures as we know them after RNA extraction are artefacts and are not concerned with RNA synthesis, which proceeds via a template to which nascent RNA strands are held by the polymerase (Fig. 8.3).

In vivo, the infecting Qβ RNA first directs the synthesis of the viral polymerase, which, together with the host component, uses the infecting RNA as a template for the synthesis of minus strands. Then the minus strand is copied into plus strands. Production of plus strands exceeds minus strands by 10-fold, possibly because plus strands are removed as a source of template by being packaged into virus particles, whereas minus strands remain available as template continuously.

The synthesis of phage RNA *in vitro*

The synthesis of Qβ RNA, in the absence of whole bacterial cells (*in vitro*), has been of use in studying the replication of RNA, since the product is infectious, the most stringent test possible for the fidelity of this system. Early studies on RNA replication *in vitro* were carried out with extracts of infected cells, but purification of an RNA-dependent RNA polymerase (also called RNA synthetase or RNA replicase) has now been achieved. The complete polymerase has an M_r of around 287 000 and con-

Table 8.1 Qβ polymerase function and structure.

Polypeptide function	Source	$M_r \times 10^{-3}$
Binding to + strand	Ribosomal protein S1	70
Initiation	Elongation factor Tu	45
Initiation	Elongation factor Ts	35
Chain elongation	Qβ-encoded	65
Minus strand synthesis	Ribosome-associated (hexameric protein)	72

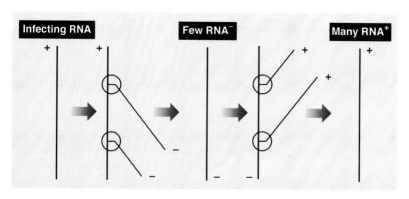

Fig. 8.3 Synthesis of Qβ RNA. O, polymerase. Note that there is a 10-fold excess of single-stranded (SS) RNA$^+$ over SS RNA$^-$.

sists of one virus-coded and four host-coded polypeptides (Table 8.1).

The Qβ polymerase shows great specificity for templates; only Qβ RNA, the strand complementary to Qβ RNA and 6S RNA present in Qβ-infected *Escherichia coli* are known to undergo replication in the reaction catalysed by this enzyme. Recognition of the template must involve more than just the primary structure of a limited region of the RNA, since fragments of RNA do not have template activity.

In vitro synthesis of the minus strand was first detected by the appearance of an RNase-resistant product after deproteinization. As would be expected, the complementary strand is synthesized early in the reaction, before progeny RNA is detected, and is correlated with the loss of infectivity of the template virus RNA molecules and the appearance of non-infectious RNase-resistant RNA in deproteinized reaction mixtures.

Although the RNA-dependent RNA polymerase is capable of synthesizing the minus strand from viral RNA, the predominant product of the *in vitro* reaction is plus strand RNA, as demonstrated by the infectivity of the product and by hybridization. Thus, the reaction is markedly asymmetric with respect to the synthesis of complementary and viral strands. The most obvious explanation — that there are two enzymes involved, one to synthesize the complementary strand and the other to use this strand as a template for synthesis of new viral strands — is not borne out in fact, as there is only one phage gene coding for an RNA polymerase and, furthermore, the phage genome is not sufficiently large to code for a second polymerase as well as the other phage-specified proteins. Recent experiments show that the synthesis of minus strands requires the viral polymerase, in conjunction with the host proteins shown in Table 8.1, whereas a tetramer of the viral polymerase directs the synthesis of plus strand RNA.

The Qβ *in vitro* system has contributed to our understanding of the roles of RF and RI *in vivo*, as explained above; for example, purified polymerase from phage-infected cells cannot use RF as template. Structures resembling the double-stranded RF and partially double-stranded RI found *in vivo* after deproteinization are also products of the *in vitro* reaction after deproteinization. Furthermore, in pulse—chase experiments, radioactivity in the RI could be chased into viral RNA. Thus, it has been suggested that the RI is a direct precursor of viral RNA *in vitro*. However, no one has yet been able to construct an active complex of purified enzyme and these double-stranded structures. The enzyme neither initiates synthesis with these molecules nor utilizes the nascent strand as primer for the continuation of synthesis. These observations support the view that the replicative template is not a double-stranded structure but rather a free complementary strand base-paired with nascent Qβ RNA solely in the region of the polymerase.

The evolution of an RNA molecule in vitro: *the 'little monster' experiment*

Regardless of the failure of studies on *in vitro* replication to provide an answer to the mechanism of *in vivo* replication, they have led to a very interesting series of experiments. Since the RNA instructs the *in vitro* replicative process, an opportunity is provided for studying the evolution of a self-replicating molecule outside a living cell. In the test-tube, the RNA molecules are liberated from many of the restrictions of the complete virus life cycle, and the situation mimics a precellular evolutionary stage, when environmental selection presumably operated directly on the genetic material rather than on the gene product. The only restraint imposed is that they retain whatever sequences are necessary for the polymerase to recognize the RNA molecule. With this in mind, Spiegelman and his collaborators designed an experiment to determine what would happen to the RNA molecules when the only demand made on them was that they replicate as fast as possible. Since the product of the reaction of polymerase with RNA in itself acts as a template and because longer RNA chains take longer to be replicated, molecules would gain an advantage by discarding any unneeded genetic information to achieve a smaller size and therefore a more rapid completion. For example, suppose the polymerase were presented with an equal mixture of whole molecules and half molecules of Qβ RNA. Because it takes half the time to replicate the half molecules compared with the whole molecules, there would soon be an excess of half molecules over whole molecules. Since there are now more half molecules than whole molecules, more half molecules would be used as templates, leading to an even greater excess of half molecules. In this way, there is rapid selection for the shorter molecules. Spiegelman and his collaborators prepared a reaction mixture containing the polymerase Qβ RNA, to act as template, and the appropriate ribonucleoside triphosphates. After incubation, the mixture was diluted into another tube containing the same amount of polymerase and triphosphate but *no* new Qβ RNA. This incubation and dilution sequence was repeated 75 times but, whereas the dilution factor was kept constant, the incubation time was reduced gradually. After the seventy-fifth transfer, an RNA molecule was isolated which retained only 17% of the original RNA sequence. Furthermore, during these experiments a 'palaeontological' record was obtained by freezing each reaction tube until there was time to examine its contents. In this way, it was observed that the size of the RNA molecule gradually decreased during the selection experiments.

The synthesis of poliovirus RNA

The mechanism by which poliovirus RNA is synthesized appears to be similar to that of Qβ, but one notable difference is that RF molecules isolated from cells infected with poliovirus are infectious whereas those from Qβ-infected cells are not. However, poliovirus RF is not infectious in enucleated cells, indicating that the nucleus is needed to separate the duplex so that the plus strand can be translated. Pulse–chase experiments with poliovirus suggest that RF is synthesized from RI. However, eukaryotes have a large pool of precursor molecules and, in order to obtain a satisfactory chase, pulse–chase experiments are usually attempted *in vitro* rather than *in vivo*. When extracts of poliovirus-infected cells are pulsed with labelled uridine and then chased, label moves from RI to single-stranded RNA and RF. Furthermore, this RF appears to be an end product, since the ratio of labelled RF to labelled single strands remains constant, i.e. RF is not a precursor of single-stranded RNA (Fig. 8.4). The RI making negative single strands has never been isolated but is assumed to exist as a means of amplifying the population of templates needed for the synthesis of RNA$^+$ virion and mRNA.

Nearly all studies on the synthesis of polio- and related viruses employ the antibiotic actinomycin D (see Appendix) to prevent the synthesis of cellular RNA, which would otherwise obscure the viral RNAs. Although actinomycin D does not inhibit the multiplication of polioviruses, one of the dangers inherent in using such antibiotics is that its variety of effects in the cell can never be fully known. Thus, what is deduced to be 'normal' viral RNA synthesis may be due to the presence of the antibiotic. In fact, RF

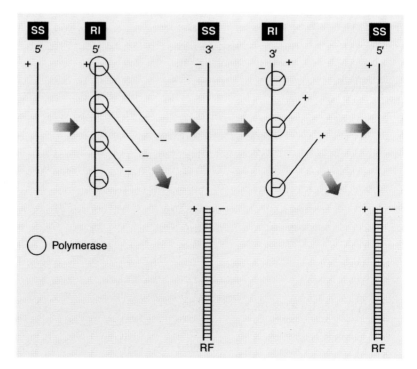

Fig. 8.4 Synthesis of poliovirus RNAs. Single-stranded RNA and RF are end-products.

is only found when cells are treated with actinomycin D and is absent in untreated cells.

SYNTHESIS OF THE RNA OF CLASS V VIRUSES

Class V consists of viruses, such as the Orthomyxo-, Paramyxo-, Arena-, Bunya- Filo- and Rhabdoviridae, which have a single-stranded negative-sense RNA genome that is complementary in base sequence to the mRNA. Thus the synthesis of mRNA involves the transfer of information from one single-stranded RNA to its complementary RNA and requires a pre-existing RNA-dependent RNA polymerase. This enzyme is present in purified preparations of class V viruses. However, activity can only be detected after partial disruption of the virus with detergent. In the presence of the four ribonucleoside triphosphates and appropriate ions, viruses 'activated' in this way in the absence of whole cells (*in vitro*), synthesize RNA *in vitro* at a linear rate for at least 2 hours. In general, the *in vitro* systems yield only RNA which is complementary and subgenomic in size and corresponds to the mRNAs made *in vivo*. It is technically much more difficult to obtain replication *in vitro* (Fig. 8.5).

When all the RNAs from an *in vitro* polymerase reaction mixture

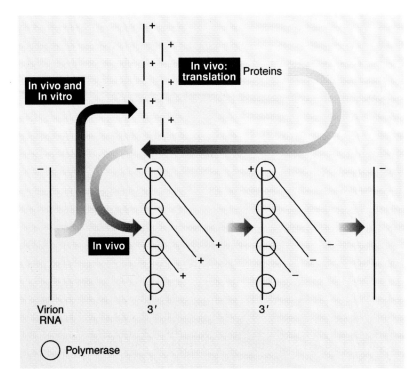

Fig. 8.5 Hypothetical scheme for the synthesis of the RNAs of vesicular stomatitis virus (Rhabdoviridae).

directed by a rhabdovirus, vesicular stomatitis virus (VSV), are extracted and annealed together, the virion template RNA is protected from RNase digestion and thus has been entirely copied into complementary RNA. However, only a small proportion of the newly synthesized RNA is protected by the virion RNA, showing that it is synthesized in excess of the template. In other words reinitiation of RNA synthesis takes place *in vitro*. The fidelity of RNA synthesis *in vitro* is elegantly shown by combining it with an *in vitro* translation system. Analysis by polyacrylamide gel electrophoresis demonstrates that all the viral proteins are synthesized from the novel mRNAs.

Synthesis of influenza virus mRNA is unique. Although the *in vitro* activity of influenza virus polymerase is not affected by actinomycin D or α-amanitin (the latter specifically inactivates eukaryotic RNA polymerase II, which synthesizes cellular mRNA; see Appendix), viral RNA synthesis *in vivo* is totally inhibited if the drugs are administered early in infection. In fact, only viral mRNA synthesis is sensitive to these inhibitors, but all viral RNA synthesis stops because mRNA-encoded proteins are required for these other processes. We now know through the work of R. M. Krug that *in vivo* viral mRNA synthesis requires the synthesis of cellular mRNA. All viral RNA synthesis take place in the nucleus of the infected cell. The viral transcriptase complex contains three proteins PB1, PB2 and PA. PB2 recognizes the methylated cap 2 structure at the 5' terminus of the cell mRNA (p. 74), reads along its sequence for 10−13 nucleotides and finally, through an exonuclease, cleaves after a purine residue. This fragment of cell mRNA is then elongated by PB1, which transcribes from the virion RNA (Fig. 8.6). The 5' terminal sequence of viral mRNA will be heterogeneous, as no specific cellular mRNA is needed. Inhibitors of cellular mRNA synthesis have their effect because very little mRNA is present in the nucleus. Although all three proteins move together along the viral RNA template, PA plays no part in transcription; rather, it is involved in replication to produce virion RNA$^-$ molecules. Recently mRNA of bunyaviruses has been shown to be synthesized in a similar manner, although synthesis is not nuclear.

SYNTHESIS OF THE RNA OF CLASS III VIRUSES

Members of the Reoviridae, reoviruses and rotaviruses of animals and wound tumour virus of plants and other families listed in Chapter 21, contain double-stranded RNA. By analogy with DNA replication, this double-stranded RNA could replicate by a *semiconservative* mechanism such that the complementary strands of the parental RNA duplex are displaced into separate progeny genomes, or the parental duplex could be conserved or degraded. In

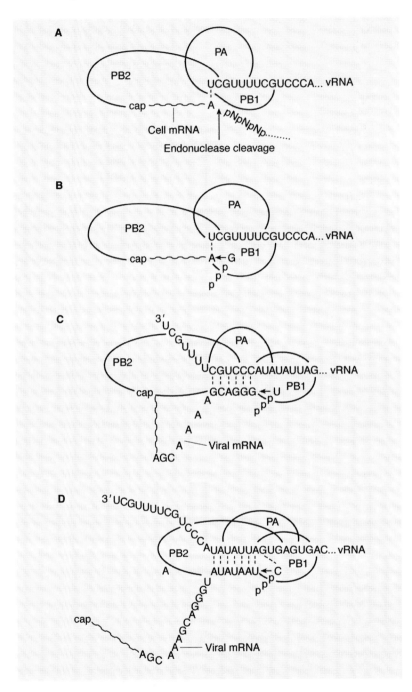

Fig. 8.6 Model of the initial stages in the synthesis of influenza virus mRNA. The virus replicative enzyme complex consists of proteins PB2, PB1 and PA. PB2 and PB1 are concerned in mRNA synthesis and PA is involved in replication. In (A) the transcriptase binds to the common 3′ terminal sequence of virion (v) RNA. Then PB2 recognizes and binds to a capped cellular mRNA. An adenine residue at position 10–13 of the mRNA hydrogen-bonds to the 3′ terminal U of the virion RNA and an endonuclease activity of PB2 cleaves the cellular mRNA on the 3′ side of the adenine as shown. In (B) PB1 begins new strand synthesis and the transcriptase complex moves down the template as elongation continues in (C) and (D).

fact, double-stranded RNA genomes are replicated *conservatively* as demonstrated by the following experiment.

When reovirus is treated with chymotrypsin, the outer capsid sub-virion is removed and cores are produced which possess RNA polymerase activity. These cores are infectious and have a density in CsCl of 1.45 g/ml, as distinct from that of mature virions

(1.37 g/ml), and this property was used to distinguish between the possible modes of replication. Cells were infected with cores which had previously been labelled with ^{3}H-uridine. About 15 hours after infection, the cells were collected and a cytoplasmic extract prepared and analysed by equilibrium centrifugation in CsCl. All the ^{3}H-labelled parental RNA remained with the cores at their original density of 1.45 g/ml, while the progeny virus, as measured by plaque-forming activity, was unlabelled and had a density of 1.37 g/ml. Thus, the parental RNA is conserved during replication. How, then, is the double-stranded RNA molecule replicated?

The RNA polymerase contained in 'cores' transcribes one strand of each of the 10 double-stranded segments of the reovirus genome (p. 75). The RNA products are single-stranded and do not anneal with each other. Since the same RNA molecules function as messengers on the polyribosomes, they have been designated as plus strands. Consequently, the complementary single strand has been designated as the minus strand. These minus strands are found exclusively in double-stranded RNA and do not appear free in the cytoplasm, suggesting that the synthesis is asymmetric. Replication takes place in subviral particles which contain single-stranded RNA$^+$ and have replicase activity. These particles have been isolated and the change from single to double-stranded RNA is observed *in vitro* (Fig. 8.7).

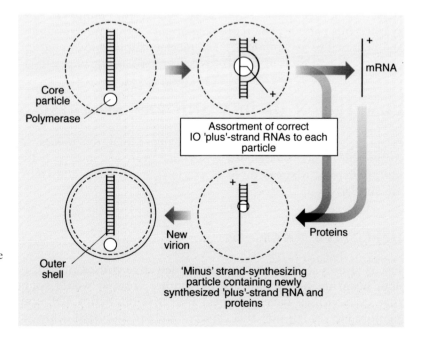

Fig. 8.7 Scheme for the synthesis of reovirus RNA. Virions contain 10 double-stranded RNA molecules but only one is shown.

If double-stranded RNA is formed by the synthesis of a complementary strand upon a preformed single-stranded RNA$^+$ template the double-stranded RNA that is formed on pulsing the cells with ^{3}H-uridine should be preferentially labelled in one of the two strands. If both strands of the newly formed duplex molecules were synthesized simultaneously, then they should be labelled equally, regardless of the length of the labelling period. To distinguish between these possibilities, cells were infected with reovirus and at various times the culture was pulsed with ^{3}H-uridine. The double-stranded RNA was extracted, denatured and then reannealed in the presence of excess unlabelled plus strands. In cultures that were continuously labelled for 9 hours or more after infection, the amount of label in the plus and minus strands of the double-stranded RNA was the same, as indicated by the fact that annealing of the denatured double-stranded RNA with excess unlabelled plus strands resulted in an annealed product containing 50% of the label present in the original double-stranded RNA. In contrast, the distribution of label in double-stranded RNA isolated from cells that had been pulse-labelled for 30 minutes, at a time when double-stranded RNA synthesis was already under way, was asymmetric with regard to the complementary strands. Over 95% of the radioactivity in the pulse-labelled double-stranded RNA was conserved in a double-stranded form after reannealing with excess unlabelled plus strands.

The packaging of the 10 segments of RNA to form a complete genome is a remarkable phenomenon but also occurs with other segmented viruses, notably influenza (eight RNAs). The observed high rate of generation of 'recombinants' occurs through the random assortment of segments in a common pool, presumably at the stage before free plus strand RNA enters the minus strand-synthesizing particle. The term 'reassortant' is thus more accurate and preferred to 'recombinant'. The mechanism by which a virion packages one of each of the 10 RNA segments is not understood.

COMPARISON OF DNA AND RNA SYNTHESIS

From the foregoing discussion, it is evident that the modes of replication of double-stranded DNA and RNA are unexpectedly quite different, whereas those of single-stranded RNA and DNA are similar.

DEFECTIVE RNAS

Just as viruses are parasitic on cells, so defective viruses are dependent upon function(s) provided by another virus. It would seem that even viruses have parasites! Defective-interfering (DI) and satellite viruses fall into this category and are discussed below.

The generation and amplification of defective–interfering (DI) virus RNA

All RNA and DNA viruses produce DI particles as the result of errors in their nucleic acid synthesis. Here we shall talk only about DI RNA viruses, about which more is known. DI viruses are deletion mutants which are unable to reproduce themselves without the assistance of the infectious parental virus (i.e. they are defective). For this reason, propagation of DI virus is optimal at a high multiplicity of infection when all cells contain an infectious virus genome. DI genomes depress (or interfere with) the yield of infectious progeny, by competing for a limited amount of some product synthesized only by the infectious parent. Interference only takes place when the ratio of DI to infectious genomes reaches a critical level. Many DI viruses synthesize no proteins and some have no open reading frame. Because they depend upon parental virus to provide those missing proteins, DI and parental viruses are composed of identical constituents, apart from their RNAs. Thus it is usually difficult to separate one from the other. A notable exception is the DI particle of the rhabdovirus, VSV, which is a truncated version of the infectious particle. When centrifuged these remain

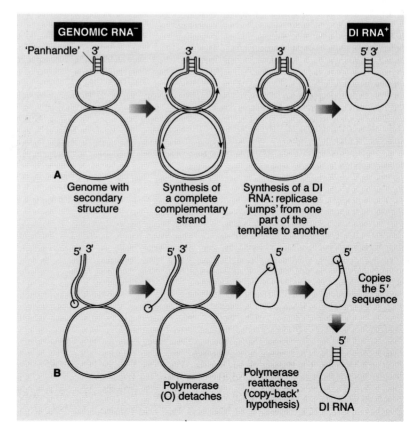

Fig. 8.8 Hypothetical schemes to explain the generation of DI RNAs having sequences identical with both the 5′ and 3′ regions of the genome (A) and with the 5′ region only (B).

at the top of sucrose velocity gradients and are thus called T particles to distinguish them from infectious B particles, which sediment to the bottom. Some biological implications of DI particles are discussed in Chapter 13.

DI RNA molecules are very variable in length and can represent as little as 5% of the infectious genome. A clue leading to one hypothesis explaining how DI RNAs are generated came from electron microscopic examination of genomic and DI RNAs from single-stranded RNA viruses. Both were found to be circularized by hydrogen bonding between short complementary sequences at the termini, forming structures called 'panhandles' or 'stems' (Fig. 8.8A). The deletion that results in DI RNA may arise when a polymerase molecule detaches from the template and reattaches at a different point or to the nascent strand (Fig. 8.8B). Thus the polymerase begins faithfully but fails to copy the entire genome. Most VSV viruses are of the latter type and lack the 3' end of the standard virus genome. There are no DI viruses known which lack the 5' terminus. In other DI viruses, parts of the genome can be duplicated, often several times over, during subsequent replication events (Semliki Forest and herpes simplex viruses), making complex structures which bear little resemblance to the standard genome from which they were derived. The three classes of DI genome are summarized in Fig. 8.9.

Usually, interference only occurs between the DI virus and its parent. This is because DI virus lacks replicative enzymes and requires those synthesized by infectious virus. Specificity resides in the enzymes, which only replicate molecules carrying certain unique nucleotide recognition sequences. Intuitively, it can be seen that in a given amount of time an enzyme will be able to make more copies of the smaller DI RNA. Thus, as time progresses, the concentration of DI RNAs increases relative to the parenteral RNA. This is the *amplification* step. Astute readers will have

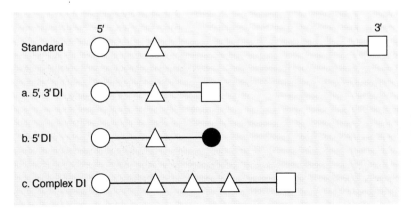

Fig. 8.9 Comparison of the genomes of standard and DI viruses. Open symbols represent common sequences, solid symbols their complement.

recognized the interesting parallel of the generation of DI RNA with the *in vitro* evolution of the small Qβ RNAs discussed above (p. 120). However, this is not the whole story, as some large DI RNAs interfere more efficiently than smaller ones. Such DI RNAs seem to have evolved a transcriptase recognition sequence which has a higher affinity for the enzyme than that of the infectious parents and hence confers a replicative advantage. All DI genomes retain a packaging or encapsidation sequence since without this they cannot recognize virion proteins and form viral particles.

RNA satellite viruses

We have already mentioned satellite DNA viruses (p. 114). There are also satellite RNA viruses, most of which occur in plants. Satellite viruses cannot multiply without a helper virus, which is specific for a particular satellite virus. Satellite viruses are distinguished from DI viruses, as their RNA has little or no homology with that of the helper. The origin of satellite viruses is not known.

Preparations of the Rothamsted strain of tobacco necrosis virus (TNV) contain particles of two sizes. Only the larger particles are infective and they are serologically unrelated to the smaller satellite particles. The larger particle (TNV) causes characteristic necrotic lesions when inoculated on tobacco leaves, but isolated satellite particles do not. Multiplication of satellite TNV reduces the size and number of lesions caused by TNV and this phenomenon is used to assay the satellite virus. The satellite virus cannot replicate by itself and, as a result of the interaction of the two viruses during replication, the helper virus is inhibited.

The reason why satellite TNV depends on TNV can be obtained from consideration of the smaller particle's structure (p. 44). It contains a single RNA molecule of 1200 nucleotides. This corresponds to a protein of 400 amino acids, which is approximately the size of the satellite virus coat protein. Thus, the helper virus is required to provide the enzymes necessary for replication of the RNA of the satellite virus.

Diversity among the satellite viruses of plants is exemplified by their presence in several different groups, including the necroviruses (TNVs), nepoviruses (tobacco black ring virus (TBRV)) and cucomoviruses (cucumber mosaic virus). Furthermore, unlike TNV, other satellites (such as TBRV satellite) have no effect upon the disease caused by the helper virus.

Hepatitis delta virus (HDV) of humans is the best known animal satellite virus. This 1700-nucleotide single-stranded circular RNA virus has as its helper hepatitis B virus (HBV), a DNA virus with an RNA intermediate (p. 137). HBV only provides HDV with its coat protein. The precise role of HDV in liver disease is not known, but it appears to exacerbate the effects of HBV.

INFECTIOUS NUCLEIC ACIDS

For a long time, virologists asked why it was that infectious nucleic acid could be extracted from some viruses but not from others. An examination of the Baltimore classification provides the answer. Viruses belonging to classes I, II, V and VI require the transcription of their genetic information to another nucleic acid upon infection of a susceptible cell. In most cases, this transfer is mediated by an enzyme associated with the virus particle, such as the DNA-dependent RNA polymerase of vaccinia virus, the RNA-dependent RNA polymerase of VSV (p. 122) and reovirus (p. 125) and the RNA-dependent DNA polymerase of tumour viruses (p. 133). Purified nucleic acid from such viruses will lack these enzymes and consequently will not be able to initiate a productive infection. Cells do not possess an enzyme capable of using the viral RNA as template.

Bacteriophage φX174 belongs to class II, indicating that viral mRNA can only be synthesized after formation of the complementary strand. Thus it might be *expected* that purified φX174 DNA would not be infectious. Since the viral DNA *is* infectious, we can conclude that the first step in infection, the synthesis of the complementary strand, is carried out by a host cell enzyme. Clearly, the possession of infectious nucleic acid should not be the sole criterion for the assignment of any virus to a particular group.

MODIFICATIONS OF RNAS

Described below are a number of interesting properties of viral RNA molecules. Functions or biological significance of these structures are not always well understood.

Modification of the 3′ terminus of viral RNA

Polyadenylation

Many cellular mRNA species isolated from eukaryotic, but not prokaryotic, cells have a homopolymer of about 100 adenosine residues, covalently bound to their 3′ terminus. This 'poly-A tract' was first discovered in RNA from poliovirus virions and is now known to be present on the mRNAs of most viruses. Such RNAs are readily isolated by affinity chromatography on a column which contains poly-dT linked to a support matrix. At relatively high ion concentration (400 mM NaCl), RNA with a poly-A tail anneals to poly-dT, while other RNA molecules pass through. Then by reducing the salt concentration to 1 mM, the hydrogen bonding breaks down and the poly-A-containing RNA can be collected.

Poly-A sequences are added to viral RNAs post-transcriptionally by the addition of adenosine 5′-triphosphate (ATP) molecules. Polio-

$$\text{pppG} + \overset{\bullet\,\bullet}{\text{ppp}}\text{N} \cdot \text{pN} \longrightarrow \text{pppG} \cdot \text{pN} \cdot \text{pN} + 2\,\text{Pi}$$

$$\overset{\bullet\,\bullet}{\text{ppp}}\text{G} + \overset{\bullet}{\text{ppp}}\text{G} \cdot \text{pN} \cdot \text{pN} \longrightarrow \text{GpppG} \cdot \text{pN} \cdot \text{pN} + 3\,\text{Pi}$$

$$2\ \text{S-adenosylmethionine} + \text{GpppG} \cdot \text{pN} \cdot \text{pN} \longrightarrow \text{m}^7\text{GpppG}^m \cdot \text{pN} \cdot \text{pN} + 2\ \text{S-adenosylhomocysteine}$$

Fig. 8.10 Reactions involved in the synthesis of the cap of reovirus mRNA. 'Deleted' phosphate groups are those lost during cap formation.

and influenza viruses have a short (~ 6 mer) oligo-uridine (oligo-U) tract on RNA$^-$ strands, which is transcribed to oligo-A and subsequently extended to poly-A.

The role of poly-A is not clear. Poliovirus RNA from which poly-A has been removed by enzymatic digestion has reduced infectivity, but Qβ and tobacco mosaic virus (TMV) RNAs are not poly-adenylated and are infectious. Poly-A does not affect translational efficiency of the message but confers stability to the message both *in vitro* and *in vivo*. However, reovirus mRNAs are entirely lacking in poly-A. In general, the length of the poly-A is not a trait of the family but is characteristic of individual viruses: poliovirus has a tract of about 80 while another picornavirus, EMCV, has about 14 adenylic acid residues. Thus, we are almost driven to the conclusion that the sole function of poly-A tracts is to help virologists to isolate polyadenylated RNAs! Hopefully, the real significance of terminal poly-A sequences will soon be discovered.

Binding of amino acids by plant virus RNAs

The RNAs of a number of plant viruses are not poly-adenylated. However, the 3' end can fold to form a structure resembling transfer RNA (tRNA), which can accept amino acids. The significance of this is not known and there is no evidence that the viral RNA functions as a tRNA in the cell. *In vitro*, the amino acid–viral RNA complex substitutes poorly for tRNA in protein synthesis. Each viral RNA has specificity for one amino acid, e.g. TMV and turnip yellow mosaic virus bind histidine and valine respectively.

Modifications of the 5' terminus of viral RNA

Capping

Eukaryotic mRNAs and most viral mRNAs have their 5' terminus modified to form a cap structure (p. 75). This is made contrary to the rules of nucleic acid synthesis (Fig. 8.10): firstly, the penultimate guanine (G) of the cap is added to the 5' terminus of the message proper and, secondly, the terminal G is joined by a 5'–5' linkage. An additional point to notice is that both Gs are methylated. This is a modification common to cellular mRNAs and is of unknown significance. The cap is synthesized by a series of five enzymes shortly after transcription has started. The cap is necessary for the formation of a stable initiation complex with a terminal AUG initiation codon and hence for protein synthesis.

Terminally linked protein

Poliovirus virion RNA is not capped but has a small protein, called VPg (M_r about 7000), covalently linked to its 5′ terminus. The protein is absent from polysomal RNA$^+$; thus it is removed before the virion RNA is translated. Nascent plus strands on the replicative intermediate also have the protein attached, which may indicate that its removal is the process which decides if newly synthesized RNA$^+$ ends up as message or in progeny virions. The function of VPg is to permit initiation of protein synthesis from an internal (i.e. non 5′-terminal) AUG codon, a unique situation that does not occur in capped mRNAs (see above). Some DNA viruses (adenoviruses, hepatitis B, phage 529 of *Bacillus subtilis*) also have a protein linked covalently to each 5′ end of the double-stranded DNA (p. 72) but this functions to circularize the genome.

FURTHER READING

Braam, J., Ulmanen, I. & Krug, R. M. (1983) Molecular model of a eukaryotic transcription complex: functions and movements of influenza P proteins during capped RNA-primed transcription. *Cell*, **34**, 609–618.

Dimmock, N. J. (1991) The biological significance of defective interfering viruses. *Reviews in Medical Virology*, **1**, 165–176.

Francki, R. I. B. (1985) Plant virus satellites. *Annual Review of Microbiology*, **39**, 151–174.

Lewin, B. (1977) RNA phages. In: *Gene expression 3*, Chapter 9 pp. 790–824. London and New York: John Wiley and Sons.

Maramarosch, K. (ed.) (1991) *Viroids and satellites: molecular parasites at the frontier of life.* Boca Raton, FL: CRC Press.

Roossinck, M. J., Sleat, D. & Palukaitis, P. (1992) Satellite RNAs of plant viruses: structures and biological effects. *Microbiological Reviews*, **56**, 265–279.

Roux, L., Simon, A. E. & Holland, J. J. (1991) Effects of defective interfering viruses on viral replication and pathogenesis *in vitro* and *in vivo*. *Advances in Virus Research*, **40**, 181–211.

Strauss, E. G. & Strauss, J. H. (1983) Replication strategies of the single stranded RNA viruses of eukaryotes. *Current Topics in Microbiology and Immunology*, **105**, 1–98.

Taylor, J. M. (1992) The structure and replication of hepatitis delta virus. *Annual Review of Microbiology*, **42**, 253–276.

Also check Chapter 21 for references specific to each family of viruses.

9 The process of infection: IID. RNA viruses with a DNA intermediate and vice versa

The synthesis of deoxyribonucleic acid (DNA) by ribonucleic acid (RNA) viruses, once regarded as heresy to the doctrine which stated that information flowed from DNA to RNA to protein, now has a place of honour in molecular biology. Such viruses belong to the family Retroviridae and include the major pathogen human immunodeficiency virus type 1 and many others (p. 349). Only some cause cancer and there is thus no link *per se* between their strategy of replication and the ability to subvert the normal functioning of a cell. Also covered in this chapter are the DNA animal viruses of the Hepadnaviridae and the DNA plant caulimoviruses. Although their genomes are DNA, we shall see that they have a very similar strategy to that of retroviruses, having an RNA$^+$ intermediate central to their replication.

RETROVIRUSES

The unexpected involvement of DNA in viral RNA replication

Early studies showed that DNA played a critical role in the multiplication of RNA tumour viruses and in their ability to transform cells. Infection and transformation by these viruses can be prevented by inhibitors of DNA synthesis added during the first 8–12 hours after exposure of the cells to the virus. Also, the formation of virions is sensitive to actinomycin D, suggesting a requirement for DNA-dependent RNA synthesis. These data led Howard Temin of the University of Wisconsin to propose his 'provirus' theory, which postulated the transfer of the information of the infecting RNA to a DNA copy, which then serves as a template for the synthesis of viral RNA.

Temin's theory required the presence in infected cells of a unique enzyme, an RNA-dependent DNA polymerase or 'reverse transcriptase'. Until 1970, no enzyme had been found in any type of cell which could synthesize DNA from an RNA template. It was clear that, if such an enzyme existed, then the RNA tumour virus must induce its synthesis soon after infection or else carry the enzyme into the cell as part of the virion. Since there was already a precedent for the occurrence of polymerases in animal viruses (p. 122) a search was begun for a reverse transcriptase in oncornaviruses. David Baltimore (of the Massachussetts Institute of Technology) and Temin reported the presence of such an enzyme simultaneously in 1970. Purified virions were disrupted with detergent and incubated with dATP, dCTP, dGTP and ^{3}H-dTTP. This resulted in a rapid incorporation of label into acid-insoluble material, i.e. DNA. By omitting one of the deoxyribonucleotide triphosphates or by pretreating the virus with ribonuclease (RNase), this incorporation could be prevented, proving that the enzyme was indeed making DNA from an RNA template.

Properties of the oncornavirus genome

Early studies showed that genomic RNA sedimented at 70S, which is equivalent to an M_r of around 7×10^6. However, when denatured, this RNA sedimented at 35S, indicating that there were two molecules of equal size. Did these contain different information or were they identical? This question was resolved at that time by *oligonucleotide mapping*. (Today the RNA would be copied into DNA and sequenced.) Here oligonucleotides derived by RNase T_1 digestion are separated in two dimensions. T_1 cuts RNA, leaving guanylate at the new 3' terminus, and the number of large oligonucleotides (runs of nucleotides not containing an internal guanosine residue) gives a measure of the size or complexity of the RNA, since the runs are more likely to occur in large molecules. Analysis of oncornavirus RNA gave a complexity expected for a molecule of around $3 \times 10^6 M_r$. Subsequently its complete sequence of nucleotides was determined. Thus each virion contains two copies of the RNA and hence is diploid. Electron microscopy shows two RNA molecules which are linked together as in Fig. 9.1.

Structurally, the viral genome resembles a typical eukaryotic messenger RNA (mRNA), having messenger polarity, a poly-adenine (poly-A) tract at the 3' terminus and a cap structure at the 5' terminus. The order and number of the major genes was first determined for Rous sarcoma virus (RSV) (Fig. 9.1). These are known by the abbreviations *gag*, encoding three group-specific *a*ntigens, or major internal virion proteins, as they are now known to be; *pol*, encoding three minor but functionally important virion proteins, including the *pol*ymerase or reverse transcriptase enzyme and integrase enzyme; and *env*, encoding the virion *env*elope protein

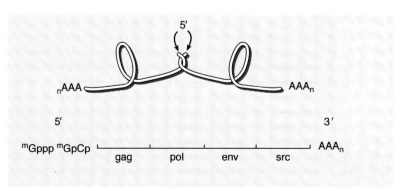

Fig. 9.1 Top: structural map of the oncornavirus genome, showing identical molecules held together at their 5' ends. Each has an intramolecular loop. Bottom: genetic map of Rous sarcoma virus (not to scale).

(see Figs 10.19 and 19.5). These are present in this order in all retroviruses. A fourth, originally cellular gene, *src*, is not needed for virus multiplication but is the key gene in transformation and carcinogenesis caused by RSV (Chapter 17).

Structure of the genome and its integration into cellular DNA

In vivo, retroviral RNA is converted into linear double-stranded DNA in the cytoplasm. This then migrates to the nucleus, circularizes and integrates into cellular DNA. Successful integration requires the five-base-pair inverted terminal repeat sequence (Fig. 9.2) and an integrase enzyme encoded by the 3' part of the *pol* gene. To be effective, this enzyme must enter the cell with the infecting particle. Evidence for these requirements comes from deletion mutagenesis, as removal of these sequences results in no integration and no transformation of the cell. However, non-integrated DNA lacking the inverted repeats still gave rise to progeny virus, from which we make the important observation that *integration is not required for retrovirus multiplication* (although under natural conditions the virus always integrates). Thus, with some justification, we can separate viral multiplication from viral carcinogenesis, which will be discussed in Chapter 17.

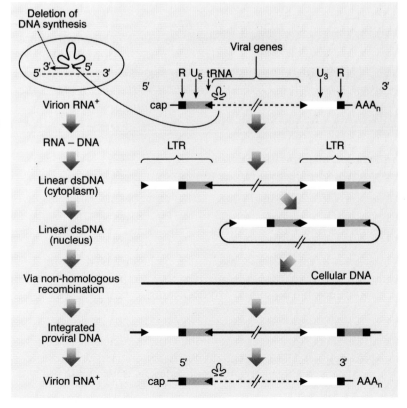

Fig. 9.2 Structure of retrovirus RNA (top panel) and an outline of the replication cycle. Transfer RNA (tRNA) is the primer for reverse transcriptase (see text and Fig. 9.3); U5 and U3 are unique sequences at the 5' and 3' ends of virion RNA; R is a short sequence common to both termini. Synthesis of the long terminal repeats (LTRs) found only in DNA, is shown in Fig. 9.3. Note the five-base-pair inverted repeat sequences (▶) which define the LTRs. Not to scale. ds, double-stranded.

Properties of reverse transcriptase

The template requirements of reverse transcriptase have been investigated. Polyribonucleotides act as efficient templates, but, like all DNA polymerases, reverse transcriptase cannot initiate DNA synthesis but requires a primer, which can be as short as four bases. Even with a primer, polydeoxyribonucleotides are much poorer templates than the corresponding homologous polyribonucleotide and in most cases yield no detectable synthesis.

Reverse transcription of viral RNA also requires a primer. This is a cellular 4S transfer RNA (tRNA) and there is one molecule per viral genome. The tRNA is intimately associated with the 5′ terminus but can be separated after denaturation. Each oncornavirus contains only one type of tRNA, e.g. RSV has tryptophan tRNA and Moloney murine leukaemia virus has proline tRNA. It is not known how these tRNAs are specifically selected.

Physically, the reverse transcriptase protein of avian oncornaviruses is composed of two subunits: α $(60\,000\,M_r)$, which has enzymatic activities, and β $(90\,000\,M_r)$, which binds to the tRNA primer and is enzymatically inactive. The α subunit has three activities, namely it, (i) synthesizes DNA from RNA; (ii) synthesizes DNA from DNA; and (iii) digests the RNA strand from RNA : DNA hybrids (ribonuclease H).

The basic problem is to explain how RNA is copied into DNA when synthesis commences with the tRNA primer at the 5′ terminus of the virion RNA (remembering that nucleic acid synthesis always proceeds from the 3′ to the 5′ terminus of the template). It would, of course, have been easier to explain if nature had put the tRNA primer at the 3′ end! However, Fig. 9.3 presents a model which suggests how DNA may be synthesized and at the same time explains the generation of proviral long terminal repeats (LTRs) and why the viruses are diploid.

Integration of retrovirus DNA and production of progeny RNA genomes

There are three steps in the integration process (Fig. 9.4). First, two bases are removed by the action of the viral integrase from both 3′ ends of the linear DNA molecule, synthesized as in Figs 9.2 and 9.3. Second, the 3′ ends are annealed to sites a few (four to six) bases apart in the host genome, which is then cleaved. Third, gaps and any mismatched bases are repaired. Cells contain 1 to 20 copies of integrated proviral DNA. There are no specific sites for integration, so this takes place by non-homologous recombination.

From integrated DNA new viral RNA genomes are synthesized. In all respects, this process resembles the production of cellular mRNA; cellular DNA-dependent RNA polymerase II synthesizes the viral capped polyadenylated molecules, as indicated in Fig. 9.2.

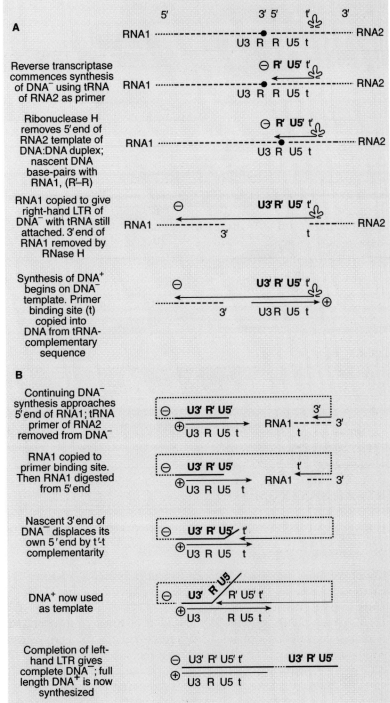

Fig. 9.3 Scheme for the synthesis of retroviral linear DNA by reverse transcriptase. (A) Synthesis of right-hand LTR of minus strand DNA, (B) minus strand DNA synthesis is completed with the left-hand LTR.

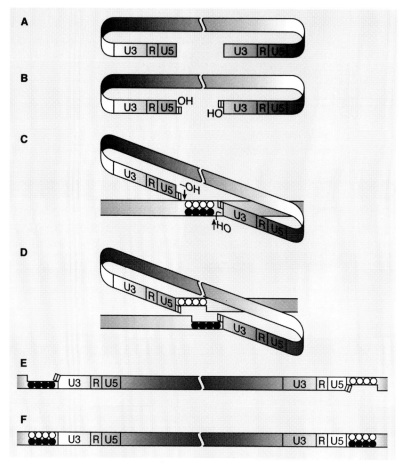

Fig. 9.4 Integration of retroviral DNA into the host genome. Linear viral DNA, synthesized as in Figs 9.2 and 9.3, is attacked by the integrase, which cleaves two bases from each 3′ end (B). This molecule migrates to the nucleus and the integrase cleaves the cellular DNA by catalysing an attack on it by the viral 3′ OH groups (C, D). DNA repair reconstitutes the double strand (E, F). From Whitcomb & Hughes (1992).

HEPADNAVIRUSES

Human hepatitis B virus (HBV) is the pre-eminent member of this group. It currently infects 200 000 000 people world-wide, causing acute liver disease, chronic disease and liver cancer. This virus is the single most important cause of cancer in humans (Chapter 17). Chapter 15 records the excellent progress that has been made in anti-HBV vaccines.

HBV has a DNA genome composed of a linear minus strand, which is covalently linked to a protein at its 5′ end, and a complementary plus strand, which is an incomplete copy (Fig. 9.5). As expected, sequence analysis of the incomplete strand shows a variable 3′ terminal nucleotide. The plus strand always overlaps the 5′−3′ junction of DNA⁻ and endows the linear DNA molecules with circularity. The virus carries its own polymerase. Virion DNA is encapsidated in core protein, which is surrounded by a lipid envelope coat containing S (surface) antigen. Coding assignments are shown in Fig. 9.5. Antibody to S antigen provides protec-

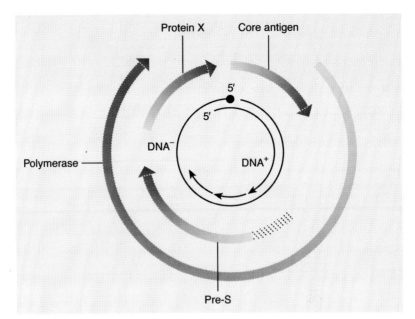

Fig. 9.5 Genome structure of hepatitis B virus showing the coding of the major proteins. There is a viral protein bound to the 5' end of DNA⁻. Pre-S is converted to S antigen by removal of hatched box region.

tive immunity and is the keystone to the immunization policy.

Molecular analysis of HBV replication was slow in coming, as the virus could not, and still cannot, be grown in culture. This frustration was partly relieved by the discovery of a very closely related virus in a colony of woodchucks in the Philadelphia Zoo (woodchuck hepatitis virus) and another in ducks (duck hepatitis virus). Data obtained from all systems were inconsistent with the requirements of semi-conservative replication of viral DNA, as the virion polymerase only synthesizes DNA$^+$ strands and in the cell only replicative intermediates making DNA$^-$ could be found. Furthermore, DNA$^+$ synthesis was blocked by actinomycin D, an antibiotic which inhibits DNA-directed *RNA* synthesis, but DNA$^-$ synthesis was not inhibited; also nascent DNA$^-$ was found to be associated with viral RNA$^+$. With these considerations, a scheme for replication was derived (Fig. 9.6), but many essential details, including the nature of the reverse transcriptase and the RNA$^+$ template, have yet to be determined.

In summary, we see that hepadnaviruses synthesize virion DNA from an RNA template, whereas with retroviruses the reverse is true (Table 9.1). Replication of these viruses is thus said to be

Table 9.1 Comparison of replication strategies of hepadnaviruses and retroviruses.

	Virion	Cell
RNA$^+$ template for replication	retrovirus	hepadnavirus
Location of dsDNA	hepadnavirus	retrovirus

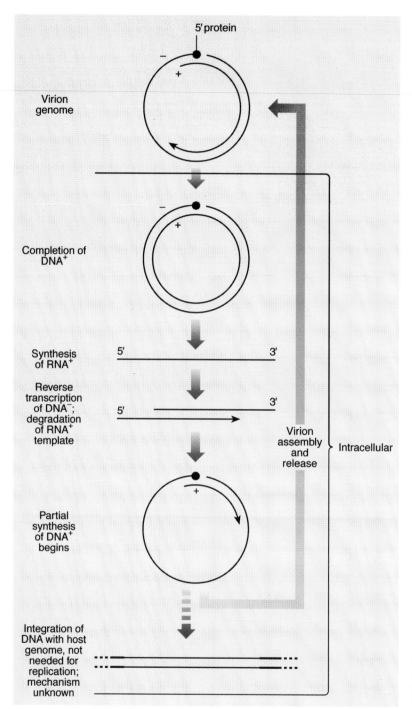

Fig. 9.6 Replication of the hepadnavirus genome by reverse transcription.

'permuted with respect to time'. Integration of hepadnaviral DNA into the cellular genome takes place infrequently and is therefore unlikely to be essential for replication. Integration of retrovirus DNA is also not essential for virus multiplication. The finding that hepadnavirus DNA integrates with cellular DNA is highly significant to the proposed involvement of HBV with liver cancer. Indeed, the majority of these tumours contain HBV sequences.

CAULIFLOWER MOSAIC VIRUS (CaMV)

CaMV, a representative of the caulimoviruses, the only truly double-stranded DNA virus family in the plant kingdom, has been viewed with considerable interest, as it is a likely vector for the genetic manipulation of plants. Investigation of its molecular biology has shown that it shares properties intermediate between retro- and hepadnaviruses (Fig. 9.6).

Firstly, all three types of virus are temporally permuted versions of each other, having an RNA$^+$ template which is central to replication. Both CaMV and retroviruses have a short repeat sequence at both termini and use as primer for DNA$^-$ synthesis a host tRNA which has a binding site close to the 5' terminus of the virion RNA. The genomes of CaMV and hepadnaviruses are similar; both are DNA, both are circular with neither strand closed, both

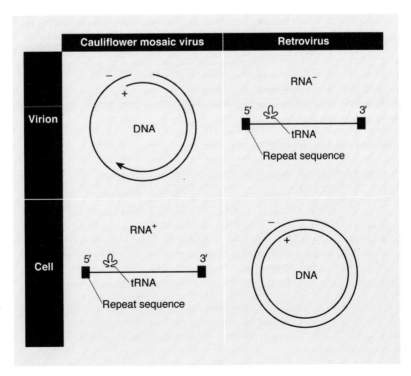

Fig. 9.7 Comparison of the replication strategies of cauliflower mosaic virus and retroviruses: an example of temporal permutation.

have a complete DNA$^-$ strand only and both have an incomplete DNA$^+$. CaMV DNA$^+$ is more complete than that of HBV and seems to result from a displacement synthesis which has been prematurely terminated. It integrates, but rarely — at a frequency of about 10^{-5}.

One should not get too carried away by the similarities in replication emphasized here, as the three groups of viruses are very different in their biology, particle formation and structure, but the thought that they all have a common type of replication makes it tempting to speculate on their evolutionary past. It could be argued that we are looking at some one-time common cellular element that has adopted different ways for its transfer to and from host cells. It is now known that some cellular DNA sequences (mammalian pseudogenes, some highly repetitive sequences and the transposable *copia* elements of *Drosophila*) involve reverse transcription and traverse the pathway of DNA → RNA → DNA.

FURTHER READING

Chen, I. S. Y. (ed.) (1990) Recent advances in retrovirology. *Seminars in Virology*, **1**, no. 3.

Cullen, B. R. (1991) Human immunodeficiency virus as a prototypic retrovirus. *Journal of Virology*, **65**, 1053–1056.

Goff, S. P. (1992) Genetics of retroviral integration. *Annual Review of Genetics*, **26**, 527–544.

Hohn, T., Hohn, B. & Pfeiffer, P. (1985) Reverse transcription in CaMV. *Trends in Biochemical Sciences*, **10**, 205–209.

Howard, C. R. (1986) The biology of hepadnaviruses. *Journal of General Virology*, **67**, 1215–1235.

Hull, R. & Covey, S. N. (1985) Cauliflower mosaic virus: pathways of infection. *Bio Essays*, **3**, 160–163.

Levy, J. A. (ed.) (1992) *The retroviridae*, Vol. 1. New York: Plenum.

Levy, J. A. (ed.) (1993) *The retroviridae*, Vol. 2. New York: Plenum.

Nassal, M. & Schaller, H. (1993) Hepatitis B virus replication. *Trends in Microbiology*, **1**, 221–228.

Tiollais, P., Pourcel, C. & Dejean, A. (1985) The hepatitis B virus. *Nature (London)*, **317**, 489–495.

Weiss, R., Teich, N., Varmus, H. & Coffin, J. (1984/5) *RNA tumour viruses* (2 vols, 2nd edn). Cold Spring Harbor Laboratory, NY. (Dated now; see Levy (1992 and 1993), if available.)

Whitcomb, J. M. & Hughes, S. H. (1992) Retroviral reverse transcription and integration: progress and problems. *Annual Review of Cell Biology*, **8**, 275–306.

Also check Chapter 21 for references specific to each family of viruses.

10 The process of infection: III. The regulation of gene expression

When a virus infects a cell, not all the genes are expressed at the same time, and this was suggested in Chapter 1 when we discussed the pulse–chase experiments of Hershey and co-workers. These workers noted the existence of 'early' and 'late' proteins, but more definitive proof of regulation comes from a systematic study of the appearance of phage T4 proteins. Infected cells were pulsed with radioactive leucine, and the radioactive proteins extracted, electrophoresed on polyacrylamide gels and detected by autoradiography. These studies showed that not all proteins are synthesized at the same time nor are many synthesized continuously. Rather, some are synthesized for only a few minutes early in infection, others are synthesized late in infection, while others are synthesized for about half the infection cycle. Initially, knowledge came from studies of bacteriophage, but there is a growing body of information on regulation in eukaryotic cells, and the mechanisms regulating the development of animal viruses are considered in the second part of this chapter. There is still relatively little knowledge about regulation and gene expression in plant viruses.

DNA BACTERIOPHAGES

During gene expression, genetic information in deoxyribonucleic acid (DNA) nucleotide sequences is transcribed into complementary ribonucleic acid (RNA) sequences and these are translated into the polypeptide chains, which form the ultimate products of the gene. Control of gene expression might occur at the level of transcription or translation, but it is becoming increasingly clear that in bacteriophage-infected cells transcriptional control is of major importance.

Transcription of DNA into RNA is carried out by the enzyme RNA polymerase which, in *Escherichia coli* at least, has a rather complicated structure. The enzyme from *E. coli* consists of five different polypeptide chains—α, β, β', ω and σ—held together by hydrogen bonding in the ratio $\alpha_2\beta\beta'\omega\sigma$ (Fig. 10.1). The polypeptide chain σ, or σ factor, can easily be separated from RNA polymerase, and the rest of the structure, the core enzyme, retains its catalytic activity. Thus, the σ factor does not contribute towards the catalytic activity of the enzyme, but it still has an important role to play in RNA synthesis, for in its absence correct initiation of RNA synthesis fails to occur *in vitro*. Thus, the most likely role of the σ factor is recognition of start signals on the DNA molecule.

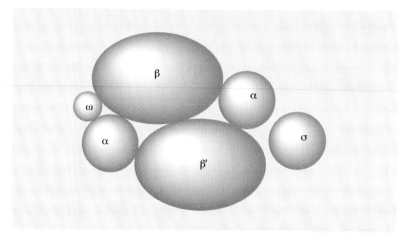

Fig. 10.1 Structure of RNA polymerase from *E. coli.*

In Chapter 1, we noted the existence of 'early' and 'late' proteins during the infection. Thus, at late times after virus infection, new sets of genes are transcribed, which were not transcribed early in infection. Based on our knowledge of transcription, how might this occur? There are at least four possible models:

1 The initiation specificity of the host RNA polymerase might be altered by synthesis of a new sigma factor or by modification of the core enzyme.

2 There could be *de novo* synthesis of a new polymerase having a new initiation specificity.

3 There could be synthesis of a protein factor which antagonizes normal termination of RNA chains (anti-termination factor) and hence allows host RNA polymerase to read through to distal genes.

4 There could be synthesis of a positive control protein, which is required for initiation of RNA synthesis at certain promoter sites but which is not a component of RNA polymerase.

These mechanisms are not necessarily exclusive and, in the case of bacteriophages T7 and T4, more than one of them is utilized. Since the pattern of T7 transcription is the simplest, we shall discuss it first.

The regulation of bacteriophage T7 multiplication

During T7 infection, a number of discrete messenger RNA (mRNA) molecules are formed and, since these are relatively stable, they can be isolated and characterized. These mRNA molecules can be divided into two classes: early mRNA, which consists of four species, and late mRNA, consisting of eight or nine discrete species. The two complementary strands of T7 DNA can be separated, and it is found that both early and late mRNA synthesized *in vivo* anneals to only one of the two strands. Thus, only one DNA strand is used as a template for mRNA synthesis.

Only early mRNA is made during infection with gene 1 conditional lethal mutants. Thus, the gene 1 product is necessary for transcription of mRNA from late T7 genes, i.e. those genes which are inaccessible to the bacterial RNA polymerase, and is in fact a new T7-specific RNA polymerase. This consists of a single polypeptide chain and consequently it is much simpler than the corresponding host enzyme. Furthermore, the T7 enzyme is resistant to rifamycin, an antibiotic inhibiting the host RNA polymerase, and to antibody directed against the host enzyme.

The current picture of T7 multiplication involves *transcription* of early genes by the host cell RNA polymerase (Fig. 10.2). Termination occurs at a specific site located between genes 1.3 and 1.7 and without the intervention of the termination factor ρ *In vitro*, *E. coli* RNA polymerase synthesizes a single mRNA species equal to the entire early region in length. However, *in vivo*, this mRNA is cut into discrete monocistronic messengers by the host enzyme ribonuclease (RNase) III. A number of discrete classes of late T7 mRNA can be isolated from infected cells. Each class is initiated at a different promoter site for T7 RNA polymerase, but all terminate at a site close to the right-hand end of the T7 DNA molecule.

The question remains as to how T7 turns off early phage and bacterial functions. The product of a late gene is required for the turn-off of host RNA synthesis and, since this protein is able to bind to bacterial RNA polymerase, it is possible that this enzyme is inactivated, thus preventing further synthesis of either host mRNA or early T7 mRNA.

The pattern of T7 regulation just described correlates nicely with genetic studies on T7. With an M_r of 25×10^6, T7 has a coding potential of approximately 30 genes. So far, 25 genes have been identified, mostly with the aid of amber mutants, and these can be ordered on a linear genetic map (Fig. 10.2) by means of

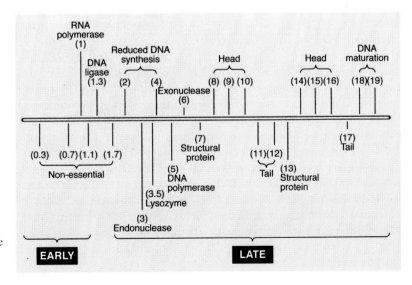

Fig. 10.2 The genetic map of bacteriophage T7. The numbers in parentheses are the gene numbers. Note the clustering of early and late genes.

standard genetic crosses and deletion mapping. The T7-specified proteins in infected cells can be analysed by electrophoresis on polyacrylamide gels. Any protein affected by an amber mutation or a deletion should disappear from its normal position in the gel, and this permits the identification of each band as the product of a particular gene. Studies on the time course of synthesis of different *gene products* shows that the proteins specified by genes 0.3 to 1.3 are synthesized early, while all the other genes are synthesized late. Thus, data from both transcription and expression studies agree that all the early genes are clustered together at the end of the genetic map, while the late genes form a larger cluster. Furthermore, since mRNA is synthesized in a 5' to 3' direction, the early genes must be clustered at the 3' end of the transcribed strand.

Late in infection, there should be sufficient amounts of the enzymes involved in DNA replication for their continued synthesis to be unnecessary. It is interesting to note that the genes determining the synthesis of these enzymes are not expressed throughout infection but are shut off sometime before lysis occurs. However, the mechanism controlling this is not known and no mutants are known in which this shut-off does not occur.

The regulation of bacteriophage T4 multiplication

The regulation of transcription in cells infected with phages T4 and λ is far more complex than that of T7. We shall postpone our discussion on regulation in λ until Chapter 12, because it is intimately tied in with the phenomenon of lysogeny, and consider only T4 regulation here. At least three classes of mRNA are found in T4-infected cells: immediate early, delayed early and late mRNA. As the name implies, immediate early mRNA is the first viral-specified RNA synthesized in the cell. Delayed early mRNA makes its appearance about 2–3 minutes after infection and late RNA is synthesized from about 10 minutes onwards. Furthermore, in contrast to T7, all T4-specific RNA synthesis is sensitive to rifamycin throughout the developmental cycle. Since the β subunit of the host polymerase is the site of rifamycin sensitivity, this subunit must be necessary for all phage RNA synthesis. Furthermore, pre-labelling of RNA polymerase before infection has shown that the α and β' subunits are also conserved, although they are modified slightly. By 2 minutes after infection, when delayed early RNA synthesis begins, 5' adenylate is covalently added to the α subunit. By 10–15 minutes after infection, when late RNA synthesis starts, the mobility of the β' subunit in electrophoretic gels has altered, suggesting that it too is modified. Furthermore, a newly synthesized polypeptide has been isolated from the core polymerase early in infection, the phage-specified ω factor. Thus, at different times, three different core polymerases are present in the infected cell (Fig. 10.3), the last two to appear being modifications of the host polymerase.

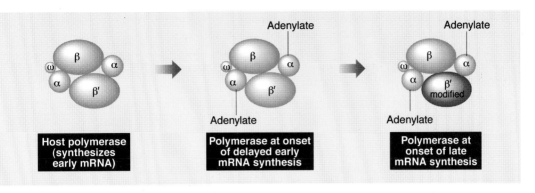

Fig. 10.3 Modification of the *E. coli* RNA polymerase following infection with bacteriophage T4.

Delayed early cistrons are contiguous with the immediate early cistrons, and both immediate early mRNA and delayed early mRNA are transcribed from the same DNA strand. To explain the synthesis of delayed early RNA, two models have been proposed. The first model suggests that delayed early transcription results from the inhibition of a chain termination factor (ρ), thus allowing polymerases which had been initiated at immediate early promoters to read through into delayed early regions. This process could occur *in vivo*, because early RNA sequences can be found in long RNA molecules which were initiated immediately after infection, when only immediate early promoters are active. The second model suggests that delayed early RNA is synthesized as a consequence of the altered initiation specificity of RNA polymerase itself. A factor has been isolated from T4-infected cells which directs both host and 5′ adenylated RNA polymerase to transcribe delayed early RNA, together with some immediate early RNA. In similar conditions, the *E. coli* σ factor directs the synthesis of 80–100% immediate early RNA. These results suggest that the early modifications to the core polymerase do not influence the accuracy of promoter selection.

Late RNA synthesis is absolutely dependent on the product of T4 gene 55. When the RNA polymerase−DNA complex is isolated from cells infected with gene 55 mutants, late RNA synthesis may be induced by addition of a crude preparation of gene 55 product. Now it is known that this is one of several RNA polymerase-binding proteins encoded also by genes 33 and 45.

In the absence of T4 DNA replication, late RNA is not synthesized unless DNA ligase is inactivated. Binding studies have shown that the late polymerase binds preferentially to denatured DNA, and it is possible to interpret these two results as follows. A specificity factor, possibly the gene 55 product, could either act as a σ factor, enabling the late polymerase to recognize certain breaks in lieu of promoters, or could itself bind to the breaks and generate a receptive structure for core polymerase.

Whereas T7 utilizes only one of the possible mechanisms of regulation listed on p. 144, namely, the formation of a new RNA polymerase, T4 uses at least three other mechanisms. Thus, studies on T4 development are more complex and, from the foregoing, it is clear that much has still to be learned.

RNA BACTERIOPHAGES

The genomes of the small RNA bacteriophages, such as MS2 and R17, code for only three proteins: phage coat protein, maturation protein (A protein) and the enzyme RNA synthetase. It might be thought that, with such limited genetic potential, these phages could reproduce without the use of any control mechanisms. This is not so. If susceptible cells are infected with these phages in the presence of radioactive amino acids and then extracted, the amount of radioactivity in each of the phage proteins can be measured after electrophoretic separation. From experiments of this type, it is clear that the coat protein is synthesized in great excess over the other two proteins. In fact, the three proteins are produced in the ratio of 20 coat proteins:5 synthetase molecules:1 maturation protein.

Gene order

Until the gene order was determined, it was not possible to explain how differential expression was achieved. Since the RNA of these phages is not known to undergo recombination, conventional genetic analysis was not possible, but a genetic map has been produced by biochemical means. Each of the three phage genes commences with an initiating sequence of nucleotides which can bind ribosomes directly, and, if a complex of ribosomes and phage RNA is digested with nucleases, these initiating regions are protected against enzymatic attack. These regions can thus be isolated and sequenced. The enzyme RNase IV from *E. coli* cleaves R17 RNA at a position 40% along the molecule from the 5' end. The initiating regions present on each of the two RNase IV fragments can also be identified, following nuclease degradation of fragment−ribosome complexes. In this way, the initiation sequence for the maturation protein and the coat protein were shown to be on the 40% fragment. Furthermore, nucleotide fingerprints corresponding to the internal parts of the gene appear in the 60% fragment. These data unambiguously show that the order of the genes in the phage must be:

5'−maturation protein−coat protein−synthetase−3'

Translational control of gene expression

It now appears that differences in the frequency of initiation of synthesis of the three proteins is the basis for regulation of RNA phage development. However, before presenting the evidence on which this statement is based, let us consider briefly how initiation of synthesis of each protein can be assayed. The simplest assay determines the amount of each initial dipeptide formed in an *in vitro* protein-synthesizing system, under conditions where chain elongation is inhibited. Since the N-terminal amino acid of each phage protein is different, the initial dipeptides are distinctive. In the case of MS2 and R17, they are formal methionine-serine for coat protein, formal methionine-alanine for synthetase and formal methionine-arginine for the maturation protein. The basis of the assay is to determine the amount of radioactivity incorporated into each fmet-labelled dipeptide, following separation by paper electrophoresis. Another initiation assay depends on the ability of ribosomes in an initiation complex to protect against RNase digestion that region of the phage RNA which is bound to the ribosome.

The effect of RNA conformation

The relative amounts of coat protein, synthetase and maturation protein initiated in *E. coli* extracts depend on the integrity and conformation of the viral RNA. With native RNA isolated from phage particles, only coat protein is initiated, as shown by the initial dipeptide assay. However, alteration of the conformation of the viral RNA by formaldehyde treatment or fragmentation of the RNA allows all three proteins to be initiated. The similar effects of these two treatments strongly suggest that specific conformational features in native RNA restrict initiation at the other two sites.

Determination of the nucleotide sequence of an extensive stretch of the coat protein and adjacent synthetase cistrons has revealed a probable region of hydrogen bonding, 21 nucleotides long, between codons 24 and 32 of the coat cistron and the synthetase initiation site. Hydrogen bonding in this region could account for the inability of ribosomes to bind to the synthetase initiation site. Furthermore, an amber mutation in any of the first 24 codons of the coat cistron would prevent ribosomes opening up this double-stranded region, thus accounting for the polar effect of some coat protein mutants on initiation of synthetase.

In contrast with synthetase formation, initiation of the maturation protein is not affected by conformational changes that occur during translation of the coat gene. Thus, amber mutations in the coat cistron have no polar effect on production of maturation protein *in vitro*. How then is the initiation of the synthetase regulated? As indicated in Chapter 8, much of the RNA in the infected cell exists as replicative intermediate (RI), and it will be recalled that this consists of an intact minus strand template, with one or more nascent single-stranded plus chains extending from it. *In vitro*, RI initiates about five times as much maturation protein, relative to the total protein initiated, as does single-stranded RNA isolated from phage particles. Presumably, the conformation of the 5' end of the single-stranded RNA chains in the RI is different from that of completed RNA molecules, allowing ribosomes access to the initiation site for the maturation protein on the nascent strands. Only a fraction of the maturation protein molecules initiated on the RI *in vitro* are actually completed, in contrast with coat protein, which is the major product directed by RI. This failure to produce complete molecules of maturation protein could be explained by assuming that only the shortest nascent strands have open initiation sites for the maturation protein, and that these strands are not long enough to contain the entire cistron. According to this hypothesis, as each nascent RNA chain is synthesized in the infected cell, a ribosome attaches to the maturation protein initiation site in the short 5' tail; as the RNA chain is elongated, that ribosomes proceeds to translate the maturation protein cistron, but folding of the elongated RNA strand (Fig. 10.4) prevents additional ribosomes from initiating at the initiation site.

Other factors affecting translational control

It appears that conformation of the phage RNA is not the only factor influencing the independent translation of the three genes, for there is some evidence indicating that the specificity of the ribosomes and initiation factors are also important. Since the ribosome binding sites differ in nucleotide sequence for each phage cistron, it is possible to imagine that the same ribosome can recognize all three initiation sites but bind with different affinity

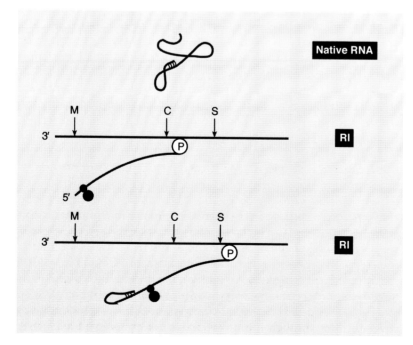

Fig. 10.4 Model to account for the linear rate of production of maturation protein. In native RNA the initiation site for maturation protein synthesis (M) is not available to the ribosome because of hydrogen bonding. During replication an RNA tail is produced at the 5' end in which the initiation site is available to the ribosome. As the nascent RNA chain grows in length, hydrogen bonding again prevents ribosome binding. P, polymerase; C, coat protein; S, synthetase.

to each. Alternatively, a separate class of ribosomes might recognize the initiation sequence of each cistron.

None of the three control systems so far described accounts for the cessation of synthesis of the RNA synthetase late in infection. This is explicable by a fourth mechanism of translational control — the specific binding of phage proteins to the messenger. When *E. coli* is infected with phage carrying mutations in the coat protein, there is an enhancement of synthetase formation, suggesting that coat protein acts as a translational repressor. Furthermore, when coat protein is incubated with phage RNA and the RNA is then used as a messenger *in vitro*, the formation of synthetase is specifically inhibited, as shown by the dipeptide assay. By way of contrast, the presence of coat protein does not affect the initiation or elongation of either the coat protein or the maturation protein. Thus the role of coat protein as a translational repressor of synthetase formation has been established.

ANIMAL VIRUSES

The regulation of gene expression by animal viruses can occur at the level of transcription or translation (or both) and to a degree which varies not only between different viruses but also during a multiplication cycle. The bewildering permutation of events which can ensue is further complicated by the possible confinement of certain events in the virus multiplication cycle to the cell nucleus, or an inhibition of host cell macromolecular synthesis by viral components. Examples showing some of the variety existing are given in Table 10.1. While it is clear that many animal viruses do indeed regulate their gene expression, it is not yet known, in many cases, how they achieve this. However, investigation of animal viruses has led to the elucidation of mechanisms totally different from those found in bacterial systems. Thus, animal viruses, as well as being of interest in their own right as infectious agents, are powerful tools for the study of the molecular biology of eukaryotic cells.

RNA viruses: class IV: diverse strategies for gene expression

Post-translational cleavage

The entire genome of picornavirus is translated as a huge polyprotein (M_r approx. 250 000), which is cleaved in a series of ordered steps to form smaller functional proteins (Fig. 10.5). Such post-translational cleavage is not found in bacteria but, after the pioneering work with poliovirus, is acknowledged as common in eukaryotic cells. Cleavage starts while the polyprotein is still being synthesized and only by inhibiting protease activity or altering the cleavage sites through the incorporation of amino acid analogues

Table 10.1 Some properties of animal virus nucleic acid and protein synthesis. (Adapted from A. E. Smith (1975) in *Society for General Microbiology Symposium*, **25**, 187.)

Class	Group	Example	Genome			No. mRNAs	Virion polymerase	Nucleus involved	Early/ late phases	Poly-A in mRNA	Poly- cistronic mRNA	Host shut-off
			Nucleic acid	M_r $\times 10^{-6}$	No. segments							
I	Papova	Polyoma	DNA	ds	5	6	–	+	+	+	–	–
I	Adeno	Adeno	DNA	ds	20–30	Several	–	+	+	+	–	+
I	Herpes	Herpes simplex	DNA	ds	80–150	Several	–	+	+	+	–	+
III	Reo	Reo	RNA	ds	15	10	+	–	+	–	–	+
IV	Picorna	Polio	RNA	ss	2.5	1	–	–	–	+	+	+
IV	Toga	Semliki Forest	RNA	ss	4	2	–	–	–	+	+	+
V	Rhabdo	VSV	RNA	ss	3.8	5	+	–	–	+	–	+
V	Paramyxo	NDV	RNA	ss	5–7	6	+	–	–	+	–	+ & –
V	Orthomyxo	Influenza	RNA	ss	4	10	+	+	+	+	–	+
VI	Retro	RSV	RNA	ss	1–3	2*	3	+	+	–	+	(+)	–

* 2 identical molecules of RNA.
poly-A, poly-adenosine; ds, double-stranded; ss, single-stranded; VSV, vesicular stomatitis virus; NDV, Newcastle disease virus; RSV, Rous sarcoma virus.

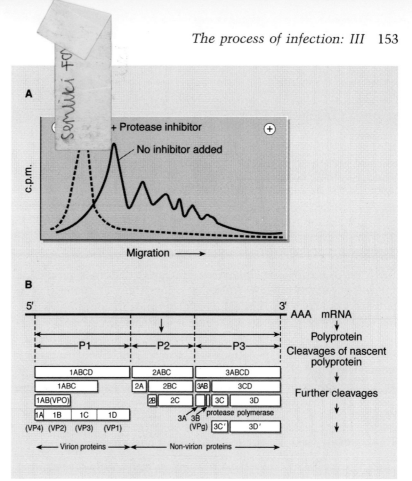

Fig. 10.5 (A) Polyacrylamide gel electrophoresis of proteins synthesized in poliovirus-infected cells in the presence and absence of protease inhibitors. (B) Representation of the cleavages of the polyprotein and the smaller products to yield mature viral proteins (see also p. 183). Note that most poliovirus mRNA lacks the 5' VPg protein present on virion RNA. 3C' and 3D' are produced by cleavage of 3CD at an alternative site to that producing 3C and 3D.

can the single polyprotein be isolated. Cleavage can also be demonstrated by pulse–chase analysis. In theory, all picornavirus proteins should be present in equimolar proportions, but this is not found in practice. It seems that a differential is introduced by degradation of some of the viral proteins; for instance, it is known that the virus-specified polymerase activity is unstable. All cleavages are carried out by virus-coded proteases. The best-known is 3C which is formed autocatalytically when P3 achieves the required confirmation.

Selective translation of virion RNA and synthesis of a subgenomic mRNA

Using cell-free translation, non-structural but not virion proteins are synthesized from the virion RNA of alphaviruses, such as Semliki Forest virus. This infers that there is an internal initiation site for the synthesis of virion proteins, which, as in RNA phage (p. 150), is hidden by the conformation of the RNA.

It is not known how transcription is controlled but, since no 26S

RNA⁻ is found in the cell we can assume that virion and 26S RNA⁺ are transcribed directly from the 42S RNA⁻ (Fig. 10.6). Despite this elaborate process, both mRNAs function simultaneously, and the 26S RNA appears to be an adaptation for making large amounts of virion proteins without an equimolar production of non-structural elements. Like the picornaviruses, the 26S RNA and non-structural region of the alphavirus genome are translated as polyproteins. Pulse–chase experiments and tryptic peptide maps show clearly that the small functional proteins are derived from larger precursor molecules.

Functionally monocistronic subgenomic mRNAs

Coronaviruses carry the alphavirus strategy to extremes, producing not two but seven mRNAs from the virion RNA. These have been called a 'nested set' as they all have 5′ termini in common, which are 'spliced' (see p. 156) to a particular downstream sequence (Fig. 10.7). Each mRNA is functionally monocistronic as it expresses only the gene which is situated at the 5′ end. Such a strategy seems wasteful, as about 60% of all mRNA sequences are not used, but perhaps compensating economy is achieved by exerting control at both transcriptional and translational levels.

RNA viruses: class V: one mRNA makes one protein

Class V viruses have a variable number of RNA segments making up their genome (Table 10.2), but gene expression is predominantly from monocistronic mRNAs. All have the same problem of tran-

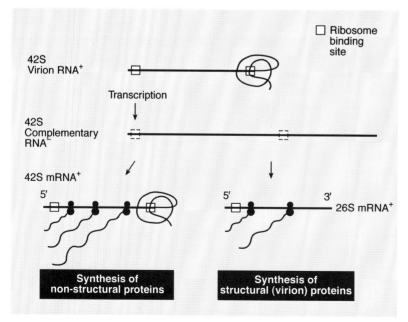

Fig. 10.6 Separate control of synthesis of structural and non-structural proteins in alphavirus-infected cells is achieved by blocking of the internal ribosome-binding site in the virion RNA and the synthesis of a separate smaller mRNA.

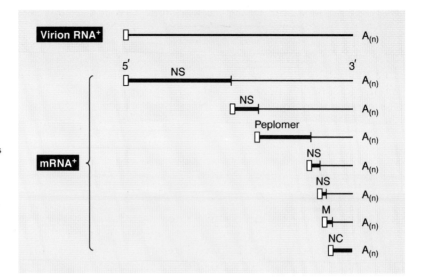

Fig. 10.7 Control of gene expression in Coronaviridae. mRNA's expressing virion peplomer (spike), matrix (M), nucleocapsid (NC) and various non-structural (NS) proteins are all transcribed from full-length virion (v) RNA⁻.

Table 10.2 The segmented class V viruses.

Family	No. of segments/genome
Orthomyxoviridae	8
Bunyaviridae	3
Arenaviridae	2
Rhabdoviridae	1
Paramyxoviridae	1
Filoviridae	1

scribing two or more monocistronic mRNAs from a polycistronic template.

Of the Orthomyxoviridae, most is known about the type A influenza viruses of humans and other animals. From each genomic RNA segment, a monocistronic mRNA is synthesized. The eight viral proteins are not present in equimolar amounts in the infected cell and the relative proportion of proteins changes during infection. Clearly, viral protein synthesis is controlled and the primary control operates at the level of transcription, although the molecular details are not known. However, the mechanism which distinguishes between mRNA⁺ and RNA⁺, which is template for virion RNA⁻, is understood. Careful annealing and sequencing studies have shown that the virus synthesizes two types of RNA⁺: template RNA⁺ is a faithful complement, while mRNA⁺ lacks some sequence from the 5′ end of the virion RNA (Fig. 10.8), as well as starting with a host mRNA sequence at its 5′ end (p. 123). Thus mRNAs cannot be templates for the synthesis of full-length genome segments.

Superimposed on this basic pattern is a control system which permits the synthesis of two proteins from influenza virus

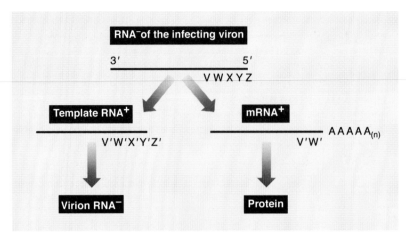

Fig. 10.8 Discrimination between mRNA$^+$ and template RNA$^+$ by the synthesis of two types of complementary RNA in influenza virus type A-infected cells. Note also that only mRNA is polyadenylated and that just one of the eight segments of the genome is shown here.

RNAs 7 and 8 by a process known as 'splicing'. Splicing is a post-transcriptional event which occurs in the nucleus. An internal section is removed from the primary transcript and the two remaining molecules are ligated. The newly spliced molecule may be read in the same reading frame to produce a truncated version of the original polypeptide or, as in the case of influenza virus, in a different reading frame (Fig. 10.9)—a nice example of biological economy. Segment 7 encodes the virion M1 (matrix) protein and

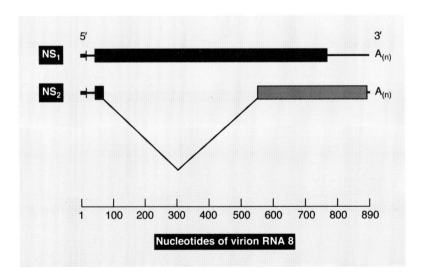

Fig. 10.9 Synthesis of mRNAs encoding NS1 and NS2 proteins, which are transcribed from RNA 8 of influenza type A. The shaded areas represent the coding regions. Note that NS1 and NS2 share a short common 5' amino acid sequence but differ in the majority of the sequence, since most of NS2 uses a different reading frame. The 'V' region is spliced out for NS2 mRNA. At the 5' end of the mRNA is the heterogeneous sequence obtained from cellular mRNA (p. 123).

M2, which is a minor virion transmembrane protein but is present in abundance in cell membranes. Segment 8 encodes the two non-structural proteins NS1 and NS2. It is galling that the NS1 protein, which was discovered by the author (NJD) in 1969 in the nucleolus of infected cells, still has no function ascribed to it. Neither has NS2.

The orthomyxoviruses are unique amongst 'true' RNA viruses (i.e. excluding the retroviruses, which have a DNA intermediate) in having an essential part of their multiplication cycle within the nucleus. For example, immediately after infection, the uncoated particle is transported to the nucleus, where viral mRNA is synthesized (p. 123), while, later in infection, certain of the newly synthesized proteins migrate from their site of synthesis in the cytoplasm into the nucleus. Since passage of molecules across the nuclear membrane is a highly selective process, this constitutes a level of control which is unique to eukaryotic cells. The mechanism is also important to those DNA viruses which replicate in the nucleus, and will be discussed later.

Of the remaining class V viruses, most is known about vesicular stomatitis virus (VSV, Rhabdoviridae). Transcription of the RNA$^-$ genome (about 3.8×10^6 M$_r$) yields separate monocistronic mRNAs, which are capped and polyadenylated (Fig. 10.10). mRNAs are synthesized sequentially from a single promoter. The proportion of mRNAs and their products decreases towards the 5' terminus, and this may reflect the probability of the transcriptase falling off before it reaches the end of the genome. Read-through from the 5' terminus of the leader RNA to the exact 3' terminus requires the nucleocapsid (N) and NS proteins and results in a complete RNA$^+$ strand, which is the template for replication. Paramyxoviruses have a similar strategy.

Although Arenaviridae appeared to be conventional segmented negative-strand viruses with just two (L and S) genomic RNAs, the smaller of their two genomic segments proved to have an unusual

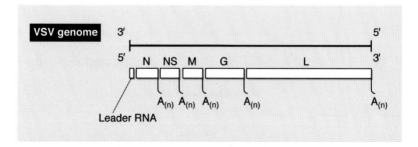

Fig. 10.10 Transcription of mRNAs of vesicular stomatitis virus encoding the nucleocapsid (N), NS, matrix (M), glycoprotein (G) and large (L) proteins (see Fig. 3.16). Note that the 48-nucleotide leader RNA is transcribed exactly from the 3' terminus of the virion RNA (see text). mRNAs are separated by non-transcribed regions of two to four nucleotides.

sequence structure. While the 3' part of S RNA is indeed RNA$^-$, the 5' region is RNA$^+$; hence its name, 'ambisense' RNA (Fig. 10.11).

All class V viruses have monocistronic mRNAs, and it might be expected that post-translational cleavage would not be involved. However, this is not always the case; for example, after its synthesis, the influenza virus haemagglutinin (HA) protein undergoes a single protease cleavage (Fig. 10.12). Uncleaved molecules occur in cells deficient in protease activity but are still incorporated into virus

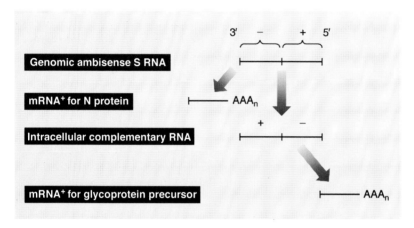

Fig. 10.11 Ambisense S RNA segment of Pichinde virus (Arenaviridae).

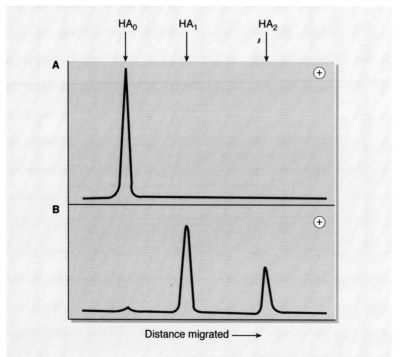

Fig. 10.12 Cleavage of influenza virus haemagglutinin. (A) Electrophoresis of an uncleaved molecule from non-infectious virus grown in cells deficient in protease activity. (B) Haemagglutinin molecules isolated from the same virus incubated with 100 μg/ml trypsin for 30 minutes at 37°C. After incubation with trypsin the virus becomes infectious.

particles which are morphologically normal and can agglutinate red blood cells. However, such virus particles are non-infective until the HA protein is cleaved. The cleaved HA changes conformation and the fusion peptide is now free at the N terminus of HA2 (Fig. 13.17). Activation of cellular proteins in eukaryotes by proteolysis is well known: for instance, conversion of inactive proinsulin to insulin and of inactive trypsinogen into trypsin.

RNA viruses: class III: viral proteins positively control mRNA synthesis

Control of protein synthesis in viruses with a segmented double-stranded RNA genome, the Reoviridae, arises from the interplay of transcriptional and translational activities. Each genomic segment is transcribed as a monocistronic mRNA, but, when protein synthesis is inhibited in the cell by the addition of cycloheximide, only four mRNAs are made, showing that there is positive control by viral proteins over the synthesis of the remaining six mRNAs. Further indication of transcriptional control is seen when the polysomes from an infected cell are analysed for their content of viral mRNAs. All 10 mRNAs are represented, but not in equimolar amounts. In addition, there is evidence of translational control, since there is more protein made from mRNAs of the $0.7 \times 10^6 \, M_r$ class than those of $0.4 \times 10^6 \, M_r$, even though more of the latter is synthesized. Interestingly, none of the reovirus mRNAs is polyadenylated, but they obviously still function effectively.

The 'DNA' viruses: classes I, II and VI: using viruses to understand nuclear controls

All DNA viruses, with the exception of the Poxviridae, synthesize their mRNA from a double-stranded DNA molecule, which is located in the cell nucleus, and thus they come closest as model systems for studying the synthesis of cellular mRNA. The finding that cell-coded histones, which form the major protein component of cellular DNA, are also complexed with papovavirus virion DNA underlines the similarity of these viral and cellular nucleic acids. The range of genome size spans two orders of magnitude, from parvovirus DNA of $1.5 \times 10^6 \, M_r$ to vaccinia virus DNA of $160 \times 10^6 \, M_r$, and the problem of analysing these genomes is in proportion. The reason why large viruses exist at all when the smallest function perfectly well is obscure. Certainly the larger viruses are less dependent upon the cell and can multiply when cells are not in S phase (synthesizing cellular DNA), whereas the smallest viruses are unable to do so. Viruses which have a genome of intermediate size have the capacity to induce the host cell to enter S phase.

Excepting poxviruses and parvoviruses, all DNA animal viruses are capable of transforming cells (Chapter 17). One can therefore speculate that transformation may be an aberrant expression of the interaction of virus with the mechanism that regulates cell division. Much viral information appears to duplicate that present in dividing cells. However, it seems likely that such functions are essential during natural infections, when the virus is found in highly differentiated cells, such as neurones, which are not dividing or are metabolically deficient in some respect.

Papovaviruses

Studies of the papovaviruses, simian virus type 40 (SV40) and polyoma, have been most revealing. Other virus groups will be mentioned only when their gene expression differs in a significant aspect. Papovaviruses have a double-stranded circular genome, which encodes early and late mRNAs (Fig. 10.13), and from which early and late proteins are translated.

Papovavirus DNA is transcribed by cellular polymerase II, which normally synthesizes cellular mRNAs. SV40 has an abundant major early protein, the large T (for tumour-specific) antigen, and a lesser protein called small t. Polyoma has three T antigens, which are closely related to those of SV40, called large, middle and small. Their roles in transformation are discussed later (p. 262).

Synthesis of early proteins

All the mRNAs are formed by splicing. This process was first

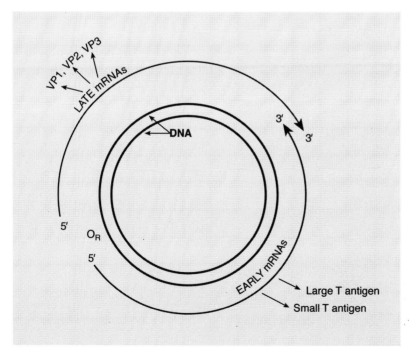

Fig. 10.13 Map of the genome of SV40 virus. VP1, VP2 and VP3 are virion proteins. O_R is the origin of replication.

discovered in adenovirus-infected cells and has since proved to be a mechanism commonly used for the production of mRNAs by eukaryotic cells and of mRNAs from RNA templates (Fig. 10.9). Usually, the DNA template consists of coding regions (exons), separated by non-coding intervening sequences (introns). These are transcribed into a pre-mRNA molecule, which is then spliced. Sometimes, an RNA is differentially spliced so that all three reading frames can be used, as in the synthesis of polyoma virus T antigens (see below).

The large T and small t antigens of SV40 have similar-sized mRNAs, but only the initial part of the small t mRNA is expressed, as there is a UAA termination codon situated about a third of the way in (Fig. 10.14A). The mRNA of large T has a 346-nucleotide intron spliced out, which voids the little t termination signal just mentioned and allows transcription to continue in a second reading frame. Synthesis of polyoma virus T antigens is similar but more complex. Although all start in the same reading frame (and hence like SV40 have the same N-terminal sequence), there are three different introns, which allow translation to continue in all three reading frames and to produce three different proteins. Nature can be wonderfully economic.

Synthesis of late proteins

The early phase is followed by induction of the synthesis of host cell enzymes and of host cell DNA. Large T antigen binds to viral DNA and permits replication to commence (Fig. 10.15). Early and late mRNAs are transcribed from different strands of DNA. How-

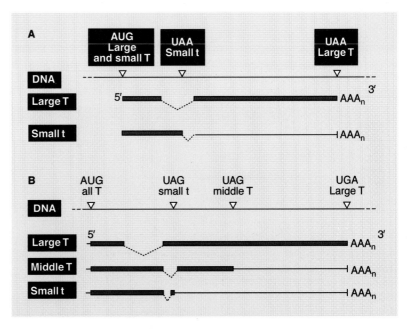

Fig. 10.14 Synthesis of early mRNAs for the T antigens of (A) SV40 and (B) polyoma virus. In SV40 splicing avoids the little t UAA termination codon and results in a frame shift. In polyoma all three reading frames are used. A solid block indicates the translated region of each mRNA, the dotted line is an intron and the single thin line is untranslated RNA. Note that the template translation initiation (AUG) and termination codons (UAA, UAG, UGA) are designated as they appear on mRNA.

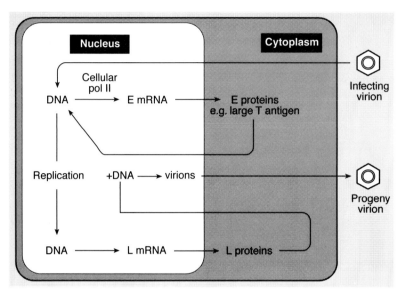

Fig. 10.15 Summary of the major early (E) and late (L) events controlling synthesis of papovavirus proteins.

ever, strand selection does not appear to represent a control process, since both early and late mRNAs are found in the nucleus late in infection. The predominant synthesis of late proteins is caused by down-regulation of the synthesis of early mRNA by the large T antigen. This is followed by the synthesis of late (virion) proteins, which are then assembled with DNA into progeny virions. Synthesis of late mRNAs uses a similar strategy to that of early mRNAs and is illustrated in Fig. 10.16.

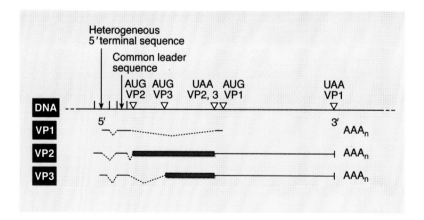

Fig. 10.16 Synthesis of late mRNAs for virion proteins (VP) of polyoma virus. The scheme is that used in Fig. 10.14. Note the untranslated heterogeneous 5′ terminal sequence, which is spliced on to a common untranslated leader sequence. In turn, this is spliced on to the coding region of the mRNA. Note also that VP2 and VP3 have the same reading frame.

Other DNA viruses

Regulation of gene expression in the Papovaviridae, described above, and all other DNA viruses does not differ in principle, and only brief mention will be made below to certain specific aspects of these other systems. For further information, the reader should consult references listed in Chapter 21.

Adenoviruses

Adenoviruses have a linear genome. As with papovaviruses, adeno-virus gene expression is divided into an early (pre-DNA synthesis) and late (post-DNA synthesis) phase (Fig. 10.17A). Early mRNA synthesis occurs from both DNA strands, which are called by convention 'left' and 'right'. Cellular polymerase II carries out both early and late transcription.

Early transcription takes place from five early (E) regions (Fig. 10.17B), each of which has its own promoter and termination sequences. Transcription is complex and, within each E region,

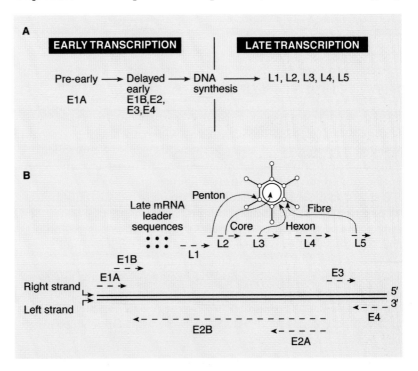

Fig. 10.17 Gene expression in adenovirus. (A) The division into 'early' (E) and 'late' (L) transcription. The numbers refer to regions of the viral genome from which transcription takes place (see below). (B) Organization of the adenovirus genome. Arrowheads show the direction of transcription. The dashed arrows are reminders that each region encodes several spliced mRNAs. E regions shown below the genome are synthesized from the left DNA strand and both E and L regions shown above the genome from the right DNA strand. The origin of some of the virion proteins is also indicated.

several different mRNAs are synthesized. This is accomplished by splicing, a strategy which is commonly used by adenoviruses (in contrast to the herpesviruses, which have very few spliced mRNAs). The E1 regions are critical for transformation, and we shall return to them in Chapter 17.

Transcription is tightly regulated and takes place in an ordered sequence; E1A is the first protein to be synthesized. DNA synthesis punctuates the change from early to late gene expression, but the molecular mechanisms of the switch are not known. Late RNAs are principally encoded by the right DNA strand and share a common leader sequence, which is formed by splicing of three disparate leader sequences (Fig. 10.17B).

Herpesviruses

Herpesviruses have a linear genome. Their gene expression departs from the scheme outlined for the papovaviruses only in that transcription of late mRNAs is more or less independent of DNA replication. Thus, when DNA synthesis is inhibited, both early and late transcripts are detected in the nucleus. However, protein synthesis in herpes simplex virus type 1 (HSV-1)-infected cells follows a system of induction and repression which occurs in three phases, called, α, β and γ.

After entry and uncoating of the outer virion structures, the virion core (DNA and associated proteins) of HSV enters the nucleus (Fig. 10.18A). The virion protein VP16 (a γ peptide) is required to initiate viral transcription, which is carried out by the cellular RNA polymerase II. Unlike adenovirus, which has a regional control system, transcription of each herpesvirus gene is controlled by individual initiation and termination controlling elements. VP16 is also important in the control of latency (p. 220).

Genetic analysis shows that α gene products induce transcription of β genes; β gene products inhibit expression of α genes and induce γ gene transcription, which in turn inhibits expression of β (Fig. 10.18B). Controls operate at both transcriptional and translational levels, but there is transcription of at least some of the α and β genes throughout infection; α and β genes are not topographically separated and are found scattered throughout the genome. DNA replication depends on the β gene products, including DNA polymerase, and takes place by a rolling circle mechanism (see p. 105).

Poxviruses

Poxviruses have a linear genome and are the only DNA viruses to multiply in the cytoplasm. As a family, the Poxviridae have the largest genomes, with a coding capacity 50-fold greater than parvo- and papovaviruses. Their site of multiplication in the cytoplasm can be likened to the establishment of a second 'nucleus'. Indeed,

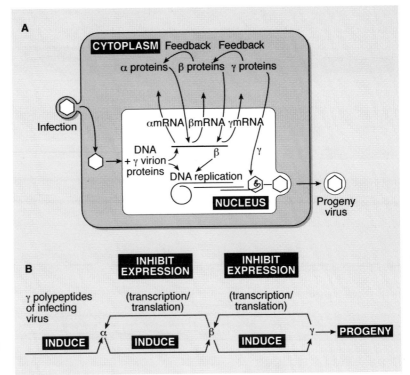

Fig. 10.18 Gene expression in herpes simplex virus type 1. (A) Shows some of the intracellular events of infection and (B) successive induction and feedback inhibition of α, β and γ gene products.

virions contain many enzymatic activities which duplicate those found in the nucleus itself. The virion has its own DNA-dependent RNA polymerase, as well as enzymes which cap, methylate and polyadenylate the resulting mRNAs.

Translational control, too, is shown during the synthesis of viral thymidine kinase, an early protein. In the cell, early viral mRNAs are very stable and yet synthesis of thymidine kinase ceases to increase after synthesis of the late class of proteins. This argues that there is direct inhibition at the translational level of the thymidine kinase mRNA.

Retroviruses

Retroviruses combine many of the properties of both RNA and DNA viruses. The DNA provirus is synthesized in the cytoplasm (Chapter 8) and is transported to the nucleus and there integrated with host DNA. mRNAs are synthesized from integrated DNA and transported to the cytoplasm. Transcriptional controls operate to synthesize three mRNAs, two of which are spliced (Fig. 10.19A). Rous sarcoma virus encodes three viral proteins common to all retroviruses in the order *gag*, *pol* and *env*, and one protein (*src*) originally from the host cell genome. There is a termination codon at the end of the *gag* gene, so usually only gag is synthesized. Translation of *pol* is permitted by occasional non-sense suppression

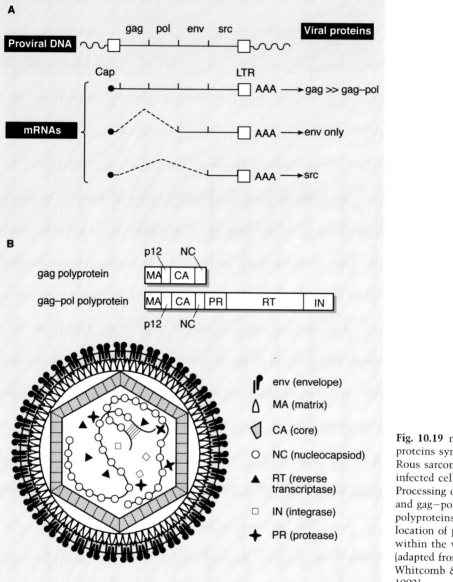

Fig. 10.19 mRNAs and proteins synthesized in Rous sarcoma virus-infected cells. (B) Processing of the gag and gag–pol polyproteins and the location of proteins within the virion (adapted from Whitcomb & Hughes, 1992).

events at the *gag–pol* boundary, producing a gag–pol fusion poly-peptide. The gag polyprotein is processed by the viral protease to make three major internal virion proteins (matrix, core and nucleo-capsid) and a p12 protein, whose function is poorly understood (Fig. 10.19B). Similarly the gag–pol polyprotein is processed to form the gag products as before, and three minor but functionally important virion proteins (protease, reverse transcriptase and inte-grase: see p. 134–6) situated within the core structure. In the cell, *gag* gene products far exceed those of *pol* in amount, presumably reflecting the relative amounts of structural protein and enzymatic activity required for successful multiplication. The structure of mRNA coding for *env* is similar, as it contains the information for *src*. However, *src* is not synthesized from this structure (which thus has a similar strategy to the alphavirus virion RNA and coronavirus mRNA) but is made on a monocistronic message. The genome of the human immunodeficiency virus (HIV-1) is more complex and has a number of regulatory genes (see Chapter 19).

FURTHER READING

This will be found in Chapter 21 listed under each family of viruses.

11 The process of infection: IV. The assembly of viruses

In an infected cell, virus-specified proteins and nucleic acid are synthesized separately and consequently must be brought together and assembled into mature virus particles. There are three ways in which this occurs: (i) viruses may *self-assemble* in a manner akin to crystallization, whereby the various components spontaneously combine to form particles which represent a minimum energy state; (ii) the viral genome may specify certain morphogenetic factors which are not structurally part of the virus but are required for normal assembly; or (iii) the particle may be assembled from precursor proteins, which are then modified to form the virion, and in such a case it is not possible to dissociate and then reassemble a mature particle from its constituent parts.

SELF-ASSEMBLY FROM MATURE VIRION COMPONENTS

Absolute proof of self-assembly requires that purified viral nucleic acid and purified structural proteins, but no other proteins, be able to combine *in vitro* to yield particles which resemble the original virus in shape, size and stability and are infectious. In practice, however, the critical step in demonstrating assembly *in vitro* is really disassembly. The virus must be disassembled in such a way that the released subunits retain their ability to reassemble in a specific manner relating to their origin. Consequently, proper assembly constitutes the test for proper disassembly! Ideally, the protein coat separates into its constituent monomers, which are not denatured.

Assembly of tobacco mosaic virus: (TMV) introduction

Although self-assembly is known to occur in the formation of other biological structures, e.g. flagella, the best studied example is undoutedly the *in vitro* reconstitution of TMV. This virus consists of a long, helically wound, molecule of ribonucleic acid (RNA), embedded in a framework of small, identical protein molecules ('A' protein)*, which are also arranged helically (Chapter 3). As already outlined in Chapter 1, TMV can be disassembled to yield protein and RNA components, which can be reassembled *in vitro* to yield active virus. However, the isolated protein, free from any RNA, can also be polymerized into a helical structure, indicating

---

* TMV protein is usually obtained by mild alkali treatment of pure virus. Alkali destroys the RNA leaving *'alkalischer'* or 'A' protein.

that bonding between the subunits is a specific property of the protein. Thus, the most likely model for the assembly of the virus would be for the protein molecules to arrange themselves like steps in a spiral staircase, enclosing the RNA as a cork-screw-like thread. Recent research, however, has indicated that the assembly of TMV is a much more complicated process than that suggested above.

In solution, TMV protein forms several distinct kinds of aggregates, depending on the environmental conditions, particularly ionic strength and pH (Fig. 11.1). Of these, the disc structure is considered the most important, since it is the dominant aggregate found under physiological conditions. A rod-like particle built of stacked rings could well arise as a variant of the normal helical structure; the protein subunits would have a similar bonding pattern to that in the virus, but in the absence of RNA−protein interactions there could be small local differences in packing so that turns of the helix are transformed into closed rings. Since there are 17 subunits per ring, which is close to the 16.34 per turn of the viral helix, the lateral bonding between the subunits is probably very similar. However, if the stacked-disc structure is a true variant of the helical structure, then successive rings would face in the same direction, as do successive turns of the helix. That is, the structure would be polar like the helix and polar aggregation of rings would be expected to continue indefinitely and not stop at the two-ring structure.

This problem was resolved by careful analysis of electron micrographs, which revealed that there is a slight axial perturbation at

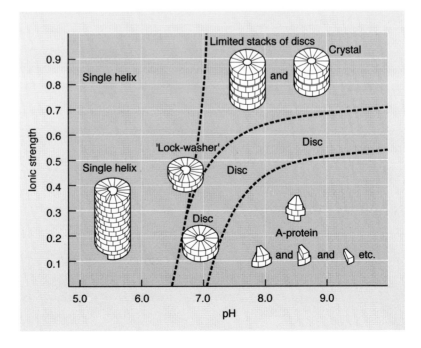

Fig. 11.1 Effect of pH and ionic strength on the formation of aggregates of TMV protein.

the edges of the subunits, such that a double disc presents a marginally different aspect to the under-side of a further disc adding to it (see Fig. 11.4). When discs are taken to a lower pH, aggregation progressively occurs, with the appearance of short rods made up of imperfectly meshed sections of two helical turns, which, after many hours, can anneal to give the regular virus-like structure (Fig. 11.1). An explanation for this comes from titration studies, which showed that TMV contains two ionizable groups per subunit titrating with a pK close to 7.0. Since the protein contains no histidine and no terminal amino groups, it has been suggested that these are carboxylic acid groups titrating abnormally. This would result from the assembled helix constraining two pairs of carboxylate groups per subunit to be close to one another. By reducing the pH, the electrostatic repulsion between the carboxylate groups is reduced, thus converting the disc into a 'lock-washer'. For the purposes of reconstitution of the virus, the conversion to the helical mode could be brought about by interaction of the protein with RNA, which provides the additional energy necessary for the stabilization of the helical form.

When A-protein subunits and small aggregates are mixed with TMV, RNA polymerization is slow and formation of virus requires about 6 hours. However, when discs are mixed with RNA under the same conditions, polymerization is rapid and mature virus forms within 5 minutes. Addition of small aggregates as well as discs to the RNA does not increase the rate of polymerization. These results strongly suggest that discs are the normal precursor for the assembly of virus. A possible model for the assembly of TMV is shown in Fig. 11.2 (but note that an alternative model is presented later). A disc is added to the growing helix (A) and this is

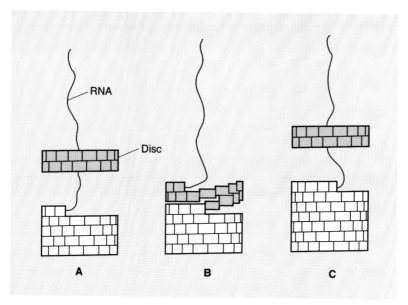

Fig. 11.2 Simple model for assembly of TMV. See text for details.

converted to the lock-washer form (B) as a result of neutralization of the juxtaposed carboxyl groups by interaction with the viral RNA. Following conversion to the lock-washer form, the subunits progressively unroll, entrapping the RNA until the structure is ready to receive another disc (C).

The involvement of discs rather than individual subunits is not too surprising, since the first turn of a helical structure is hard to build. One solution is to have a firmly bonded pre-existing ring, which could also be transformed into the first turn of the helix. A single ring tends to be an unstable structure, because there is only a single bond between adjacent subunits. A disc made of two rings is considerably more stable, because the ratio of bonds to subunits is much greater. Furthermore, when discs are used rather than single subunits, the assembly process is less likely to be affected by short unfavourable nucleotide sequences in the RNA.

The assembly of TMV: bidirectional growth

When purified TMV RNA is mixed with limited amounts of viral coat protein in the form of the disc aggregate, a unique region of the RNA becomes protected from nuclease digestion. The protected RNA consists of fragments up to 500 nucleotides long, which are found in nucleoprotein particles having a protein–nucleic acid ratio similar to that of the mature virus. The shortest fragments define a core about 100 residues long common to all the fragments, while larger ones are covalently extended by up to 400 nucleotides in one direction and up to 30 in the other. These data are interpreted as showing that assembly is initiated at a unique internal packaging site on the RNA, and that growth occurs bidirectionally but at greatly unequal rates. The major direction of assembly would be 3' to 5', and consistent with this view is the finding, from sequence analysis of the protected fragments, that the packaging site is close to the 3' end of TMV RNA. Furthermore, sequence analysis of the packaging site suggests strongly that it exhibits a hairpin configuration (Fig. 11.3).

Assembly of TMV: the 'travelling loop' model

This model suggests that the RNA hairpin inserts itself through the central hole of the disc into the 'jaws' between the rings of subunits. The nucleotides in the double-stranded stem then unpair and more of the RNA is bound within the jaws. As a consequence of this interaction, the disc becomes converted to a lock-washer structure trapping the RNA (Fig. 11.4). The special configuration generated by the insertion of the RNA into the central hole of the initiating disc could subsequently be repeated during the addition of further discs on top of the growing helix; the loop could be perpetuated by drawing more of the longer tail of the RNA up through the central hole of the growing virus particle. Hence the

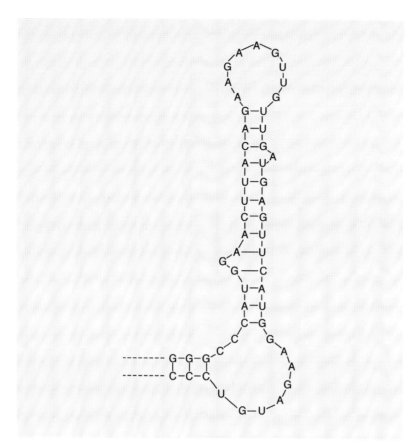

Fig. 11.3 The packaging site of TMV RNA. The loop probably binds to the first protein disc to begin assembly. The fact that guanine is present in every third position in the loop and adjacent stem may be important in this respect.

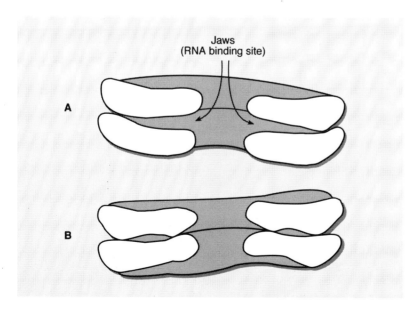

Jaws
(RNA binding site)

A

B

Fig. 11.4 Diagrammatic representation of a cross-section through a disc (A) and a lock-washer (B). The arrangement of the two layers of subunits in the disc can be likened to a pair of nutcrackers. The 'jaws' are the RNA binding site, and on conversion of the disc to a lock-washer the jaws close, trapping the RNA.

particle could elongate by a mechanism similar to initiation of packaging, only now instead of the specific packaging loop there would be a 'travelling loop' of RNA at the main growing end of the virus particle (Fig. 11.5). This loop would insert itself into the central hole of the next incoming disc, causing its conversion to the lock-washer form and continuing the growth of the virus particle.

A disturbing feature of the earlier model shown in Fig. 11.2 is that discs have to be threaded on to the RNA chain and this would obviously be the rate-limiting step. However, the 'travelling loop' model shown in Fig. 11.5 overcomes this problem as far as growth in the 5' direction is concerned, for incoming discs would add directly on to the growing protein rod. Discs would still have to be threaded on to the 3' end of the RNA and thus elongation in this direction would be much slower, as has been experimentally observed. One prediction of the 'travelling loop' model is that both the 5' and 3' tails of the RNA should protrude from one end of partially assembled TMV particles. Electron micrographs of such structures have now been observed.

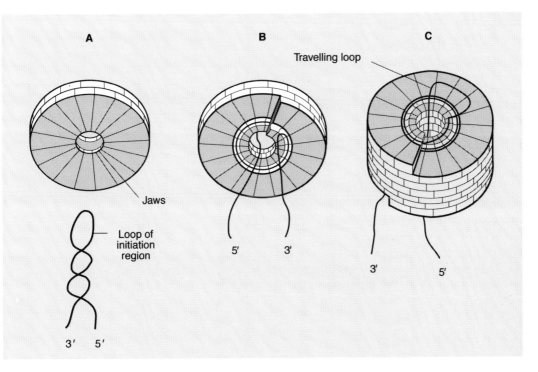

Fig. 11.5 The 'travelling loop' model for TMV assembly. Nucleation begins with the insertion of the hairpin loop of the packaging region of TMV RNA into the central hole of the first protein disc (A). The loop intercalates between the two layers of subunits and binds around the first turn of the disc. On conversion to the lock-washer (B) the RNA is trapped. As a result of the mode of initiation the longer RNA tail is doubled back through the central hole of the rod (C), forming a travelling loop to which additional discs add rapidly.

The spherical plant and bacterial viruses

The most detailed study so far reported on the self-assembly and reconstitution of a spherical virus is that of cowpea chlorotic mottle virus (CCMV). The formation, in a stoichiometric mixture of initially separated CCMV RNA and protein, of a nucleoprotein with the infectivity of CCMV affords proof of the ability of this virus to self-assemble.

The first reconstitution experiments with spherical bacteriophages, such as R17, yielded particles which had the same size, density, RNA content and electron microscopic appearance as the original phage. However, they sedimented more slowly than phages formed *in vivo* and could not adsorb to bacteria. These defective particles formed *in vitro* resemble those formed in restrictive hosts infected with phages carrying an amber mutation in the A-protein cistron. Both types of particles lack the A protein. When the A protein is added in stoichiometric amounts to the reconstitution mixture, infectious virus is formed.

Assembly pathway for spherical viruses

The two groups of spherical viruses just discussed exist as a complex of RNA with 180 coat protein molecules, and this obviously cannot be synthesized in a single step. Intermediates in virus assembly must exist. The RNA appears to form an initiation complex by reacting with a few protein subunits or, in the case of the RNA phages, with the A protein. RNA condensation occurs either during this step or in the subsequent aggregation of additional subunits that yields the capsid. Consistent with this model is the isolation of complexes consisting of a few protein subunits and one RNA strand.

Assembly *in vivo* is probably different in some respects from the occurring *in vitro*, especially with regard to local concentrations of the components and their nascent states. Thus, single-stranded RNA is used in the *in vitro* reconstitution of bacteriophages, while *in vivo* the pool of single-stranded RNA is very small, if it exists at all. Thus translation and self-assembly might be coupled and regulate each other. A model for such a process is shown in Fig. 11.6. In it RNA has been transcribed to the extent that the A-protein cistron and the coat protein cistron are available for translation. One molecule of A protein and a few molecules of coat protein have already been synthesized (pp. 148–50) and are bound to the RNA. The ribosomes continue to translate the RNA. The coat protein molecules situated close to the RNA replication fork delay further RNA synthesis, and RNA replicase production is thus repressed at the level of RNA transcription. More coat protein is synthesized, which binds to the RNA on different sites, except for the coat protein cistron, which is occupied by ribosomes. Then the RNA replicase resumes its activity and releases the completed

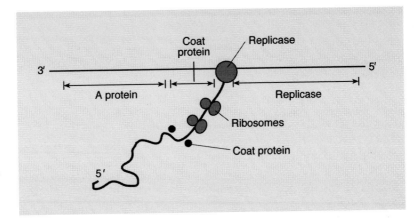

Fig. 11.6 A model for coupling of transcription, translation and protein synthesis in the regulation of RNA phage development.

RNA⁺ strand. However, the completed plus strand remains re-pressed for RNA replicase synthesis, but this time at the level of translation. Finally, the ribosomes fall off and the capsid is completed.

DIRECTED ASSEMBLY

Assembly of bacteriophage T4

As we saw in Chapter 3, the structure of the tailed bacteriophages is considerably more complex than that of the viruses we have just been considering. Extensive studies on these bacteriophages, particularly T4 and λ, have revealed that non-structural proteins can direct assembly of viruses. The results obtained from the study of assembly of both phages are analogous, but the greatest body of information exists for T4.

The first advance in elucidating the morphogenesis of T4 came with the isolation of conditional lethal mutants (see p. 11). When non-permissive *Escherichia coli* cells are infected with phage carrying amber mutations in any of the structural genes, phage particles are not produced. However, electron microscopic exam-ination of lysates of these infected cells reveals the presence of structures readily recognizable as phage components. Thus, cells infected with mutants in genes 34, 35, 36, 37 or 38 accumulate phage particles which appear normal except for the absence of tail fibres. We can thus conclude that these genes code for proteins involved in tail fibre assembly and that attachment of tail fibres occurs late in the assembly process. Since tail fibres also accumulate in cells infected with mutants in genes 34, 35, 36 and 38, but not gene 37, this latter gene must be the structural gene for the major fibre protein. Extension of this technique to cells infected with other mutants led to the identification of the function of many T4 genes (Table 11.1), as well as suggesting that the head, tail and tail fibres were all synthesized independently of each other.

The next advance came a few years later, when it was discovered that the morphogenesis of T4 could be made to occur *in vitro*. In one experiment, purified fibreless phage isolated from cells infected with a tail fibre mutant (gene 34) were mixed with an extract of cells infected with a mutant (gene 23) which cannot synthesize heads. Infectivity rapidly increased by several orders of magnitude, indicating that the gene 23 mutant extract was acting as a tail fibre donor, whilst the other extract supplied the heads (Fig. 11.7). This type of experiment is analogous to a genetic complementation test but is different in that it occurs *in vitro*.

The foregoing analysis provides only limited information concerning the sequence of assembly reactions. To determine the exact sequence, the precursor structures have to be isolated prior to the complementation tests. For example, free base plates can be isolated from cells infected with T4 which has a defective gene 19, gene 48 or gene 54 by zone sedimentation, and after isolation they still retain their *in vitro* complementation activity. By determining the ability of each to complement the other two unfractionated extracts, it is possible to establish the sequence in which the three gene products interact with the base plate. If isolated gene 19-defective base plates complement a gene 54-defective extract, the

Table 11.1 Phenotype of T4 mutants as determined by electron microscopy of infected cell lysates.

Genes	Phenotype	Conclusions
Mutant group Y 20, 21, 22, 23, 24, 31, 40	Fail to produce heads or generate aberrant heads	These genes involved in determination of head shape
Mutant group X 2, 4, 16, 17, 49, 50, 64, 65	Produce heads which appear normal by electron microscopy but extracts inactive when tested by *in vitro* complementation	These genes involved in head maturation subsequent to shape acquisition
5, 6, 7, 8, 10, 11, 25, 26, 27, 28, 29, 51, 53	Fail to produce base plates	Required for synthesis of base plates
3, 15, 18, 19	Fail to produce complete tails	Required for assembly of tail
34, 35, 36, 37, 38, 57	Fail to produce tail fibres or produce only half-fibres	Required for assembly of tail fibres
13, 14	Normal heads and tail produced but no mature particles	Required for head–tail joining

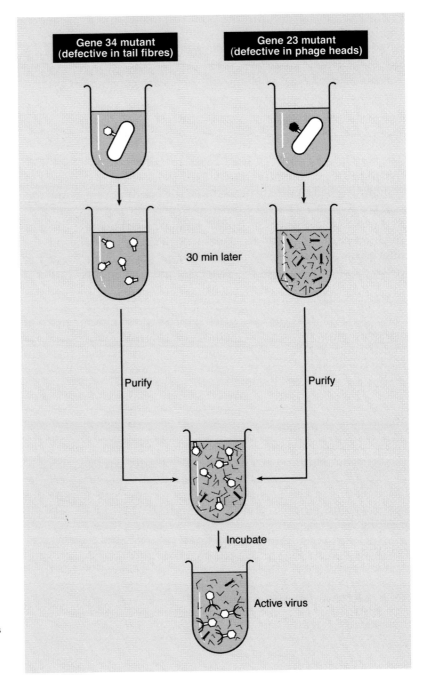

Fig. 11.7 *In vitro* complementation between two mutants which are blocked at different stages in morphogenesis.

gene 54 product must have carried out its function by interacting with the base plate in the absence of the gene 19 product, i.e. the gene 54 protein acts prior to the gene 19 protein. If, moreover, in the converse experiment, isolated gene 54-defective base plates do not complement a gene 19-defective extract, then the gene 19 product cannot act prior to the gene 54 product and the sequence 54, 19 is established.

Using these techniques in conjunction with analysis of polypeptides in polyacrylamide gels, it has been possible to determine the way in which heads, tails and tail fibres are constructed prior to their assembly into mature virions.

Head assembly

The way in which the T4 head is assembled can be deduced from the properties of mutants classified in groups X and Y (Table 11.1). A more detailed summary of the properties of the group Y mutants is given in Table 11.2. The order in which the Y group genes act during head morphogenesis can be defined by considering the relative complexities of these aberrant structures in terms of the information needed to specify them. The fewer genes needed to specify a given structure, the simpler it must be and the earlier it must lie relative to the assembly pathway. Based upon this concept, the first stage of head morphogenesis is as shown in Fig. 11.8. Note, however, that the structures shown are *not* intermediates in assembly but represent the products of abortive assembly. Since multilayered polyheads would appear to be more complex than single-layered polyheads the reader can be excused for thinking that genes 20 and 40 should act before gene 22, but this is not so.

Maturation of the head

When infected cells are given a pulse of ^{3}H-leucine at 13–14 minutes after infection and the phage precursors extracted, label is first found in particles sedimenting at 400S. This label can be chased into particles sedimenting at 350S, then 550S and finally enters mature heads, sedimenting at 1100S. The particles sedimenting at 400S, called proheads I, resemble tau particles and contain largely gene 23 protein. Proheads I are converted to proheads II, which contain P23* instead of P23. The difference in the rates

Table 11.2 Phenotypes of T4 mutants belonging to group Y.

Mutant gene	Phenotype
23	No head structure
31	Produce lumps (aggregates of head protein associated with inner cell membrane)
22	Formation of multilayered polyheads consisting of several concentric tubes approx. 20 times length of normal head
20, 40	Produce single-layered polyheads
21, 24	Produce tau particles (head-like structures which lack sharp edges and vertices)

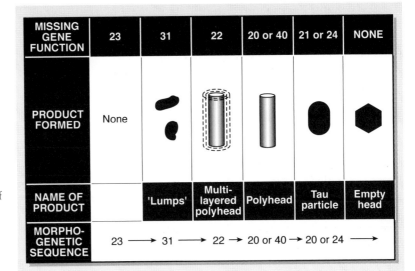

MISSING GENE FUNCTION	23	31	22	20 or 40	21 or 24	NONE
PRODUCT FORMED	None					
NAME OF PRODUCT		'Lumps'	Multi-layered polyhead	Polyhead	Tau particle	Empty head
MORPHO-GENETIC SEQUENCE	23 →	31 →	22 →	20 or 40 →	20 or 24 →	

Fig. 11.8 Functions of morphogenetic genes of group Y. Note the structures formed are *not* intermediates in assembly but are products of abortive assembly.

of sedimentation of prohead I and prohead II is about what would be expected from a reduction in mass of the particle when P23 is cleaved to P23*. The 550S particle, termed prohead III, is about half-full of deoxyribonucleic acid (DNA) and appears to represent an intermediate between the 350S particle and the mature head. Its isolation as such implies that head filling may take place in two stages: first, half of the DNA is inserted rather rapidly and, then, the second half is inserted. Thus packaging does not take place continuously since this would generate precursor heads at all stages of fullness. These stages are shown diagrammatically in Fig. 11.9.

The dimensions of the T4 head (100×65 nm) correspond to an internal volume of $2.5 \times 10^{-4}\,\mu m^3$, assuming the head proteins themselves occupy no space. Since the genome of T4 is $56\,\mu m$ long, it probably occupies a minimum volume of approx. $1.8 \times 10^{-4}\,\mu m^3$. Clearly, the T4 genome must be compactly packed into the T4 head. Exactly how this is achieved is not yet clear. Three genes, 16, 17 and 49, have been implicated in DNA packaging.

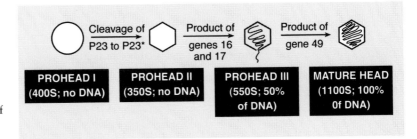

Fig. 11.9 Maturation of the head of bacteriophage T4.

Genes 16 and 17 are required to initiate packaging, i.e. the conversion of proheads II to proheads III, and gene 49 to complete packaging. However, the mode of action of these proteins is not known.

Assembly of the tail

The T4 tail includes the head–tail connector, the core to which it is attached, the surrounding sheath and the base plate and tail fibres. These components control the initial stages of adsorption outlined on pp. 88–90. The pathway of tail morphogenesis is shown in Fig. 11.10.

Assembly of tail fibres

There are a series of mutations which disrupt the formation of tail fibres (Table 11.1). Mutants requiring cofactors for attachment map in gene 34. Since, in the absence of cofactors, the fibres remain bound to the sheath, the product of gene 34 is probably responsible for the interaction of the fibre with the particle. Host-range mutations map in gene 37. Since only half-fibres containing antigen C can bind to T4 receptors isolated from sensitive bacteria, antigen C must be responsible for making bacterial contact. The function of gene 57 is not clear, but it may play a regulatory role, since mutants in gene 57 produce fibreless particles on infection.

The overall assembly process

The way in which the different components of T4 virion are assembled into infectious particles is shown in Fig. 11.11. Although the products of genes 13 and 14 must be functional to permit heads to join to tails, they are not involved in the joining process itself. Heads spontaneously join to tails, so genes 13 and 14 (Table 11.1) must be involved in some modification of mature heads which is a necessary prerequisite to tail addition. In contrast, the tail fibres do not spontaneously join to the base plate but require the active participation of the product of gene 63.

The way in which a head vertex with fivefold symmetry stably and spontaneously unites with a tail with sixfold symmetry is not clear. Nor is it clear why tails only attach to one vertex. It is possible that one end of the DNA protrudes through a vertex, disturbing the fivefold symmetry, and that this promotes tail addition. Indeed, protrusion of the DNA a short way into the tail may be a necessary structural feature for successful injection following contact of the phage with a susceptible bacterium.

Scaffolding proteins

The correct assembly of many spherical viruses, and in some cases the heads of tailed phages, appears to require the help of proteins

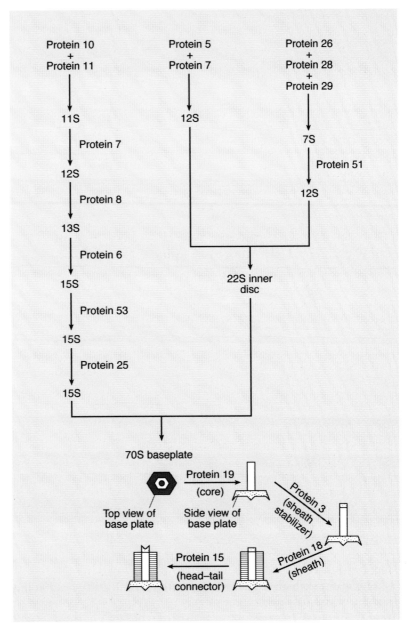

Fig. 11.10 The pathway of tail assembly. The precursors of the base plate are not individually named and only the sedimentation coefficient is given. Note that the proteins involved in base plate formation are not necessarily structural proteins.

not found in the mature virion. Such proteins are known as scaffolding proteins. During the morphogenesis of phage P22, about 250 molecules of scaffolding protein catalyse the assembly of 420 coat protein molecules into a double-shell prohead containing both proteins. Upon encapsidation of the DNA, all the scaffolding protein molecules are ejected from the prohead and these then take part in further rounds of prohead assembly. The way in which DNA replaces the scaffolding protein is not known but represents

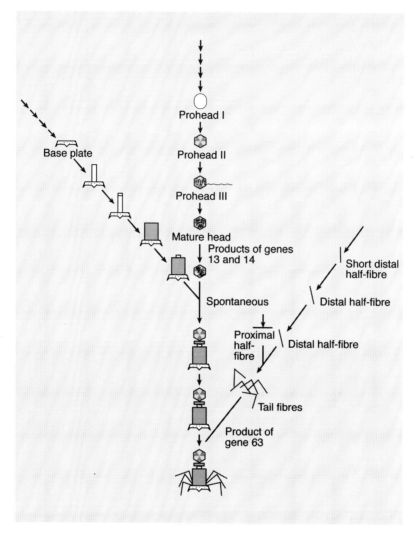

Fig. 11.11 Assembly of bacteriophage T4.

a fascinating exchange reaction. In the case of T4, the scaffolding protein, which is the product of gene 22 (Table 11.1), is removed from the prohead by proteolysis at the time of DNA encapsidation. It seems that eukaryotes also use a similar mechanism and during adenovirus assembly scaffolding proteins are lost at the time of the DNA entry.

Protein cleavage

With most animal and bacterial viruses, formation of mature virions requires the cleavage of larger precursor protein molecules after they have aggregated into proheads or provirions. In this context the conversion of P23 to P23* prior to DNA encapsidation in T4 heads has already been discussed (pp. 178–9). Possible explanations for the widespread occurrence of post-translational protein cleavage in the assembly of viruses are, firstly, that in the absence of DNA or RNA the uncleaved proteins may form more stable aggregates than cleaved proteins. Secondly, protein cleavage may represent a way of restricting the number of minor structural components which have to be added during morphogenesis.

In cells infected with picornaviruses, two kinds of post-translational protein cleavage occur. These have been termed *formative* and *morphogenetic* and the distinction between the two is best illustrated with poliovirus. The entire genome of poliovirus is translated as a single giant polypeptide, which is cleaved as translation proceeds into three smaller polypeptides (see p. 153). One of these is P1, which is the precursor to all four virion proteins. P1 is derived from the 5' end of the genome and is synthesized completely before being cleaved, suggesting that folding is necessary for cleavage. Secondary cleavage of P1 gives rise to VP0, VP1 and VP3, which are found associated with each other in infected cells as 5S and 14S aggregates and empty capsids. The 5S complex is a precursor of the 14S complex, which in turn is a precursor of empty capsids. After RNA is encapsidated, VP0 undergoes cleavage to yield the virion proteins VP2 and VP4. These steps are summarized in Fig. 11.12.

The initial cleavage of the primary translational product into three smaller peptides is considered to be 'formative' cleavage. The cleavage of P1 to VP0, VP1 and VP3 and the later cleavage of VP0 to VP2 and VP4 are examples of 'morphogenetic' cleavage.

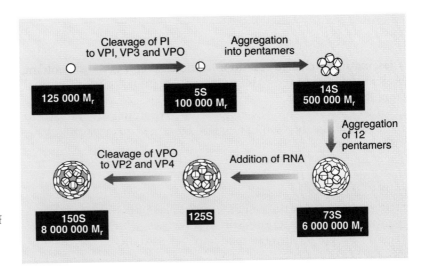

Fig. 11.12 Summary of the steps involved in the assembly of poliovirus.

The assembly of enveloped viruses

A large number of viruses, particularly viruses infecting animals, have a lipid envelope as an integral part of their structure. These include many animal viruses (herpes-, toga-, oncorna-, orthomyxo-, paramyxo-, corona-, arena-, pox- and irido-), the tomato spotted wilt virus group and rhabdoviruses of plants and three families of bacteriophage (cortico-, plasma- and cystoviruses) (see Table 3.2 and Chapter 21). Herpesviruses replicate in the cell nucleus, although the viral proteins are synthesized in the cytoplasm and transported back into the nucleus. After assembly of the nucleo-capsid, the virus buds off from the nuclear membrane and thus becomes enveloped. Prior to the budding process, the membrane is modified by incorporation of viral-specified proteins, which are subsequently glycosylated. Very little is known about the assembly of the lipid-containing phages and the iridoviruses, but it appears that the envelope is *not* incorporated by a budding process. In this respect, they resemble poxviruses, whose morphogenesis has been studied extensively by electron microscopy of infected cells. In thin sections, particles initially appear as crescent-shaped objects within specific areas of cytoplasm, called 'factories', and even at this stage they appear to contain the trilaminar membrane which forms the envelope. The crescents are then completed into spherical structures. DNA is added, and then the external surface undergoes a number of modifications to yield mature virions.

The majority of enveloped viruses acquire their envelope by budding from the plasma membrane or one of the internal cyto-plasmic membranes (Fig. 11.13). Four events leading to the matur-ation have been identified. First, the nucleocapsids form in the cytoplasm. Secondly, patches of cellular membrane incorporate viral glycoproteins, which are transmembrane proteins. Thirdly, these are focused by interaction of their cytoplasmic tails with the nucleocapsid (or intermediary matrix protein), which becomes aligned along the inner surface of the modified membrane, and finally the virion is formed by budding. It is remarkable that during the budding process host membrane proteins are excluded from viral particles (retroviruses are the exception), although the bulk of the lipid in the envelope is derived from the host's normal complement of lipid.

Viruses budding into the endoplasmic reticulum are released to the exterior via the Golgi complex. First, the virus particle buds into a vesicle, which moves to and fuses with Golgi complex. Then, following the normal route of transport of cellular proteins, the virus moves through the Golgi and buds from the concave surface into another vesicle, which is transported to the plasma membrane. Fusion of the vesicle with the plasma membrane releases the particle to the exterior of the cell. Viruses which bud from the plasma membrane are automatically released when the budding process is complete.

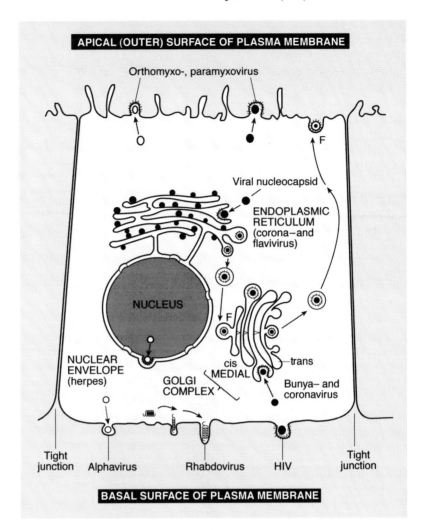

Fig. 11.13 Sites of maturation of various enveloped viruses. F, fusion of a vesicle with a membrane.

One complication is that many differentiated cells *in vivo* are polarized, meaning that they carry out different functions with their outer (apical) surface and their inner (basal) surface. For instance, cells lining kidney tubules are responsible for regulating Na^+ ion concentration. They transport Na^+ ions via their apical surface from blood to the cytoplasm and then expel them from the basal surface into urine. Some cell lines, such as Madin–Darby canine kidney (MDCK) cells, retain this property in culture, but this can only be demonstrated when they form confluent mono-layers and tight junctions between cells. The latter serve to separate and define properties of the apical and basal surfaces. These proper-ties reside in cellular proteins which have migrated directionally to one or other surface. This happens too with some viral proteins. The lipid composition of a polarized cell is also distributed asymmetrically between apical and basal surfaces. For instance, if a cell is dually infected with a rhabdovirus and an orthomyxovirus, electron microscopy will show the rhabdovirus budding from the basal surface, which is in contact with the substratum, and the orthomyxovirus from the apical surface (Fig. 11.13). This property is determined by a molecular signal on the viral envelope glycoproteins which directs them to one surface or the other; tight junctions are needed or proteins diffuse laterally and viruses can then bud from either surface.

FURTHER READING

Butler, P. J. G. (1984) The current picture of the structure and assembly of tobacco mosaic virus. *Journal of General Virology*, **65**, 253–279.

Casjens, S. & King, J. (1975) Virus assembly. *Annual Review of Biochemistry*, **44**, 555–611.

Compans, R. W. (1991) Protein traffic in eukaryotic cells. *Current Topics in Microbiology and Immunology*, **170**, 1–186.

Earnshaw, W. C. & Casjens, S. R. (1980) DNA packaging by the double-stranded DNA bacteriophages. *Cell*, **21**, 319–331.

Harrison, S. C. (1984) Structure of viruses. In: *'The Microbe 1984'*, *Society for General Microbiology Symposium*, **36** (Pt 1), 29–73.

Kaper, J. M. (1975) *Chemical basis of virus structures, dissociation and reassembly*. Amsterdam: North-Holland.

Lewin, B. (1977) *Gene expression — Vol. 3. Plasmids and phages*. London and New York: John Wiley and Sons. (Contains several lucidly written chapters with a wealth of information on the assembly of bacteriophages.)

Tucker, S. P. & Compans, R. W. (1993) Virus infection of polarized epithelial cells. *Advance in Virus Research* **42**, 187–247.

Wood, W. B. & King, J. (1979) Genetic control of complex bacteriophage assembly. In: *Comprehensive virology*, Vol. 13, pp. 581–633. H. Fraenkel-Conrat & R. R. Wagner (eds). New York: Plenum Press.

Also check Chapter 21 for references specific to each family of viruses.

12 Lysogeny

Shortly after the discovery of bacteriophages by Twort (1915) and d'Hérelle (1917), strains of bacteria were isolated which appeared to 'carry' phage in the sense that phages were always present in culture fluids of these strains. The 'carrying' strain itself was never sensitive to the 'carried' phage, but many related strains were sensitive. Although repeated single-colony isolation failed to free the carrying strain of phage, as did heating or treatment with antiphage serum, many workers refused to believe that *lysogens*, as these bacteria are called, represented anything other than contamination. By 1950, two more properties of lysogens had been discovered. Firstly, lysogenic bacteria are able to adsorb the phage which they 'carry', but lysis does not ensue. Neither does lysozyme lysis liberate phages from lysogens. Secondly, if the phage from a lysogen is plated on a sensitive strain, these produce progeny phage in the normal way.

In 1950, André Lwoff ended all controversy by a simple but elegant study with a lysogenic strain of *Bacillus megaterium*. He cultivated a single cell of his lysogen in a microdrop of culture fluid and watched its division under the microscope. When the cell divided, he removed one of the daughter cells as well as some culture fluid. Observation on the remaining daughter cell was continued, and, each time the cell divided, one of the progeny and some more of the culture were removed. Each of the cells which had been removed was plated on agar to determine if a lysogenic population grew, and the culture fluid was tested for the presence of phage. Repeated experiments of this kind showed that a lysogen could grow and divide without releasing phage. Nevertheless, a filtrate of a culture of a lysogen always contains some bacteriophages. How, then, do the phages arise in a lysogenic population? Lwoff reasoned that only a small percentage of bacteria release bacteriophages. This was confirmed when he observed spontaneous lysis of some of the cells in microdrops and that this was always correlated with the presence of free phage. It will be recalled, however, that artificial lysis of lysogens does not release phage. Thus, in lysogenic strains of bacteria there must be maintained a non-infective precursor of phage, called *prophage*, which endows the cell with the ability to give rise to infective phage without the intervention of exogenous phage particles. Lysis only ensues when some of the cells are stimulated to produce phage.

Lwoff and his two students, Siminovitch and Kjeldgaard, believed that the presence of free phage in cultures of lysogens was due to the induction of prophage by external factors. They began a long search for conditions that would increase the frequency of induction. Their search was ended when they found that small doses of ultraviolet (UV) irradiation induced phage production in the majority of cells of a lysogenic population. After irradiation, the lysogenic cells continue to grow and divide for some time and then suddenly the turbidity of the culture drops and there is a rapid increase in the number of free phages in the culture.

Thus, not only can a lysogenic bacterium pass on the ability for producing phage to all of its descendants, a few of which might spontaneously lyse and liberate infective particles, but it can also be induced to lyse at will by the application of UV treatment. Bacteriophages which are capable of existing as prophages within a host cell are said to be *temperate* bacteriophages. After a lysogen is induced with UV light, there is initially an eclipse period, during which no infectious phage can be detected inside the cell. However, phage-specific deoxyribonucleic acid (DNA) and protein can be detected, and shortly before lysis this is assembled into mature phages. Thus, the events following induction resemble those occurring in cells infected with virulent phages, such as T4.

Although a lysogenic strain of *B. megaterium* was used by Lwoff, most of the subsequent studies have utilized another temperate phage, λ. In 1951, Esther Lederberg accidentally discovered that *Escherichia coli* K12, the strain in which Tatum and J. Lederberg had discovered bacterial conjugation, was also lysogenic. Non-lysogenic derivatives had unknowingly been prepared and, when these were mixed with the original K12 strain, plaques were produced, and the phage isolated from them was called *lambda* (λ). A wealth of information on this phage has been produced in the last two decades, so much, in fact, that entire books have been written about it! Consequently, we shall cover only a few of the salient features here.

THE INTEGRATION OF λ DNA

Since lysogenic and non-lysogenic derivatives of the sexually fertile *E. coli* K12 were available, it became possible to determine how the prophage behaved in genetic crosses. Early studies showed that lysogeny is closely linked to galactose fermentation. Following the isolation of Hfr bacteria (which undergo *high frequency recombination*) and the discovery that such bacteria transfer their genetic material in an orderly fashion, it became possible to map genetic loci by measuring their times of entry into the recipient, and in this way λ prophage was located between the galactose and tryptophan genes. In genetic crosses, the character of lysogeny behaves just like any other marker, and it was assumed that during lysogeny the prophage DNA is in some way integrated into the host chromosome.

How, in the course of eliciting the lysogenic response, does the phage genome manage to become integrated into the bacterial chromosome? The first clue came with the isolation of λ mutants which had a deletion in the b2 region of their DNA. These mutants were incapable of lysogenizing their host, although they were perfectly capable of undergoing lytic growth. Thus, the b2 sector is involved in the genetic exchange between the λ genome and the region of the *E. coli* genome between the galactose and tryptophan genes, which is referred to as the lambda attachment (*att* λ) locus.

In 1962, Campbell, with remarkable foresight, suggested that crossing-over between phage and host genomes results in insertion of the entire phage genome into the bacterial genome. He further suggested that the linear genome of the phage becomes circular after infection and that at the moment of lysogenization there is recognition between the b2 region of the phage and the *att* λ region of the bacteria. Thus, the phage genome is inserted as a linear structure into the host genome by reciprocal recombination (Fig. 12.1).

Genetic studies on λ had yielded a linear map in which the b2 region and genes c and h were midway between genes mi and m6, which were located at either terminus (Fig. 12.2). If Campbell's hypothesis was correct, then the genome must assume a circular configuration prior to lysogenization, followed by a reordering of the genes after insertion (Fig. 12.3). When crosses were made between Hfr and F⁻ bacteria lysogenic for different genetically marked strains, a map was obtained in which c and h were at opposite ends and mi and m6 were closely linked. Thus, the predictions of the Campbell model were borne out. About this time, Hershey and his colleagues had shown that the DNA of λ has 'sticky ends' (see p. 66), thus providing a mechanism whereby circular molecules might be formed in the infected cell. Furthermore, when *E. coli* is infected with labelled phage and the DNA extracted, some of the label can be isolated as covalently closed circular molecules.

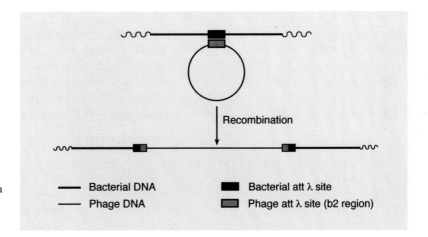

Fig. 12.1 Campbell's model for the insertion of phage λ DNA into the bacterial genome.

—— Bacterial DNA   ▮ Bacterial att λ site
—— Phage DNA   ▨ Phage att λ site (b2 region)

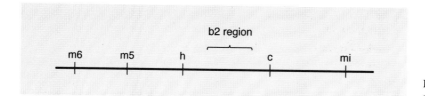

Fig. 12.2 Early genetic map of bacteriophage λ.

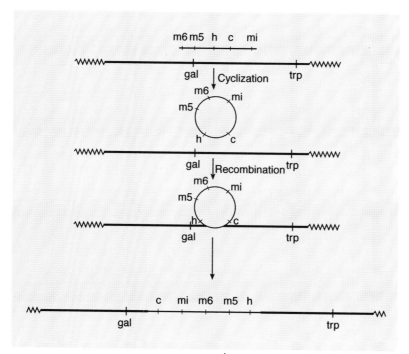

Fig. 12.3 Reordering of the genes of bacteriophage λ following insertion of the phage DNA into the bacterial chromosome. The symbols *gal* and *trp* represent the galactose and tryptophan operons respectively.

STRUCTURE OF THE ATTACHMENT SITES

Fifteen years after Campbell proposed his model for λ integration and excision, the bacterial and phage attachment sites were sequenced and shown to contain identical tracts of 15 base pairs (Fig. 12.4). The actual site of the crossover for both integration and excision must take place within or at the boundaries of this common sequence. Thus recombination λ and *E. coli* genomes is said to be *homologous*. A number of phage genes, such as *int* and its product, the integration protein, are involved in this process.

THE EXCISION OF λ DNA

Just as recombination between the phage and host genomes results in lysogeny, reversal of the process results in the regeneration of the circular phage genome and terminates in lysis. Genetic studies

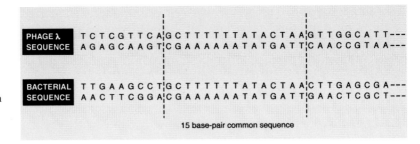

Fig. 12.4 The common sequence in the bacterial and λ attachment sites.

have revealed the presence of a battery of phage genes, such as the *xis* gene and its product, the excision protein, which specifically control excision of the prophage DNA.

Occasional errors can occur in the process of excision, such that the region excised includes a small amount of bacterial DNA, with a corresponding piece of phage DNA deleted. If the portion of bacterial DNA is located to the left of the prophage, it may include some or all of the genes of the galactose operon, with a compensatory loss of some of the genes at the right end of the prophage map. This incomplete phage chromosome then serves as the template for replication, such that essentially all of the phage progeny issuing from that cell bear the genes for galactose utilization and have lost some of the phage genes. Reinfection of a Gal⁻ cell with such a phage, under conditions allowing lysogenization, can confer upon that cell the ability to ferment galactose, since the necessary genes become inserted into the bacterial chromosome. Such a *transducing* phage is defective in its own replication, since it lacks some essential phage genes, and can only be propagated if a normal ('helper') phage is present to supply the missing gene functions.

By the time Lwoff had confirmed the existence of a prophage in lysogens, it had already been established that superinfecting phage DNA does not replicate in a lysogen. Thus the idea arose that in a lysogen an 'immunity substance' is synthesized by the prophage which not only prevents the expression of most of the prophage genes but is also capable of diffusing throughout the cell and inhibiting the expression of superinfecting phage genomes of the same origin. Among the earliest mutants of λ which were isolated was a series which gave rise to clear plaques, and these mutants could be divided into three classes (cI, cII and cIII). cI mutants were virulent and incapable of lysogenizing a host except in the presence of phages carrying a functional cI gene product. However, cI mutants failed to grow in lysogens which had a functional cI gene. Thus, the cI gene is responsible for the production of the immunity substance.

IMMUNITY TO SUPERINFECTION

Immunity in bacteria is a specific property and completely different from immunity in eukaryotes. A lysogen is immune to the phage of the same type that it carries, but is usually not immune to another, independently isolated temperate phage. Phages that induce immunity to one another are termed *homoimmune* and, if they do not, they are called *heteroimmune*. A number of temperate phages which are related to λ but are heteroimmune have been isolated and these have permitted a detailed analysis of the genetic factors involved in the specificity of immunity. In genetic crosses between two heteroimmune phages, the genetic determinants responsible for the specificity of immunity are found to be closely linked to the cI gene. In fact, for each pair of phages tested there is a region of non-homology comprising gene cI and segments of various lengths on either side (Fig. 12.5).

The production of the immunity substance by the cI gene is analogous to the production of a repressor protein for β-galactosidase synthesis by the *i* gene of *E. coli*. Carrying the analogy further, it should be possible to find phage mutants which correspond to the o^c mutants of the *E. coli lac* operon. Such mutants were isolated very soon after the discovery of λ, and, as expected, they are virulent. These grow on λ lysogens, despite the synthesis of the cI gene product, and have mutations which define two operators, O_L and O_R, to the left and right of cI respectively. Immunity acts at the level of transcription since the only λ-specific mRNA synthesized in a lysogen hybridizes with the cI region. This fact was confirmed by the isolation of the λ immunity substance (the 'λ repressor') by Mark Ptashne.

In summary the cI gene produces the λ repressor, which acts on two operator sites, O_L and O_R, and so prevents the transcription of the genes required for lytic infection (Fig. 12.6). Thus the cI gene product acts as a negative control element.

THE FUNCTION OF GENES N AND Q

Many prophage genes operate when lysogens are superinfected with closely related heteroimmune phages, as superinfecting phages defective for these genes are complemented by the prophage. Thus,

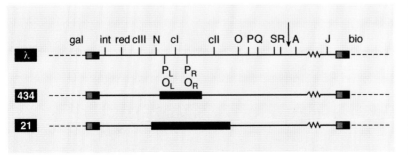

Fig. 12.5 Simplified genetic map of prophages λ, 434 and 21 (not to scale). The arrow indicates the ends of the genome in the mature phage particle and the dotted lines represent bacterial DNA. The area of non-homology is shown by the solid bar.

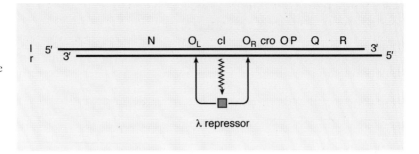

Fig. 12.6 Binding of the cI gene product, the λ repressor, to the right (O$_R$) and left (O$_L$) operators. The DNA strands are denoted as left (l) and right (r).

prophage genes can be switched on in spite of fully persisting immunity. It is not immunity *per se* which prevents expression of these genes in lysogens; rather, immunity prevents the prophage from producing diffusible molecules which switch on genes, i.e. immunity prevents positive control.

Which λ genes exert positive control? Mutations in such genes would be expected to display a recessive, pleiotropic phenotype. Mutants displaying such a phenotype have been mapped in genes N and Q (Fig. 12.7). Mutants of gene Q replicate normally, but they express all the late functions at a low rate. Since Q mutants produce very little mRNA hybridizable with the right arm of the phage chromosome, the Q gene product must act at the level of transcription. N⁻ mutants are exceedingly pleiotropic and most λ functions are not expressed in the absence of the N gene product. The N gene product also acts at the level of transcription, since very little of the mRNA produced following induction of N⁻ lysogens is hybridizable with λ DNA. The pleiotropic character of N mutants is due to the absence of the N gene product rather than to a polar effect. However, the N and Q genes do not act independently of each other. The late gene R codes for an endolysin whose expression is dependent upon a Q gene product (a protein) and this, in turn, is dependent upon the expression of N, as shown in Fig. 12.7.

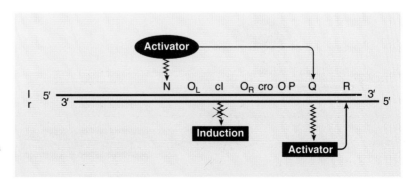

Fig. 12.7 Dependence of gene Q expression on gene N expression following induction.

EXPRESSION OF THE GENES BETWEEN cI AND Q (THE X OPERON)

The region to the right of cI comprises genes with very different functions: *cro*, which has a regulatory role (see later), cII, which takes part in immunity, and the replication genes O and P; *cro* is transcribed independently of N. Genes cII, O and P can be transcribed in the absence of N. However, their gene products are synthesized in much greater amounts in the presence of N than in its absence. Presumably, there is a weak termination signal between *cro* and cII. Since the gene to the right of P is Q, whose transcription is dependent on the presence of N, we must postulate the presence of a second terminator between P and Q. This aspect of λ control is summarized in Fig. 12.8.

SUMMARY OF EVENTS DURING THE LYTIC CYCLE OF λ

1 Following infection, transcription results in the formation of the N and *cro* gene products, as well as a little of the cII, O and P gene products.

2 N exerts positive control on the transcription of Q and some other genes.

3 Q exerts positive control on transcription of the endolysin gene R and also the structural genes A to J (remember that λ DNA circularizes following infection, so genes R and A are linked (Fig. 12.8)).

4 The *cro* gene product exerts negative control by counteracting the effect of N (see below).

5 Transcription of O and P results in DNA synthesis. As a result of the effect of *cro*, almost all the gene products synthesized will be specified by late genes, e.g. the endolysin produced by R and structural genes (A to J).

6 Phage assembly and lysis occur.

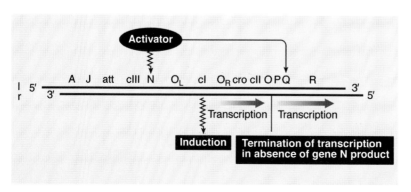

Fig. 12.8 Dependence of gene Q transcription on presence of gene N product.

THE EVENTS LEADING TO LYSOGENY

Earlier (p. 191) we alluded to the fact that clear plaque mutants of λ can belong to any one of three genes cI, cII and cIII. Since the λ repressor is produced by the cI gene, what are the functions of the cII and cIII genes? It appears that the cI gene is transcribed in two modes. In the lysogenic state, transcription of the λ repressor starts from the promoter for repressor maintenance, called P_{RM} (Fig. 12.9). In contrast, upon infection *de novo*, cI transcription begins about 1000 bases to the right of cI at a promoter for repressor establishment, called P_{RE} (Fig. 12.9). P_{RE} directs the synthesis of 5 to 10 times more repressor, per genome, than does P_{RM} and provides the large burst of repressor necessary to *establish* lysogeny. Once established, lysogeny can be *maintained* by the lower level of repressor synthesis directed by P_{RM}. When λ infects a cell, transcription of the N gene results in the formation, among other things, of some of the cII and cIII gene products. Both these proteins are positive regulators necessary for transcription from P_{RE} and hence their absence leads to the clear plaque phenotype. Once lysogeny is established, the cI protein will prevent transcription from O_L and O_R and synthesis of the cII and cIII gene products will cease, as will transcription from P_{RE}.

THE SWITCH FROM LYSOGENY TO LYTIC GROWTH

The switch from lysogeny to lytic growth depends on the products of genes cI and *cro*. Each turns off the other gene (Table 12.1). Also, each binds to two operators but most important is the right operator (O_R), which is the primary switch of lysogeny to lytic growth. O_R has three sites (O_R1, O_R2, O_R3) which control the two

Table 12.1 Phage λ gene products that control its alternative life styles.

	λ Repressor	*cro*
Lysogeny	+	−
Lytic growth	−	+

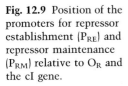

Fig. 12.9 Position of the promoters for repressor establishment (P_{RE}) and repressor maintenance (P_{RM}) relative to O_R and the cI gene.

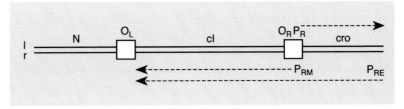

divergent promoters P_R and P_{RM}. When O_R1 and O_R2 are bound to repressor, transcription from P_R is turned off and transcription from P_{RM} is turned on (Fig. 12.9). Thus genes required for lytic growth (including *cro*) are not transcribed and genes required for lysogeny (including cI) are active.

When a lysogenized *E. coli* cell is treated with a variety of mutagens, including UV light, a number of genes are turned on. These 'SOS functions' include the *recA* gene and its product. The recA protein promotes recombination, but is also a protease which cleaves the λ repressor and inactivates it. As a result there is no repressor to bind to P_R and it is derepressed. Resulting transcription includes the *cro* gene. The *cro* protein binds to the same promoters as the repressor, but note that only *cro* or the repressor can bind, not both together. The P_{RM} is actively repressed by *cro*, completing its switch-off; P_R is now functional and the proteins necessary for lytic growth are synthesized.

For phage λ, the lysogenic state is the norm. Lytic growth occurs rarely by spontaneous mutation — about once every 10^5 generations.

COMPARISON OF λ WITH OTHER TEMPERATE PHAGES

Too few temperate phages (those capable of lysogeny) have been studied sufficiently to make detailed comparisons. However, mechanisms of prophage maintenance and immunity differ in coliphages P2 and λ and the *Salmonella* phage P22. Phage P2 seems to have only one gene that controls lysogenization and the few clear-plaque mutants that have been studied all belong to one complementation group. In addition, P2 immunity seems to involve only one operator locus, and virulent mutants thus occur at a perceptible frequency. The only P2 gene, other than the repressor gene itself, which is not repressed in lysogens is the *int* gene, but it is probably made quiescent when integrational recombination severs its transcription unit. However, it seems that phages λ and P22 have features in common which control the switch from lysogeny to lytic growth, including a complex tripartite operator and functionally similar repressor and *cro* proteins.

For P2, P22 and λ, stable lysogeny requires prophage insertion, and in each case one gene for insertion has been found. Whereas P22 and λ integrate at a single site on the host chromosome, P2 can insert into at least nine different sites in the chromosome of *E. coli* C. The extreme example of low insertion specificity is offered by coliphage Mu, which can apparently insert at any point in the chromosome. The reason why Mu has this exceptional ability is not known. It does, however, have one obvious side-effect; since insertion frequently occurs in the middle of a gene, it leads to gene inactivation.

THE BENEFITS OF LYSOGENY

What is the basis of the natural selection favouring lysogeny? The fact that temperate phages carry so many genetic determinants for lysogeny indicates that lysogeny does confer some advantages on the phages. One advantage would be to provide a mode of persistence which does not deplete the supply of hosts for such an obligate parasite. Another would be the opportunity for extended growth under non-selective conditions, in which multiple genetic variations can occur.

Lysogeny also carries a strong selective value for the host cell, since temperate phages frequently confer new characteristics on the host cell. This phenomenon manifests itself in many ways and is referred to as *lysogenic conversion*. For λ, the only known lysogenic conversion is the capacity of λ lysogens to block the multiplication of a particular class of mutant (the rII mutant) of bacteriophage T4. This block involves the product of the *rex* gene and perhaps also the repressor. Since rII mutants may not be very common in the natural environment of λ, the *rex* gene probably has other, more significant, roles to play.

Lysogenic conversion often involves exclusion of superinfecting phages. Exclusion acts *after* phage attachment and, in a severe form, involves breakdown of the superinfecting phage DNA. Lysogens of P2 and of P22 exclude many other phages: P22 exclusion even acts against P22. In a different way, phage P1 protects the lysogen from further phage infection, by conferring a new restriction and modification phenotype on the host. Thus, phages grown on hosts lacking P1 cannot successfully infect a P1 lysogen.

Perhaps the most interesting example of lysogenic conversion is that observed in *Corynebacterium diphtheriae*. Strains of this bacterium which cause the serious childhood disease diphtheria carry the diphtheria toxin (and are called toxigenic). In 1951, Freeman found that, if non-toxigenic strains of *Corynebacterium* were infected with a phage, now called β, from virulent, toxigenic bacilli of the same species, a proportion of the survivors acquired the ability to synthesize toxin. The conversion of the non-toxigenic strain to toxigenicity was due to the establishment of lysogeny, and loss of the prophage resulted in loss of toxin production. It was possible that a phage gene coded for the toxin itself or that the product of a phage gene could control the expression of a host gene which coded for the toxin. Recent work has established beyond any doubt that the structural gene for toxin is carried by the phage itself.

FURTHER READING

Hendrix, R. W., Roberts, J. W., Stahl, F. W. & Weisberg, R. A. (1983) *Lambda II*. Cold Spring Harbor Laboratory, NY.

Howe, M. M. & Bade, E. G. (1975) Molecular biology of bacteriophage Mu. *Science,* **190,** 624–632.

Johnson, A. D., Poteete, A. R., Lauer, G., Sauer, R. T., Ackers, G. K. & Ptashne, M. (1981) λ repressor and *cro* – components of an efficient molecular switch. *Nature (London)* **294,** 217–223.

Landy, A. & Ross, W. (1977) Viral integration and excision: structure of the lambda *att* sites. *Science,* **197,** 1147–1160.

Lewin, B. (1977) *Gene expression – Vol. 3. Plasmids and phages.* London and New York: John Wiley and Son. (Contains an excellent account of all aspects of the biology of λ known prior to 1977.)

Ptashne, M. (1986) *A genetic switch.* Palo Alto: Blackwell Scientific Publications & Cell Press (a distillation of the principles of gene regulation using the phage λ system).

Ptashne, M., Jeffrey, A., Johnson, A. D., Maurer, R., Meyer, B. J., Pabo, C. O., Roberts, T. M. & Sauer, R. T. (1980) How the λ repressor and *cro* work. *Cell,* **19,** 1–11.

Scott, J. R. (1980) Immunity and repression in bacteriophages P1 and P7. *Current Topics in Microbiology and Immunology,* **90,** 49–65.

Also check Chapter 21 for references specific to each family of viruses.

13 Interactions between viruses and eukaryotic cells

So far we have dealt with viruses as if they were able only to produce a lytic infection or to undergo lysogeny. In both cases, we discussed the interactions of bacteriophages and their unicellular hosts. In this chapter we shall be concerned with the various types of interactions which occur between viruses of eukaryotes and cells in culture. We shall deal exclusively with animal systems, as plant cell culture is technically more difficult and still not widely used. Virus−cell interactions are classified here into lytic, persistent, latent, transforming and abortive infections. They have all been studied in the laboratory to determine the molecular events involved and to pave the way for our eventual understanding of the process of infection of the whole organism (Chapter 15). Two points should be borne in mind: firstly, that a prerequisite of any of these types of infection is the initial interaction between a virus and its receptor on the surface of the host cell and hence any cell lacking the receptor is automatically resistant to infection; and secondly that both cell and virus are important in establishing the interactions described below and a virus may exhibit, for example, a lytic infection in one cell and latency in another.

LYTIC INFECTIONS

Lytic infections are the ones most commonly studied in the laboratory, because cell killing is the easiest effect to observe and production of infectious progeny can usually be monitored without difficulty. The one-step growth curve (Chapter 1) describes the essential features of any eukaryotic or prokaryotic virus−host interaction which results in lysis. Viruses can inhibit cellular deoxyribonucleic acid (DNA), ribonucleic acid (RNA) or protein synthesis, but frequently death occurs earlier than can be accounted for by any of these events (see p. 203). Lysis by lysozyme is a property of some phages only.

PERSISTENT INFECTIONS

Persistent infections result in the continuous production of infectious virus and this is achieved (as will be described below) either by the survival of the infected cell or by a situation in which a minority of cells are initially infected and the spread of virus is limited, so that cell death is counterbalanced by new cells produced by division, i.e. no net loss. Persistent infections result from a balance struck between the virus and its host either: (i) through

the interaction of virus and cells alone; (ii) with the help of antibody or interferon; (iii) by the production of defective–interfering
(DI) virus (p. 126); or (iv) by a combination of these events.

Virus + cells alone

An outstanding example of persistence of this type is seen in the
infection of monolayers of monkey kidney cells with a simian
virus (SV5, Paramyxoviridae, class V). The virus multiplies with
a classical one-step growth curve (Fig. 13.1A), but the cells do
not die, and they continue to produce virus for over 30 days
(Fig. 13.1B). Infection by SV5 does not damage the cell, in the
sense that it does not perturb cellular DNA, RNA or protein
synthesis in monkey cells. In fact, virus infection makes little
demand on the host's resources; for example, the amount of viral
RNA synthesis is <1% of cellular synthesis (even though each cell
is producing about 150 000 particles/day!). An important point to
note is that the same virus in a different cell line (baby hamster
kidney (BHK) cells) is lytic. Thus, here, the outcome of infection
depends on the properties of the host cell.

Virus + cells + antibody, or virus + cells + interferon

This is a situation in which the virus would otherwise be lytic.
Usually, only a few cells are initially infected and the addition of
small amounts of specific neutralizing antibody (often low-avidity
antibody is most effective) decreases the amount of progeny virus
available to reinfect cells. Alternatively, the addition of interferon
(see p. 209) depresses the extent of multiplication, so that the cell
population continues to survive even if a small proportion of cells
die. It is a question of establishing a dynamic equilibrium. This
situation is thought to mimic certain sorts of persistent infections
in the whole animal (Chapter 15).

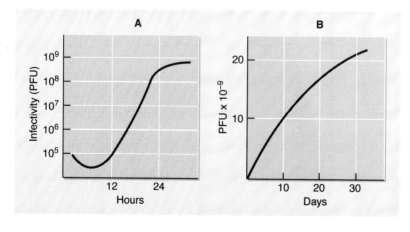

Fig. 13.1 Persistent
infection by SV5 of
monkey kidney cells in
culture. (A) Initial one-
step growth curve and
(B) the cumulative yield
from (A) over 30 days.

Virus + cells + DI virus

DI viruses are probably produced by all viruses as a result of mistakes in replication which delete part of the viral genome (p. 126). Interference comes about because DI viruses depend absolutely upon infectious virus for replicative enzymes and structural proteins and deprive infectious virus of sufficient components for its own multiplication. The DI genome is replicated preferentially since its small size allow more rounds of replication in unit time. The yield of infectious progeny is thus diminished and the cell population is more likely to survive. However, the generation of DI virus is very much dependent on the type of cell infected, so that both cell and virus contribute to a balanced situation.

An infection can be carefully engineered with critical proportions of infectious virus, cells and DI virus, so that a persistent infection is achieved. In the initial stages, increase of DI virus is seen to follow that of infectious virus, upon which it is dependent (Fig. 13.2). As DI virus increases, there is a progressive interference with multiplication of infectious virus; its numbers decrease, which in turn results in a concomitant decrease in the dependent DI virus. This cycle of events continues until the levels of virus produced fall and a persistent infection results, in which there are low levels of both infectious and DI virus.

LATENT INFECTIONS

Latent means existing but not exhibited. In a latent infection, some virus-coded products may be expressed but infectious virus is not formed. Lysogeny by temperate phages (Chapter 12) is clearly a latent infection, and in animal cells, for example, herpesviruses and some tumour viruses can exhibit latency (Chapters 15 and 17).

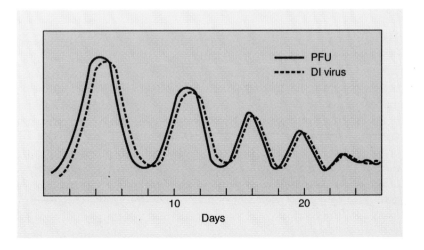

Fig. 13.2 A persistent infection established with defective-interfering virus.

However, compared with phage λ, our understanding of the molecular control processes is still at an early stage. Bacterial viruses and the animal retroviruses which achieve latency do so by integrating a DNA copy of their genome into the host's DNA. This ensures that the viral genome will be replicated, together with host chromosomal DNA, and transmitted to daughter cells, and will be protected from degradation by nucleases. On the other hand, the DNA of herpes simplex virus type 1 remains nonintegrated, although the normally linear molecule is circularized. This virus is latent in neurones, which are non-dividing cells.

Delicate molecular controls operate to maintain the latent state, since latent infections can be converted into lytic infections by a number of external stimuli. It is apparent that the latency in bacteriophage is fundamentally different from that of animal viruses, the former being maintained primarily by virally encoded repressors of lytic replication (p. 195) and the latter by the influence of host factors which are necessary for the expression of early virus gene products (p. 220).

The presence of latent virus can sometimes be shown by using labelled antibody specific for viral proteins or, better, by polymerase chain reaction amplification of virus genome sequences.

TRANSFORMING INFECTIONS

As a result of infection with a variety of DNA viruses or with RNA tumour viruses, a cell may undergo more rapid multiplication than its fellows, concomitant with a change in a wide variety of other properties, i.e. it is transformed. This is accompanied by integration of the viral and host genomes, as mentioned above. Chapter 17 deals in detail with transformation and other aspects of tumour viruses.

ABORTIVE INFECTIONS

Viruses do not grow equally efficiently in all cells which possess the appropriate receptor. There may be a reduction in the total yield of virus particles (sometimes to zero), or in the particle: infectivity ratio. Both of these reflect a defect in the production or processing of some components necessary for multiplication, be it DNA, RNA or protein. One example is avian influenza virus growing in a line of mouse L cells, where both the amount of progeny and its specific infectivity are reduced, probably due to the synthesis of insufficient virion RNA. Another is infection of other *non-permissive* cells with human influenza viruses which gives rise to normal yields of non-infectious progeny. This results from a failure to cleave the haemagglutinin protein proteolytically and can be completely reversed by adding small amounts of trypsin to the culture or to released virus (p. 158). Abortive infections present difficulties to virologists trying to propagate viruses but have been used to advantage when research into the nature of the defect has furthered understanding of productive infections. In natural infections, abortiveness contributes to tissue specificity, by permitting invasion only of those cells in which a productive infection can take place.

HOW DO VIRUSES KILL CELLS?

The recognition that viruses are often toxic for the cells in which they multiply gives the simplest criterion of infectivity, that of viral cytopathic effects (cpe). However, toxicity depends on the precise relationship between a virus and a host cell and, as exemplified by SV5 above, a virus does not necessarily kill the cell in which it multiplies (see below). Exactly how a cell is killed is by no means clear, except that in all cases viruses need to undergo at least part of the multiplication cycle. Thus, it seems that a toxic product is produced by the virus and that viruses with different replication strategies are likely to invoke different mechanisms of toxicity. One problem in these studies is in distinguishing between an effect on cell function which operates early enough to be responsible for toxicity and those which appear late on and are a consequence of those toxic effects. Viruses can inhibit synthesis of host proteins, RNA and DNA, but here we shall deal only with the proteins, about which slightly more is known (Table 13.1).

Studies of the inhibition of host cell protein synthesis by poliovirus suggest that this results from inactivation of those translation initiation factors which are responsible for recognizing capped messenger RNAs (mRNAs). Poliovirus mRNA is not capped and so its translation is unaffected. In Semliki Forest virus-infected cells, there is evidence that the virus inhibits host protein synthesis

Table 13.1 Some suggested mechanisms of viral cytopathology.

Mechanism	Virus
Loss of ability to initiate translation of cellular mRNA	Polio, reo, influenza
Imbalance in intracellular Na^+/K^+ ratio	Semliki Forest
Competing out of cellular mRNA by excess viral mRNA	Vesicular stomatitis, Semliki Forest
Degradation of cellular mRNA	Influenza
Failure to transport mRNA out of the nucleus	Adeno
Apoptosis	Sindbis, influenza

by affecting the plasma membrane Na^+/K^+ pump, which controls ion balance. As a result, intracellular Na^+ concentration increases to a level where viral but not cellular mRNA is translated. In cells infected with vesicular stomatitis virus (VSV), it is suggested that there is so much viral mRNA that it outcompetes cellular mRNA for the translational machinery. However, there is evidence against this hypothesis, as a DI VSV, which is only able to undergo primary transcription, also kills cells. Here, it is thought that an early virus product is toxic. A recent very credible suggestion is that some viruses kill cells by triggering apoptosis, which is the process of self-programmed cell death. Apoptosis is part of normal development, familiar to us as the loss of the tail of the amphibian tadpole or the removal of self-reactive T cells.

Evidently, viruses do not kill cells by any one simple process, and we are far from understanding the complex mechanisms involved. However, cells do not die immediately their macro-molecular synthesis is switched off, unless there was rapid turnover of some vital molecule. Thus, the mechanisms discussed above, with the exception of the upset in Na^+/K^+ balance and apoptosis, seem more akin to death by slow starvation than to acute poisoning. Lastly, it is by no means clear what advantage acrues to the virus in killing its host cell, and this situation may represent a poorly evolved virus–cell relationship or virus in the 'wrong' host cell.

FURTHER READING

Carrasco, L., Otero, M. J. & Castrillo, J. L. (1989) Modification of membrane permeability by animal viruses. *Pharmacology and Therapeutics A*, **40**, 171–212.

Collins, M. K. L. & Rivas, A. L. (1993) The control of apoptosis in mammalian cells. *Trends in Biochemical Sciences*, **18**, 307–309.

Fenner, F., McAuslan, B. R., Mims, C. A., Sambrook, J. & White, D. O. (1974) *The biology of animal viruses* (2nd edn). London: Academic Press. (Still a good treatment of the biology.)

Joklik, W. K. (1985) The effect of virus infection on the host cell. In: *Virology.* W. K. Joklik (ed.). Norwalk: Appleton-Century-Crofts.

Kääriäinen, L. & Ranki, M. (1984) Inhibition of cell functions by RNA-virus infections. *Annual Review of Microbiology*, **38**, 91–109.

Levine, B., Huang, Q., Isaacs, J. T., Reed, J. C., Griffin, D. E. & Hardwick, J. M. (1993) Conversion of lytic to persistent alphavirus infection by the *bcl-2* cellular oncogene. *Nature (London)*, **361**, 739–742.

Shatkin, A. J. (1983) Molecular mechanisms of virus mediated cyto-pathology. *Philosophical Transactions of the Royal Society of London Series B*, **303**, 167–176.

Also check Chapter 21 for references specific to each family of viruses.

14 The immune system and interferon

Infections of multicellular plants and animals are complicated by the variety of cell types present in an individual and the possession, in higher animals, of an elaborate defence against infection. It may be helpful to summarize the responses of the latter to virus infections. Broadly speaking, there are *adaptive responses*, which are specific for certain epitopes (see below) and are amplified by contact with those epitopes. These are brought about by B lymphocytes, which develop into cells synthesizing immunoglobulins (antibody), or T lymphocytes, which develop into cells responsible for cell-mediated immune responses. Each T and B cell has in its plasma membrane many copies of a unique receptor molecule—the T cell receptor (TCR) and the B cell receptor, respectively. The latter is identical to the antibody which is secreted when the B cell is activated, except that it is synthesized with an additional trans-membrane anchoring sequence. Both receptors recognize a small region (the epitope) of a foreign molecule (the antigen), such as a viral protein. The B cell receptor interacts with about 17 amino acids on the surface of the viral protein, and the TCR with a peptide of around 10–20 amino acids derived by proteolysis from the viral protein, as described below. Any one B cell or T cell and their clonal descendants recognize only one epitope. The second type of response is *non-adaptive responses*, brought about by macrophages, polymorphonuclear leucocytes (PMNL), natural killer (NK) cells, the complement system and interferons, which act non-specifically, although their activity may be enhanced during infection. Both adaptive and non-adaptive responses act in an interlocking and orchestrated fashion, and Fig. 14.1 tells only a small part of this complex story. It is no comfort to the student to learn that often each infection is successfully combated by only one of the elements shown in the scheme, even though many respond, and there is no way of predicting which element of the immune system will overcome a particular virus infection. It can act either directly on virus particles (antibody) or on infected cells which carry foreign antigens on their surface (cell-mediated immunity or antibody plus accessory factors).

THE IMMUNE SYSTEM

Antibodies (immunoglobulins) are made by plasma cells. Their immediate progenitors are B lymphocytes, which react with foreign molecules (antigens) and are stimulated to divide and differentiate to form a large number of plasma cells and memory cells. The plasma cells synthesize and secrete free molecules of antibody,

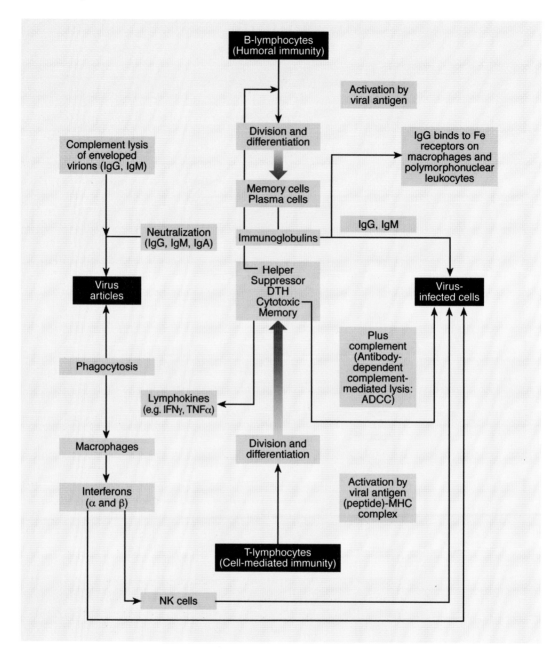

Fig. 14.1 Summary of responses of the immune system and associated cells to viruses and virus-infected cells. DTH, delayed-type hypersensitivity; TNFα, tumour necrosis factor alpha.

which react specifically with the stimulating antigen. Memory cells have the same specificity but are lymphocytes and do not secrete antibody. Because many are made as a result of the initial reaction with antigen, memory cells enable an animal to give an enhanced response to antigen when it is encountered for a second

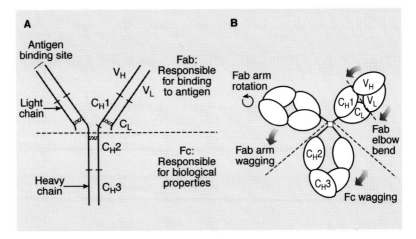

Fig. 14.2 Generalized immunoglobulin molecule, showing (A) the outline structure, consisting of two identical dimers formed of H–L polypeptides. Note the sequence-variable (V) region of the H and the L chains, which together form the unique antigen-binding site. The remaining H and L sequences are relatively constant (C). (B) The globular domains. The arrows indicate the flexibility which allows the molecule to bind to two (identical) epitopes, which can be different distances apart in all three dimensions. SS, disulphide bond.

time. They then develop into plasma cells and prevent reinfection.

Immunoglobulins (Ig) have the basic pattern seen in Fig. 14.2. The five types are distinguished by having different heavy chains called α, γ, δ, ε and μ, which characterize IgA, IgG, IgD, IgE and IgM, respectively. Neutralizing and non-neutralizing immunoglobulins bind to viruses (p. 27) or to antigens expressed on the surface of virus-infected cells. This allows the virus to react with accessory elements of the immune system, such as complement (see below) and phagocytic cells which have receptors for the Fc portion of immunoglobulin in their plasma membranes (see below). Killing of virus-infected cells by phagocytes in this way is called antibody-dependent cellular cytotoxicity (ADCC).

Complement is a nine-component enzyme system whose activation is initiated by IgG and IgM. The first activated component activates the next by proteolytic cleavage and so on, with progressive amplication. Usually, each component is cleaved into a part which adheres to the antigen–antibody complex or the cell to which the antibody has bound, and a diffusible part, which forms a chemical gradient and attracts cells of the non-adaptive immune system to the vicinity of antigen. The final stage is the insertion of pore structures (the membrane attack complex) into the cell membrane, which, in sufficient number, create an ionic imbalance and lyse that cell.

The other arm of the immune system consists of T lymphocytes. Each T cell has many copies of a receptor, the TCR (see above), which recognizes a foreign peptide. Unlike B cells, the T cells do not secrete their antigen receptors but react only with foreign peptides displayed on the surface of the other cells. These peptides arise by intracellular proteolytic *processing* of the foreign protein. They then complex intracellularly with host proteins called class I or class II major histocompatibility complex (MHC) proteins, rather like an egg in an eggcup. MHC class I proteins complex with a peptide of 8–10 amino acids and MHC class II proteins with peptides of 17–22 amino acids. The complex is transported to the plasma membrane, where the peptide is displayed on the outside of the cell (antigen *presentation*). The TCR recognizes an infected cell bearing a foreign peptide held by an MHC molecule. This results in the activation of the T lymphocyte and its clonal expansion and differentiation to become a functional T cell. Mature T cells fall into two categories, which express either CD8 or CD4 proteins in their plasma membranes and hence are referred to as $CD8^+$ or $CD4^+$ T cells. The ligand for CD8 is the MHC class I protein and for CD4 it is the MHC class II protein. This recognition is additional to that of the TCR for the foreign peptide. All cells of the body have class I proteins and hence are under immune surveillance by $CD8^+$ T cells. Normally, only immune cells carry MHC class II protein constitutively, but these can be induced in many other types of cell during an infection. Thus, T cells recognize only cells carrying a foreign peptide and cannot react with foreign antigen which is not complexed with an MHC protein. The functions of fully activated T cells can be classified as either effector or regulatory (Table 14.1). T cells have many different, sometimes contradictory, functions. These may be carried out by subpopulations, or one T cell may respond differently according to the situation it finds itself in. The major T cell functions are:

1 *Cytotoxicity*, in which cells carrying processed foreign (e.g. viral) antigens are destroyed. This is mostly, but not exclusively, carried out by $CD8^+$ T cells.

2 *Delayed-type hypersensitivity*, which is the result of the secretion of cytokines (such as γ-interferon and interleukin-2), which are soluble proteins having a wide variety of effects on other cells. Other cytokines recruit macrophages and PMNL to the site of antigenic stimulation and alter the permeability of blood-vessels in that area. These all contribute to inflammation—a swelling and reddening at that site due to an influx of fluids containing antibody

Table 14.1 Some functions of fully activated T cells.

| | Effector activity | | Regulatory activity | |
	Cytotoxicity	DTH	Help	Suppression
$CD4^+$	+	++	++	+
$CD8^+$	++	+	−	++

++ indicates a major activity.
DTH, delayed-type hypersensitivity.

and immune cells. In moderation this is helpful but in excess can cause tissue damage — like any immune response.

3 *Help* is a positive regulatory function which assists B and T lymphocytes to develop in response to foreign antigen. Thus, the entire immune response depends on helper T cells. Only CD4$^+$ T cells can provide help. We shall see later that this is the main type of cell infected by human immunodeficiency virus (HIV) and so why their destruction is so devastating.

4 *Suppression* is a negative regulatory function which controls immune responsiveness of other cells in opposition to helper T cells, but this is poorly understood. CD8$^+$ and, to a lesser extent, CD4$^+$ T cells are thought to exert suppression.

Associated cells

These cells are not derived from B or T lymphocytes. The best known are the macrophages, which can engulf (phagocytose or endocytose) particles of foreign material. Unlike B or T lymphocytes, their activity is not confined to a specific antigen. However, as mentioned above, macrophages have receptors for the Fc region of IgG molecules, and such 'armed' macrophages have antibody specificity conferred upon them. Provided the antibody is non-neutralizing, some viruses can use it as a receptor to facilitate the infection of an otherwise non-infectable cell. Macrophages are found mainly in the body cavities, but another family of phagocytes, the PMNL, occur in the bloodstream. PMNL and macrophages have Fc receptors and kill infected cells to which IgG has bound.

The last cells to have been recognized in playing a role against virus infections are the NK cells. These can arise early, within 2 days of infection, and their activity against virus-infected cells is stimulated by interferons. NK cells are not armed with antibody but are cytotoxic.

INTERFERONS

The interferons are host cell proteins with antiviral and other activities, which were discovered in 1957 by Alick Isaacs and Jean Lindenmann. They incubated the chorio-allantoic membrane from embryonated chicken eggs with heat-killed influenza virus and then washed the membrane thoroughly and continued the incubation for a further 24 hours. The membranes were then discarded and the buffer tested for antiviral activity. This was done by placing a fresh membrane in the buffer and inoculating with infectious influenza virus. They found that membranes so treated did not support the growth of active virus, in contrast to untreated membranes. It was concluded that an extracellular product had been liberated in response to the heat-killed virus, and this substance was named *interferon*.

We now know that there are three types of interferon. α-Interferon and β-interferon are similar and are synthesized by most cells. Both are stimulated by replicating or abortively replicating virus. γ-Interferon is released by stimulation of T cells by the antigen for which they are specific, i.e. it is a cytokine. The major problem of studying interferons was that they, like many other chemical messengers of the body, were present in very small amounts. Much effort was devoted to their purification and they were demonstrated to be glycoproteins with carbohydrate groups which were essential for activity. Now, since the advent of genetic engineering, milligrams of interferon can be produced by expressing cloned interferon genes in eukaryotic cells, usually yeast. The antiviral activity of interferons can be measured by inhibition of the incorporation of radioactive uridine into viral ribonucleic acid (RNA) in cells infected by a togavirus or any other interferon-sensitive virus. Activity is expressed as the amount of interferon required to reduce the normal level of viral RNA by 50% and this is arbitrarily defined as one unit. Purified interferon has a high activity of around 10^9 units per mg of protein (which is of the same order as the hormones).

Molecular cloning has allowed the determination of the sequences of α-, β- and γ-interferons. This has confirmed the amino acid sequence homology (35%) of α- and β-interferon and their difference from γ-interferon. While β- and γ-interferons are each represented by a single sequence in the human genome, there is a family of about 12 α-interferons, which differ slightly in sequence but not in function. Now that pure interferon is available, it is clear that the antiviral activity is only one of several physiological effects so far discovered. Interferon enhances the activation of NK cells (see above), inhibits cell division of tumour cells and is also a macrophage-activating factor (MAF). In addition, interferons can serve as both positive and negative regulatory controls in the expression of immune responses. One very important effect is to up-regulate the expression of MHC class I proteins (α- and β-interferons) and of both MHC class I and class II proteins (γ-interferon). γ-Interferon also causes the expression of MHC class II proteins on epithelial cells, fibroblastic cells, endothelial cells (which line blood-vessels) and astrocytes (which provide nutrition for neurones in the central nervous system). The appearance of MHC proteins allows these cells to interact for the first time with $CD8^+$ and $CD4^+$ T cells, and up-regulation increases that interaction.

The ready availability of cloned interferons has permitted clinical trials to investigate their antiviral activity under natural conditions. Such trials are conducted under 'double-blind' conditions, in which half the participants are given a placebo, an innocuous substance without known therapeutic effect. Neither they nor the clinicians who monitor the symptomatology of the disease know who has received interferon or who placebo, and thus there can be no subjective bias either from patient or doctor. Similar trials based on the ability of interferon to inhibit cell division of tumour cells have also investigated its anti-cancer effects; unfortunately, results have generally been disappointing (but see p. 252). Treatment of persistent viral infections diminishes virus replication, but these effects are transitory and virus reappears when treatment is withdrawn. Similarly, although interferons do slow down the spread of some cancers, they are considered less effective than more conventional chemotherapy with antimitotic compounds. Future work entails testing the 12 cloned α-, one β- and one γ-interferon in various permutations, chemical modification to enhance activity and much more fundamental research directed towards increasing understanding of their properties.

In the latter regard, use has been made of new 'transgenic' techniques which specifically disrupt genes in mouse oocytes. The modified oocytes are then fertilized and implanted into the uterus of females. The resulting babies are born and grow up with a specific gene function knocked out. Mice have been produced with no functional γ-interferon gene or no functional γ-interferon receptor gene. These mice developed normally but had impaired macrophage and NK cell function, and reduced amounts of macrophage MHC class II protein. CTL, T helper cell and antibody responses were normal but mice showed increased susceptibility to vaccinia virus, and the bacteria *Mycobacterium boris* and *Listeria monocytogenes*. Thus γ-interferon is important in the defence against these micro-organisms.

α- and β-interferons

A great variety of viruses can induce interferon and, once induced, this is active against the whole spectrum of viruses, not just the virus responsible for induction. The mode of action of interferon can be divided into two stages (Fig. 14.3): *induction*, which results in the derepression of the interferon gene(s), the release of the interferon protein and production of an *antiviral state* in other cells by the released interferon. The full antiviral state is not achieved until the cell is infected. Induction of interferon in human cells is controlled by chromosome 9. All multiplying viruses induce interferon and it is believed that double-stranded RNA (dsRNA) is the specific inducer. If this is so, then it means that even deoxyribonucleic acid (DNA) viruses must synthesize RNA molecules with some double-stranded regions.

Free interferon initiates an antiviral state in other cells by binding to a receptor on their cell surface. In humans, the receptor is encoded by chromosome 21. Induction of the antiviral state by interferon allows the cell to respond in two ways which curtail virus replication. Firstly, dsRNA, synthesized by the infecting virus, stimulates the phosphorylation of certain cellular proteins, notably the eukaryotic initiation factor E1F2α which impairs their

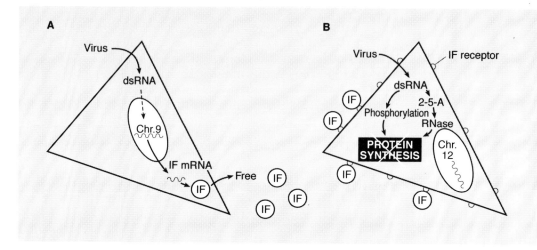

Fig. 14.3 (A) Induction of α/β interferon (IF) by viral double-stranded (ds)RNA. The dashed line indicates that we do not know if dsRNA is required to enter the nucleus to initiate interferon mRNA synthesis or whether it acts through an intermediate. (B) Establishing an antiviral state through inhibition of protein synthesis. RNase, ribonuclease.

function in the initiation of protein synthesis. Secondly, dsRNA activates a ribonuclease (RNase L) which degrades messenger RNA (mRNA) and hence also stops protein synthesis. In fact, the mechanism is more complex: interferon stimulates 2',5'-A synthetase but this is inactive without dsRNA, which is produced when the cells are infected. The dsRNA-activated 2'5'-A synthetase then synthesizes 2',5'-A (a trinucleotide pppA2'p5'A2'p5'A), which, in turn, activates RNase L. These mechanisms are not specific to the virus and, if continued, would lead to the death of the cell. However, if the virus infection is aborted, synthesis of dsRNA ceases, the inhibitory mechanisms are no longer brought into play and the cell returns to its normal state.

γ-Interferon

When infected, cells of the immune system synthesize α/β-interferon, like any other cells, but they release γ-interferon upon reaction of their antigen receptors with viral (or other foreign) antigens for which they have specificity. This can be demonstrated by infecting an animal with a particular virus and then, some days later, reacting its T cells *in vitro* with antigen purified free of nucleic acid from the same virus.

γ-Interferon has antiviral properties like α- and β-interferons but is in sequence and unstable different at pH 2. Its induction is not dependent upon the presence of dsRNA. γ-Interferon is of advantage to the host during infections by viruses which switch off host RNA and protein synthesis before α- and β-interferons can be made; it also stimulates the immune response, as already described. However, as it is part of the adaptive immune response, γ-interferon is only made on secondary stimulation of the immune system.

FURTHER READING

Arai, K.-I., Lee, F., Miyajima, A., Miyatake, S., Arai, N. & Yokota, T. (1990) Cytokines: coordinators of immune and inflammatory responses. *Annual Review of Biochemistry*, **59**, 783–836.

Lengyel, P. (1982) Biochemistry of interferons and their actions. *Annual Review of Biochemistry*, **51**, 251–282.

McMahon, M. & Kerr, I. M. (1983) The biochemistry of the antiviral state. In: *'Interferons'. Society for General Microbiology Symposium*, **35**, 89–108.

Mims, C. A. & White, D. O. (1984) *Viral pathogenesis and immunology.* Oxford: Blackwell Scientific Publications.

Moore, M. (1983) Interferon and the immune system 2: effect of IFN on the immune system. In: *'Interferons'. Society for General Microbiology Symposium*, **35**, 181–209.

Roitt, I. M., Brostoff, J. & Male, D. K. (eds) (1993) *Immunology* (3rd edn). London: Mosby.

Samuel, C. E. (1991) Antiviral actions of interferon: interferon-regulated cellular proteins and their surprisingly selective antiviral activities. *Virology*, **183**, 1–11.

Wilkinson, M. & Morris, A. G. (1983) Interferon and the immune system 1: induction of IFN by stimulation of the immune system. In: *'Interferons'. Society for General Microbiology Symposium*, **35**, 149–179.

Also check Chapter 21 for references specific to each family of viruses.

15 Animal virus–host interactions

It is useful to remember that viruses are parasites and that the biological success of a virus depends absolutely upon the success of the host species. Hence the strategy of a virus in nature must take into account that it is disadvantageous to kill the host or impair its reproductive ability. In this chapter, we shall be concerned with viruses of eukaryotes and their host cells, drawing largely upon the animal kingdom for examples. We shall lay particular emphasis upon the complexity of these interactions and the multitude of factors which share responsibility for the final outcome of infection. These studies represent one of the most important frontiers of modern virology and probably the first without a precedent in the bacteriophage systems.

CAUSE AND EFFECT: KOCH'S POSTULATES

Over a century ago, the bacteriologist Robert Koch enunciated criteria for deciding if an infectious agent was responsible for causing a particular disease. With the continuing evolution of new diseases, these are equally relevant (with some modifications) today. In essence, Koch postulated that:

1 The suspected agent must be present in particular tissues in every case of the disease.

2 The agent must be isolated and grown in pure culture.

3 Pure preparations of the agent must cause the same disease when introduced into healthy subjects.

As understanding of pathogenesis has increased the following modifications have proved necessary:

Postulate **1** Koch originally postulated that the agent should not be found in the absence of the disease, but this he abandoned with the realization that there were asymptomatic carriers of cholera and typhoid bacteria. As will be seen below, this is also the case with many viruses.

Postulate **2** Viruses were not known in Koch's time and some still cannot be grown in culture today, so this postulate is modified to say that it is sufficient to demonstrate that bacteria-free filtrates induce disease and/or stimulate the synthesis of specific antibodies.

Postulate **3** Clearly, it is impossible to deliberately fulfil the third postulate when dealing with serious disease in humans, although accidental infection can sometimes provide the necessary evidence. At the time of writing, failure to be able to demonstrate fulfilment of Koch's postulates in relation to human immunodeficiency virus (HIV) and acquired immune deficiency syndrome (AIDS) is causing some controversy (Chapter 19).

A CLASSIFICATION OF VIRUS–HOST INTERACTIONS

A classification of the various types of virus–host interactions is given in Table 15.1 and each category is dealt with later. However, this is only intended as a guide, as there is really a spectrum of virus–host interactions and the divisions are imposed for convenience. It will also become apparent that a single virus may appear in several of the categories, depending on the nature of its interaction with the host.

Acute infections

Acute infections are analogous to the classical one-step growth curve of the T-even bacteriophages, except, of course, that the time-scale is measured in days rather than minutes. Infection of organisms can be described in terms of clinical signs (which are objectively assessed) and symptoms (which are a subjective impression — in humans only) and by a variety of laboratory tests. Without the latter, no identification of the causative agent is complete. Laboratory tests include isolation and titration of infectious virus, detection of viral antigens in the blood by a variety of immunological assays, detection of viral antigens in cells obtained by biopsy, by antibody tagged with fluorescent or enzyme markers, or of viral nucleic acids, by hybridization with molecular probes or by the polymerase chain reaction, the direct identification of virus, using the electron microscope, possibly in conjunction with antibody which will agglutinate homologous virus particles, and the characterization of viral deoxyribonucleic acid (DNA) genomes by restriction endonuclease mapping (Chapter 2). In an acute infection, infectious progeny are produced and infected cells die. Further cycles of multiplication ensue and eventually the first signs and symptoms appear. Thus, the infection has been progressing for several days before we are even aware of the fact (Fig. 15.1 and

Table 15.1 A classification of virus infections at the level of the whole organism.

	Infectious progeny	Cell death (lysis)	Signs and symptoms	Duration of infection[*]
Acute	+	+	+	S
Inapparent	+	+	−	S
Chronic	+	+	+ or −	L
Persistent	≪+	−	−	L
Latent	−	−	−	L
Slowly progressive	+	+	eventually +	L
Tumorigenic: productive	+	−	+	L
defective	−	−	+	L

[*]S, short (<3 weeks); L, long (up to a lifetime).

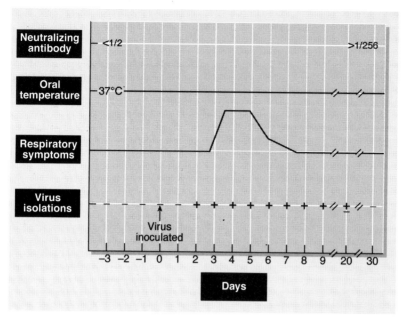

Fig. 15.1 Virus infections are measured both by symptoms and by laboratory tests for the presence of virus. The figure illustrates a 'common cold' in man caused by intranasal inoculation of a rhinovirus.

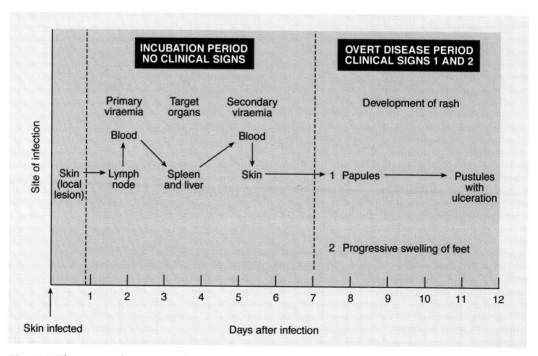

Fig. 15.2 The course of an acute infection: mouse pox virus in mice.

particularly 15.2). Fortunately, most people recover from acute virus infections within 3 weeks, and it is appropriate at this point to consider which mechanisms are responsible for allowing us to survive our first encounter with a particular virus, i.e. a primary infection. It is difficult to obtain unequivocal data from such a complex situation, far removed from the study of cells in culture, and the only major point to emerge is that it is impossible to generalize about recovery mechanisms, since different viruses are susceptible to different parts of the immune system (Table 15.2). However, it is likely that interferons, which are synthesized as soon as multiplication commences, play a major role in limiting the spread of infection. Antibody appears much earlier than is generally recognized and is found 2—3 days after infection. Initially this is complexed with viral antigen, and free antibody is not seen until it is present in excess over the antigen. Recovery is assisted by the destruction of infected cells, which have viral antigens on their surface, by cytotoxic T cells. Natural killer (NK) cells appear very shortly after infection and they too directly destroy the infected cells (see Fig. 14.1). We suffer no harm from losing infected cells in this way, providing that they represent a small proportion of the total present in a tissue. At the end of an infection, free virus is removed by interaction with specific antibody and macrophages. Recovery may be summarized in military terms:

1 Stand by to repel attack (interferon).
2 Search and destroy, counterattack (cytotoxic T cells; antibody + complement; NK cells; armed macrophages and killer cells).
3 Mopping up (macrophages; antibody).

A well-studied example of an acute infection is the disease caused in mice by mouse pox (ectromelia) virus (Poxviridae; class I) (Fig. 15.2). To obtain such data, mice are infected and killed at intervals and various organs are dissected out and ground up, so that the amount of virus there can be determined. It is important to remember that, in all infections, extensive multiplication takes place during the clinical sign-free incubation period; that any multicellular organism consists of many types of differentiated cells organized into tissues, which go towards forming a highly integrated system of organs, and that viruses home in on and multiply in certain 'target' organs; and, finally, that expression of the disease is the outcome of the combined properties of the virus and its host.

Table 15.2 Recovery from primary viral infections.

Immunity responsible for recovery	Virus-group
Activated T lymphocytes	Herpes Pox
Antibody from activated B lymphocytes (plasma cells)	Picorna Orthomyxo Paramyxo

Inapparent infections

These are the commonest infections and, as their name implies, there are no signs or symptoms of disease. In all other regards, these are the same as acute infections. Evidence of infection comes only from laboratory isolation of the virus or by a post-infection rise in a specific antibody. A virus which causes an inapparent infection has evolved to a favourable equilibrium with its host: the enteroviruses (class IV, Picornaviridae), which multiply in the gut, are one such group. The classic example is poliovirus, which causes no symptoms in over 90% of infections. An inapparent infection is arguably the expression of a highly evolved relationship between a virus and its *natural* host, since there are many examples of the virus causing lethal disease when it infects a different host, e.g. yellow fever virus (class IV, Flaviviridae) causes an inapparent infection in Old World monkeys but results in a severe infection in humans and is fatal in some New World monkeys. Inapparent infections have the same duration and are cleared by the same means as acute infections.

The following chronic, persistent and latent infections have in common the feature that the immune system cannot get rid of the viruses responsible. We shall discuss various ways in which this may come about.

Chronic/persistent infections

We have defined a chronic infection in Table 15.1 as if it were an acute or inapparent infection which is not terminated by defence processes. The main difference from a persistent infection is that large amounts of infectious virus or viral antigen are produced throughout. The pertinent question is why the combination of interferon and immune response is ineffective. In general, it appears that the infection has an inhibitory effect upon some aspect of the immune response itself, or an expression of major histocompatibility complex (MHC) antigens (Chapter 14), so that virus multiplication continues (Table 15.3). At the moment, it is not known why interferon is ineffective in curtailing infection. However,

Table 15.3 Ways in which viruses may evade immune responses and establish chronic infections.

Immunosuppression by infection and inactivation of macrophages, B cells, T cells, etc.

Stimulation of 'blocking' antibody which interferes with the action of neutralizing antibody or immune cells

Failure to synthesize or display viral antigens on the cell surface

Evolution of viral antigens within an individual host

Infection of fetus or neonate before it has a fully competent immune system viruses can both inhibit host cell macromolecular synthesis non-specifically and inhibit the functioning of individual cellular genes (p. 317), and it is possible that interferon synthesis is directly inhibited by the infecting virus. Even if it is synthesized, interferon must be present at the right place, at the right time and in a sufficiently high concentration to be effective. Perhaps viruses can in some way upset this formula.

A classic example of chronic infection occurs when lymphocytic choriomeningitis virus (class V, Arenaviridae) is inoculated into new-born mice. The animals are immunologically tolerant to the viral antigens, and virus can be found in large quantities in the circulation and tissues. One of the unexplained features is that tolerance is rarely complete, but presumably the reduced immune response is unable to cope with the infection. Similarly, little interferon is made, although the viruses are sensitive to its action. Neonatally infected animals remain healthy, whereas animals infected as adults suffer an acute infection. They mount an extreme T cell response to the virus which is lethal—a form of 'immunological suicide'.

In humans, a chronic infection results from neonatal contact with rubella virus (class IV, Togaviridae). Another outstanding example occurs when adults are infected with hepatitis B virus (HBV) (class I, Hepadnaviridae), the causative agent of serum hepatitis. After an initial acute infection, virus may continue to be present in the liver in about 10% of cases for a lifetime; however, this incidence rises to over 90% when infection occurs at a very young age (see Fig. 16.7). The virus persists by down-regulating MHC class I proteins on the surface of infected liver cells (hepatocytes), with the result that CD8$^+$ cytotoxic T cells are impotent to act. Up-regulation of MHC class I proteins by interferon can effect a cure (p. 252). During the chronic infection, cocirculation of viral antigens and antibody leads to the deposition of the resulting antigen–antibody complexes and may lead to an immune complex disease. Hence, the appearance of disease in this instance depends not on the cytolytic effects of virus but upon the relative proportions of viral antigens and antibody.

Latent infections

By formal definition, no infectious virus is present in animals with latent infections. The Herpesviridae (class I) are a diverse and ubiquitous group which normally cause latent infections, although they always initially start as, or subsequently develop into, an acute or inapparent infection. Latency is maintained by a balance between the virus and the host's defence mechanisms. Breakdown of latency leads to the activation of an acute infection, which occurs frequently when people are immunosuppressed, and activation of human cytomegalovirus is sometimes a life-threatening complication of transplant surgery or cancer chemotherapy, where immune responses are deliberately suppressed as part of the treatment. Natural activation is seen, for example, with herpes simplex virus type 1 (HSV-1) and with varicella-zoster virus (Herpesviridae) (Table 15.4). After the initial acute infection, the virus is maintained in the dorsal root ganglia of spinal or cranial nerves supplying the original area of infection. Upon activation, the virus descends the sensory nerve and erupts into an acute infection in that area of the skin supplied by this nerve. HSV-1 infection is maintained in the presence of large amounts of antibody and a T cell response, both of which probably contribute to keeping the virus in its latent state. HSV-1 has been isolated by cocultivation of the trigeminal ganglia from infected mice or humans with cells from a sensitive line, showing that virus is activated under the conditions of culture.

Molecular understanding of the distinction between acute and latent HSV infection is shown in Fig. 15.3. In an acute infection, activation of transcription depends on the interaction of a virion protein, VP16, with cellular protein(s). There is then the progressive flow of transcription of α, β and γ genes and the production of progeny virus. Latency arises when the activating cellular proteins are absent or repressor proteins bind to VP16 and/or to the activating cellular proteins, with the result that there is no α gene transcription. During latency, there is very limited transcription of latency-associated transcripts (LATs). Latency is broken by the transcription of sufficient LATs and/or the expression of ICP0 (intracellular protein (0)), an α gene product, which up-regulates its own expression and leads to an acute infection.

Is HSV DNA integrated with that of the host cell? This question was technically difficult to answer because such a small amount of DNA is present (about 0.15 to 0.015 genomes per cell). DNA can be demonstrated both in dorsal root ganglia and brain. This cell-associated DNA differed from the linear virion DNA since it had no free ends, meaning either that it was integrated or that it had circularized (Fig. 15.4). Careful work showed that it is not integrated and that latent HSV-1 DNA is episomal, existing as a free, circular molecule.

Table 15.4 The establishment of latent infections in an individual and the breakdown of latency with herpes simplex virus type 1 (HSV-1) or varicella-zoster virus (VZV).

	Acute infection	→ Latent infection	+ immunosuppression or 'physiological factors'	→ Acute infection
HSV-1	Fever blisters or cold sores around and inside the mouth	In dorsal root ganglion of trigeminal nerve	e.g. Strong sunlight, menstruation, tension	Fever blisters or cold sores
VZV	Chicken-pox	In dorsal root ganglia	'Ageing'	Shingles/zoster

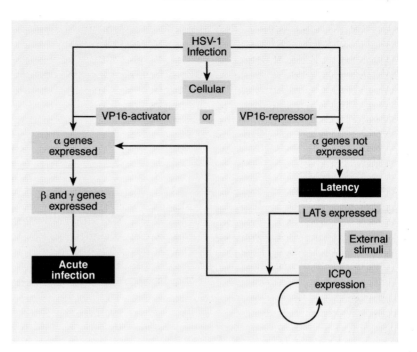

Fig. 15.3 Control of herpes simplex virus latency. Only the viral latency-associated transcripts (LATs) are expressed during latency. The viral non-structural protein, ICP0, is induced by external factors and is self-enhancing. Together with the LATs it activates expression of α genes and unlocks the pathway to acute infection. From Garcia-Blanco & Cullen (1991).

Slowly progressive infections

There are two categories: slowly progressive diseases caused by viruses, and spongiform encephalopathies, which are caused by as yet unidentified infectious agents. However, both have the central nervous system as the target organ and an incubation period of up to half a lifetime.

Slowly progressive virus infections

The classic example is a rare sequel of infection with measles virus. Before a vaccine was available (see Fig. 16.3), most children contracted an acute measles virus infection, which was cleared in around 3 weeks. However, in a very small proportion (6–22 per million cases of measles), the virus establishes itself in the brain and, after a long incubation period of 2–6 years, causes a pattern of degenerative changes in brain function, including loss of higher brain activity, and inevitable progression to death. The disease is called subacute sclerosing panencephalitis (SSPE), and at post-mortem the areas of 'hardening' or 'sclerosing' of brain tissue can be seen which give the disease its name. Why only certain individuals contract the disease is not understood but a predisposing factor is infection early in life—usually before 2 years of age. Infectious viruses cannot be isolated, although cultured cells can be infected, still non-productively, by cocultivation with brain extracts. Genomes of SSPE-measles viruses obtained from these cells have now been sequenced, and it turns out that they have

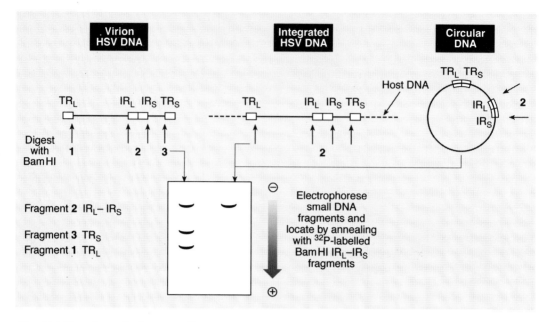

Fig. 15.4 Hypothetical scheme to illustrate the qualitative change in the structure of HSV DNA isolated from the central nervous system (CNS) of latently infected mice, which results from the loss of the free termini of the linear virion DNA. DNA from virions (A) or CNS (B, C) is digested with the restriction endonuclease *Bam* H1, which cuts at the position of the arrows. Small DNA fragments are electrophoresed in agar and annealed to a radiolabelled HSV DNA *Bam* H1 fragment from the IR_L-IR_S junction. Because of the reiterated sequences at the ends of the long and short unique regions (p. 78), the probe shows up the presence of both the IR_L-IR_S junction and the terminal fragments in the virion DNA. If the HSV DNA was integrated with host DNA the IR_L-IR_S segment would appear as before but the terminal segments would now be attached to host DNA and be too large to enter the agar gel. Other data show that (C) is the form adopted during latent infection.

accumulated many mutations and produce a defective internal (M) protein. The nature of the presumed defect in the immune response which fails to clear virus is not known.

The spongiform encephalopathies

This name refers to the vacuolating changes found in brain tissues after the disease becomes apparent. As with SSPE, there are a variety of behavioural changes which signal its onset. The best known of these infectious agents is scrapie of sheep, which causes the animal to scrape or scratch itself against obstacles. Brain suspensions from infected animals can be used as a source of infectious material and titrated in the normal way in susceptible animals. However, all attempts to purify a scrapie particle or nucleic acid have failed, so it does not appear to be a conventional virus, and is called an 'agent'.

The relationship between multiplication of the scrapie agent (inoculated peripherally with the mouse) and disease is illustrated

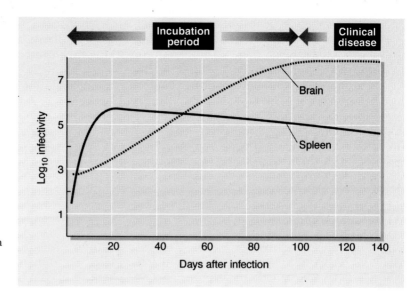

Fig. 15.5 Multiplication of the scrapie agent in mice and the time course of the disease.

in Fig. 15.5. Scrapie agent multiplies initially outside the central nervous system, mainly in organs such as the spleen or lymph nodes. Peak titres are reached in the spleen in about 20 days, which is a rate comparable to some acute virus infections. Infectivity accumulates in the brain over a longer period and reaches a higher concentration than elsewhere. Clinical disease becomes apparent when brain infectivity attains a plateau, at 3–4 months after infection. This is the best of scrapie experimental systems, for clinical disease in its natural host, the sheep, does not occur for about 2 years. If it is true that the scrapie agent codes for no proteins, there are no antigens against which the immune system can respond. Hence, this is a unique situation, where the immune response plays no part in the manifestation or suppression of the disease.

There have been many wild and wonderful theories about the nature of the scrapie agent, but there is no doubt that it has heritable properties, as exemplified by the constancy of the pattern of disease and the remarkably defined incubation period, which varies by only 1–3 days per 100 days. However, scrapie agent infectivity is remarkably resistant to inactivation by phenol and ultraviolet (UV) irradiation, which has led to suggestions that the agent is not a nucleic acid. Recently, Stanley Prusiner has isolated a protein from scrapie-infected brain which he suggests is the infectious agent and calls a 'prion'. This will be discussed later, in Chapter 20.

Recently, bovine spongiform encephalopathy (BSE) was diagnosed in dairy cattle in the UK. Colourfully called 'mad cow disease' by the media, remarkable scientific detective work showed that BSE was caused by scrapie agent contained in artificial cattle food containing sheep residues, which was fed to intensively reared cattle. We shall return to BSE in Chapter 20.

Creutzfeldt–Jakob is a distressing slowly progressive disease of the central nervous system in humans and bears many similarities to scrapie. Like scrapie, there is an infectious agent, which can be transmitted to higher primates and, by accidental contamination of neurosurgical instruments, to other people. It is a rare condition, occurring at the rate of 1 per million per year. The agent has not been isolated.

The slowly progressive diseases have initial similarities with inapparent infections and later with persistent infections, but the late development of the disease symptoms gives them a unique place in Table 15.1.

Virus-induced tumours

All viruses that cause tumours have DNA as their primary genetic material in the cell but some, the ribonucleic acid (RNA) tumour viruses, have RNA in their virions, i.e. they belong to class VI. DNA 'tumour' viruses normally cause an acute infection and it is rare that a tumour results, e.g. most young people are infected by Epstein–Barr (EB) virus (class 1, Herpesviridae), which causes infectious mononucleosis (glandular fever), and yet occurrence of the tumours (Burkitt lymphoma, nasopharyngeal carcinoma) with which the virus has been associated is very rare. Demonstration of tumorigenicity in the laboratory is frequently made of necessity under 'unnatural' circumstances which are known to be favourable to the development of the tumours. Important factors are the genetic attributes and age of the infected animal: certain inbred lines develop tumours more readily than others, and young animals are more susceptible, because their immune system is not fully mature. If no tumours result, it may be necessary to transfer cells transformed by the virus *in vitro* into the animal to demonstrate viral tumorigenicity.

The diagnostic criterion for a tumour virus in the laboratory is the conversion of a normal cell into a transformed cell. This involves many complex changes in cellular properties, which will be discussed more fully in Chapter 17. However, not all transformed cells are tumorigenic when transplanted to appropriate animals, and other steps in addition to transformation are needed to produce a cancer cell. Thus, transforming viruses that cause cells to divide more rapidly than normal are at one end of a spectrum and at the other are those causing cellular destruction. Tumours fall into two groups: those that produce infectious virus and those that do not. The latter are the more common: they contain viruses which are defective and unable to multiply or viruses in which multiplication is in some way repressed (like lysogeny). The methods described above for detecting viral components in latently infected cells may be used successfully on tumour cells. However, it is suspected that some tumours are caused by a 'hit-and-run' infection. This suspicion was strengthened by finding that cells transformed by adenoviruses in culture contained as little as 7% of the genome of the original infecting virus.

Like other virus—host interactions which have been discussed above, the induction of tumours depends upon the balance of a complex situation. This can be viewed as shown in Fig. 15.6, where, initially, infection may be acute or in a persistent or latent form. The transformation event may be enhanced by or require additional factors (such as chemical carcinogens in Fig. 15.6). How often transformation of cells takes place in the animal we do not know, but it is likely that this is frequent and that on most occasions the immune system recognizes and destroys the transformed cell. Tumour formation probably requires that the immune system is in some way deficient, a state perhaps induced by infection or by ageing.

INTERACTIONS WITH VIRUSES WHICH TAKE PLACE OUTSIDE THE HOST

In this section, we shall continue with virus—host interactions by considering how viruses are transmitted from an infected to a susceptible host. Since most viruses are inherently unstable and, when outside the host, lose infectivity rapidly, this is an important step in their life cycle. Knowledge of how viruses are transmitted may enable us to break the cycle at this stage and thus prevent further infections. We shall deal with transmission of infection by the respiratory route, by the faecal—oral route, by the urinogenital tract and by 'mechanical' means. The reader will have noted that these strategies of transmission have in common the ability to circumvent the dead, impermeable outer layer of the skin and to bring virus into contact with the naked cell surface. It is no coincidence that most viruses are spread by the respiratory and faecal—oral routes, since, in order to carry out their normal physiological functions, the lungs and small intestine each have a surface

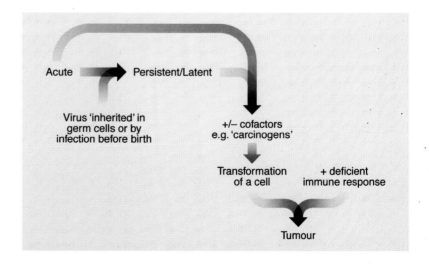

Fig. 15.6 The relationship between virus infection and tumorigenesis.

area equivalent to three tennis-courts and are lined with living cells.

Transmission via the respiratory tract

Not only viruses which cause respiratory infections but also some viruses causing generalized infections, such as measles and small-pox, are contracted by this route. After virus has multiplied, it either reinfects host cells or escapes from the respiratory tract in the aerosol which results from our normal activities, such as talking, coughing and sneezing (and particularly singing!). These aerosols are inhaled and give rise to a 'droplet infection' of a susceptible individual. The size of droplets is important, as those of large (>10 μm) diameter rapidly fall to the ground and the smallest (<0.3 μm) dry very quickly, resulting in accelerated inac-tivation of virus contained therein. Thus, the middle-sized range of droplets are those that transmit infection. The size of these will determine where the droplets are entrapped by the respiratory system of the recipient, since the 'baffles' lining the nasal cavities remove the larger airborne particles, while the smaller may pen-etrate deep into the lung. The lungs have an enormous area, equivalent to three tennis-courts (about 400 m^2), which is required for gaseous exchange. The entire surface consists of naked cells, which can potentially be infected by viruses.

It seems that the increase in nasal secretions which accompanies many respiratory infections favours the dispersal of the viruses responsible and the increase in coughing and sneezing increases the production of infected aerosols. However, transmission exper-iments from people infected with a rhinovirus to susceptibles sitting opposite at a table proved singularly unsuccessful. Equally unsuccessful was the transmission of influenza from a naturally infected husband/wife to his/her spouse. This has led to the suggestions that only particular individuals may shed sufficient virus, produce excess nasal secretion and/or aerosols containing optimum-sized particles and act as efficient spreaders of infection or that, in order to be infected, one has to be in a certain physio-logical state. Apart from the traditional stories of wet feet pre-disposing you to catch a cold (not true!), it has been shown that recently bereaved people are particularly susceptible to infectious diseases. Thus resistance is influenced by one's state of mind. Neuroimmunology is a developing field of study (see Zwilling, 1992).

In addition to direct aerosol transmission, it is now believed that some respiratory viruses are spread by physical contact, for example, large droplets which land on solid surfaces can be transferred from hand to nose. The relative importance of airborne to physically borne virus transmission remains to be determined.

Linked with transmission of respiratory infections are environ-mental factors which result in seasonal variation in the amount of illness or frequency of isolation of a virus. Influenza, for example, is a winter disease, occurring around January in the Northern Hemisphere, June in the Southern Hemisphere, but nearly all the year round close to the Equator. One can point to variations both in the environment (e.g. temperature and humidity) and in social behaviour (crowding together in winter with poor ventilation) which could well affect virus stability and transmission and hence the seasonal incidence of virus diseases, but a full explanation of these complex phenomena is not yet available. Since viruses can survive for only a limited time outside the cell, it is fair to assume that infected individuals are present continuously somewhere in the population.

Transmission by the faecal—oral route

Many viruses are ingested, and infect and multiply in the alimentary tract. The small intestine alone has the same area as a tennis-court (about $400\,m^2$). Its entire surface consists of naked cells, which normally function to adsorb nutrients and water, and can potentially be infected. One virus which invades by this route is poliovirus, and there are many similar enteroviruses spread in an identical manner; another is hepatitis A virus (Picornaviridae, class IV). All these viruses are excreted in faeces, so their spread is favoured by poor sanitation and poor personal hygiene. Not surprisingly, when one remembers how small children investigate strange objects by putting them in their mouth, many of these viruses cause infections in early childhood.

One would expect to see a reduction of enteric viruses when sanitation is improved, and basically this has been the experience. However, there have been some unpleasant surprises. In conditions of poor sanitation, poliovirus infects young children and results usually in an inapparent gut infection. With improved sanitation, poliovirus is not contracted until adolescence, and associated with this shift in age distribution is an increase in the incidence of paralytic poliomyelitis. This is another example of increased severity of a disease seen when a virus infects an 'unnatural' host, and in this case age is the important difference.

Transmission via the urinogenital tract

A few viruses, such as Lassa fever (Arenaviridae, class V), are excreted in urine. Of greater importance are those that are spread by sexual contact or are transmitted to the offspring by the mother. HSV types 1 and 2, hepatitis B virus (HBV) human papilloma viruses (p. 252; p. 263) and HIV are all spread sexually. HIV (Retroviridae) has only been recognized as a serious cause of disease since 1983 and is responsible for AIDS. AIDS predisposes to both viral and non-viral infections of exaggerated severity and to cancer, and has a mortality rate approaching 100% (Chapter 19). HBV is linked with the commonest human cancer, primary hepatocellular carcinoma (Chapter 17). All three viruses are transmitted sexually, both by heterosexual (male to female, and female to male) and by homosexual behaviour. The vaginal wall is thicker than the rectal wall and its blood-vessels are less inclined to rupture and may provide better protection from infection, but human ejaculate contains about 10^6 lymphocytes, in addition to sperm, which if carrying virus could well contribute to infection. Spread of such viruses is affected by socio-sexual behaviour and it is probably no coincidence that their spread was increased by the promiscuity of the 1960s and 1970s. There are as yet few signs that this is being curtailed by public awareness of the new wave of sexually transmitted virus diseases.

The transmission of virus from mother to offspring is called 'vertical' transmission, in contrast to the 'horizontal' transmission between other individuals. Such transmission of HBV and HIV is common. Rubella virus (Togaviridae) is also transmitted vertically and is responsible for congenital malformation if the fetus is infected in the first 3 months of pregnancy. It is impossible to decide with certainty whether a virus is transmitted to the zygote from an infected oocyte or sperm, or whether the zygote is infected from virus present in cells of the uterus, or via the bloodstream in placental mammals. Any virus that infects the mother while she is producing eggs or offspring can in theory be vertically transmitted, while those that have genomes integrated with those of the reproductive cells of the host cannot help but be transmitted by this route. Infection can also result from passage of the ovum down an infected oviduct or by contact with infected cells on its way to the exterior. Although it does not belong in this section, other means of transmission of certain viruses to the new-born animal include infected milk, or direct infection from mother to offspring, possibly via saliva.

Transmission by 'mechanical' means

Under this umbrella are included animal viruses which infect their hosts by means which puncture the impermeable outer layer of skin. This route introduces for the first time the phenomenon of *virus vectors*, usually biting arthropods which feed by piercing the skin with their mouthparts. These become contaminated with virus, which is introduced into the body of their next meal.

Animal viruses that are spread by mosquitoes are predominantly Alpha-, Flavi- and Bunyaviridae. Transmission is limited to certain species. These viruses have a complex life cycle, multiplying in the tissues of the insect vector as well as the vertebrate host. Other arthropod-borne animal viruses do not multiply in their vector, examples being myxomavirus (Poxviridae), which is spread between rabbits by passive transfer on the mouthparts of infected mosquitoes in Australia and rabbit fleas in the UK, and HBV, which is transmitted by the bedbug.

There is also transmission of serious infections (HBV and HIV) between intravenous drug abusers. Virus can be carried from one person to another by using non-sterilized hypodermic syringes. However, there is risk of transmission through any shared device which causes cuts or abrasions, such as combs, razors, and needles used for acupuncture, tattooing or ear-piercing.

Although not covered in this book, most plant viruses are spread by mechanical means. They are transmitted into plant cells by animal vectors, such as aphids, leafhoppers, beetles and nematodes, which feed on plants, by fungi, or by non-specific abrasions made to the plant tissue made by the wind etc. which expose cells sufficiently for them to be infected. Transmission by plant-feeding animals is a specific process and only certain species are implicated in the transmission of a particular virus (see Chapter 21). Like their animal virus cousins, some plant viruses multiply in the vector, while others are passively transmitted.

RECAPITULATION

It is worth reiterating here the major features of virus—host interactions:

1 The majority of infections result in a satisfactory end for both the host population (if not the individual) and the virus. In other words, infections are dealt with by the host's defence processes so that the host recovers and can reproduce itself, but not before the virus has time to multiply and perpetuate itself by being transmitted to other susceptible individuals.

2 Infections that kill (or reproductively incapacitate) large sections of a population are rare. When they occur, one should suspect that there has been a change in the virus—host interaction, i.e. the 'wrong' host has been infected. Although we are far from understanding the nature of this genetic compatibility, it is inevitably the product of long-term evolution.

FURTHER READING

An enormous area to cover, but the following provide a way in and references to further reading.

ASM News (1991) Mother-to-newborn infections pose problems. *American Society for Microbiology News*, **57**, 619.

Braciale, T. J. (ed.) (1993) Viruses and the immune system. *Seminars in Virology*, **4**, no. 2.

Dimmock, N. J. & Minor, P. D. (eds) (1989) *Immune responses, virus infections and disease*. Oxford: IRL Press.

Dimmock, N. J., Griffiths, P. D. & Madeley, C. R. (eds) (1990) *Control of virus diseases* (Society for General Microbiology Symposium 45). Cambridge: Cambridge University Press.

Fenner, F., McAuslan, B. R., Mims, C. A., Sambrook, J. & White, D. O. (1974) *The biology of animal viruses* (2nd edn). London: Academic Press. (This is still a good treatment and provides many useful references.)

Fraser, N. W., Block, T. M. & Spivack, J. G. (1992) The latency-associated transcripts of herpes simplex virus: RNA in search of a function. *Virology*, **191**, 1–8.

Fujinami, R. S. (ed.) (1990) Mechanisms of viral pathogenicity. *Seminars in Virology*, **1**, no. 4.

Garcia-Blanco, M. A. & Cullen, B. R. (1991) Molecular basis of latency in pathogenic human viruses. *Science*, **254**, 815–820.

Gibbs, C. J. & Gajdusek, D. C. (1978) Atypical viruses as the cause of sporadic, epidemic and familial chronic diseases in man: slow viruses and human diseases. In: *Perspective in virology*, Vol. 10, pp. 161–194. Pollard, M. (ed.). New York: Raven Press.

Kimberlin, R. H. (1984) Scrapie: the disease and the infectious agent. *Trends in Neurosciences*, **7**, 312–316.

Kimberlin, R. H. (1990) Unconventional slow viruses. In: *Topley and Wilson's Principles of bacteriology, virology and immunity* (8th edn), Vol. 4, pp. 671–693. Collier, L. H. & Timbury, M. C. (eds). London: Edward Arnold.

Mahy, B. W. J., Minson, A. C. & Darby, G. K. (eds) (1982) 'Virus Persistence'. *Society for General Microbiology Symposium*, **33**. (13 chapters on different aspects of persistence.)

Mellerick, D. M. & Fraser, N. W. (1987) Physical state of the latent herpes simplex virus genome in a mouse model system: evidence suggesting an episomal state. *Virology*, **158**, 265–275.

Mims, C. A. (ed.) (1985) Virus immunity and pathogenesis. *British Medical Bulletin*, **41**, 1–102.

Mims, C. A. (1989) The pathogenetic mechanism of viral tropism. *American Journal of Pathology*, **135**, 447–455.

Mims, C. A. & White, D. O. (1984) *Viral pathogenesis and immunology*. Oxford: Blackwell Scientific Publications.

Notkins, A. L. & Oldstone, M. B. A. (1984, 1985, 1986) *Concepts in viral pathogenesis*. New York: Springer-Verlag. (Very useful mini-reviews.)

Rouse, B. T. (1992) Herpes simplex virus: pathogenesis, immunobiology and control. *Current Topics in Microbiology and Immunology*, **179**, 1–179.

Sharpe, A. H. & Fields, B. N. (1985) Pathogenesis of viral infections: basic concepts derived from the reovirus model. *New England Journal of Medicine*, **312**, 486–497.

Stevens, J. G. (1989) Human herpesviruses: a consideration of the latent state. *Microbiological Reviews*, **53**, 318–332.

Stroop, W. G. & Baringer, J. R. (1982) Persistent, slow and latent viral infections. *Progress in Medical Virology*, **28**, 1–43.

ter Meulen, V. & Hall, W. W. (1978) Slow virus infections of the nervous system: virological, immunological and pathogenetic considerations. *Journal of General Virology*, **41**, 1–25.

Timbury, M. C. (1983) *Notes on medical virology* (7th edn). Edinburgh: Churchill Livingstone.

Zwilling, B. S. (1992) Stress affects disease outcomes. *American Society for Microbiology News*, **58**, 23–25.

Also check Chapter 21 for references specific to each family of viruses.

16 Vaccines and chemotherapy: the prevention and treatment of virus diseases

Humans are concerned to protect themselves, their plants and animals against death, disease and economic loss caused by virus infections. In animals (including humans), immunization with vaccines has been far more effective than chemotherapy, but vaccines against some viruses are less than ideal. With other viruses it has not yet been possible to produce a vaccine which gives any protection. These problems will be discussed below. The rationale of immunization is to raise a specific immunity without an individual experiencing the disease. Since plants do not have an immune system, prevention of plant virus diseases has relied upon other means, such as breeding plants that are genetically resistant to the virus or its vector, or by control of the vector. Control of animal virus diseases through improvements in nutrition, public health, personal hygiene, and education is a vitally important aspect of prevention.

PRINCIPAL REQUIREMENTS OF A VACCINE

Our discussion will deal initially with conventional vaccines and then take a forward look at new approaches to immunization. Conventional vaccines comprise either infectious ('live') or non-infectious ('killed') virus particles (Table 16.1). Upon administration, all vaccines should have the following properties: (i) cause less severe disease than the natural infection; (ii) stimulate effective and long-lasting immunity, and (iii) be genetically stable.

Cause less severe disease than the natural infection

In other words, a vaccine causing some illness or side-effects can be tolerated when the alternative is more serious illness or death, but obviously vaccine manufacturers strive for a product which causes the minimum of discomfort. The process of producing a virus strain which causes a reduced amount of disease is called *attenuation*. The disease-causing virus is referred to as the *virulent* strain and the attenuated strain as *avirulent*. These are relative, not absolute, terms.

Avirulent strains have been obtained by selecting and enriching for such variants that occur naturally. This is achieved empirically, and experience has shown that it can be helped by multiplication in cells unrelated to those of the normal host, by multiplication at subphysiological temperature or by recombination with an avirulent

Table 16.1 Some human and veterinary virus vaccines.

Virus	Family	Live or killed vaccine	Disease (if named differently from the virus)
Vaccinia	Pox	l	Smallpox caused by variola virus
Polio	Picorna	l, k*	Poliomyelitis
Measles	Paramyxo	l	
Mumps	Paramyxo	l	
Rubella	Toga	l	German measles
Yellow fever	Flavi	l	
Rabies	Rhabdo	k	
Influenza A and B	Orthomyxo	k†	
Hepatitis B (HBV)	Hepadna	k†	Serum hepatitis (jaundice)
Foot-and-mouth	Picorna	k, l	Affects cattle and pigs
Newcastle disease	Paramyxo	l	Fowl pest
Canine distemper	Paramyxo	l	
Equine influenza	Orthomyxo	k	
Parvo	Parvo	k, l	Causes death in young dogs and cats
Turkey herpes	Herpes	l	Lymphoma in chickens‡
Pseudorabies	Herpes	k, l	Aujeszky's disease of pigs

* Killed vaccine is used successfully in Scandinavian countries.
† A subunit vaccine is used for human HBV (surface (S) antigen) and influenza (haemagglutinin and neuraminidase antigens).
‡ Caused by Marek's disease virus, another herpes virus. Thus a virus with some related antigens can provide protection, cf. vaccinia virus against smallpox virus. This system is also notable as the first vaccine against cancer. The hepatitis B vaccine is the first anti-cancer vaccine to be used in humans.

laboratory strain. The use of strains which are antigenically related to the virulent strain and cause less disease in that host has long been known, since Edward Jenner in 1798 used cowpox virus to vaccinate against smallpox. Until recently, we used a descendant of cowpox virus, vaccinia, for the same purpose, and the term 'vaccine' has become synonymous with any immunogen used against an infectious disease. Naturally enough, killed vaccines should cause no disease at all. However, since the vaccine is made from the virulent strain, it is essential to kill *every* infectious particle present (see p. 235). An advantage of killed preparations is that other unknown contaminating viruses may also be killed by the same process. However, a severe disadvantage is that the killed virus does not multiply, so that an immunizing dose has to contain far more virus than a dose of live vaccine. This increases both the cost and the amount of impurities present; the latter may result in hypersensitivity reactions when these substances are experienced again.

Effective and long-lasting immunity

The requirement for these properties is both scientific and sociological. As explained in Chapter 15, different viruses are susceptible to different parts of the immune system; hence it is necessary for the vaccine to stimulate immunity of the correct type, in the correct location and in sufficient magnitude for it to be effective. This is a problem with the killed influenza virus vaccines currently in use, which are administered by injection and do not raise immunoglobulin A (IgA) antibody at the respiratory epithelial surface. However, sufficient IgG to be protective is thought to diffuse from the bloodstream to the respiratory surface in about 70% of individuals immunized. If an immunity is not effective, then the virus being immunized against may still be able to multiply. This can occur with foot-and-mouth disease virus of cattle and the partial immunity provides a selection pressure which favours the multiplication of new antigenic variants of the virus, which in time replace the pre-existing strain. Even animals with effective immunity against the original strain are not protected from the new strain; hence the long, expensive process of vaccine development and immunization has to begin again.

Public faith in any preventive medicine is rapidly lost if that measure is not effective in the majority of cases. Consequently, to avoid bad publicity, a vaccine must protect the majority of individuals who have received it. Immunization should result in long-lasting immunity, as it is surprisingly difficult to persuade people to come forward to be immunized. Hence a 'single-shot' vaccine, which requires only one visit to the surgery, would be ideal but has not yet been realized. Some employers have circumvented this problem by offering immunization at the place of work. Vigorous advertising campaigns are widely used also (Fig. 16.1) to overcome public apathy. In tropical countries, where immunization is often most needed, there is the problem of ensuring that a vaccine has a good shelf-life (i.e. that its infectivity or immunogenicity is stable for periods when refrigeration is not available), or immunization will be ineffective.

Genetic stability

An attenuated live virus vaccine must not revert to virulence when it multiplies in the immunized individual, but even the excellent poliovirus vaccine is not perfect in this regard (p. 235). Vaccines which are formed by 'killing' virulent strains of viruses must not be restored to infectivity by genetic interactions with each other or viruses occurring naturally in the recipient.

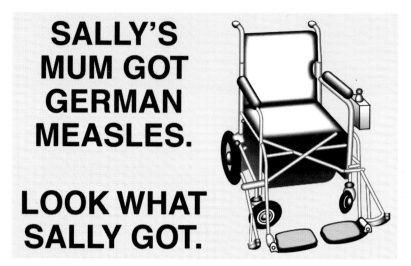

SALLY'S MUM GOT GERMAN MEASLES.

LOOK WHAT SALLY GOT.

Fig. 16.1 Newspaper advertisement aimed at increasing public awareness and acceptance of immunization against rubella virus.

ADVANTAGES, DISADVANTAGES AND DIFFICULTIES ASSOCIATED WITH LIVE AND KILLED VACCINES

Inactivation

A killed vaccine must have no infectivity and yet be sufficiently immunogenic to provoke a protective immunity. Inactivating agents should inactivate viral nucleic acid function and not merely attack the outer virion proteins, since infectious nucleic acid could be released by the action of the cell on the coat. Secondly, inactivating agents must lend themselves to the industrial scale of manufacturing processes. Formaldehyde and β-propiolactone are two which have been used. The former, reacts with the amino groups of nucleotides and cross-links proteins through ε-amino groups of lysine residues. β-propiolactone inactivates viruses by alkylation of nucleic acids and proteins. The basic problem of preparing a killed vaccine is that every one of the 1 000 000 000 or so infectious particles which are contained in an immunizing dose of virus must be rendered non-infectious, and this was the problem which faced Jonas Salk before the first poliovirus vaccine could be presented to the general public in 1953. Inactivation of the infectivity of polio- and other viruses is exponential and has the kinetics shown in Fig. 16.2. Frequently, it is found that a small fraction of the population is inactivated far more slowly than the majority, so that the whole virus population has to be kept in contact with the inactivating agent for much longer than is predicted by the initial rate of inactivation. This has the concomitant danger that immunogenicity will be destroyed. Examination of the 'resistant' fraction shows that it results from inefficient inactivation of particles trapped in aggregates or clumps. Despite these difficulties, the inacti-

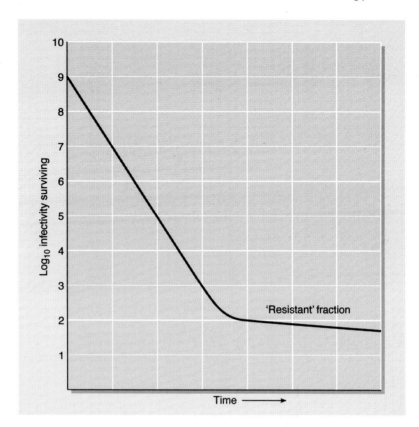

Fig. 16.2 Exponential kinetics of virus inactivation. Note the 'resistant' fraction which is inactivated more slowly.

vated Salk vaccine has been used very successfully, mainly by Scandinavian countries, to this day. Live vaccines present none of these problems, but the difficulties of producing a genetically stable attenuated strain are at least as great. For poliovirus, these were overcome by Albert Sabin, who attenuated all three serotypes to produce the successful live vaccine, which has been in use since 1957. None the less, the type 3 vaccine reverts to virulence at the rate of one case of poliomyelitis per 10^8 doses of vaccine. This is an extremely low figure but is very serious for the child involved. The molecular explanation and ways of improving the vaccine are discussed later (p. 315).

Routes of administration

The unavoidable difficulty of killed vaccines is that they do not multiply. Hence they do not reach and stimulate immunity in those areas of the body where infectious virus would normally be found. However, providing that the killed vaccine is a potent immunogen, the high levels of serum IgG which result from injection of the vaccine can often serve in place of local immunity, presumably because there is a sufficient concentration to diffuse to the extremities.

The means of administration is an important factor in persuading people to accept a vaccine. Injections are painful and unpopular, particularly with small children and their mothers. A compressed-air device which also injects vaccine through the skin has been used for mass immunization, mainly of military personnel, in the USA, in place of the traditional hypodermic syringe and needle. A disadvantage of killed viruses is that two or more injections are required to build up a sufficient secondary immune response. This could in theory be avoided by emulsifying vaccine with an adjuvant. When injected, this remains as a lump and releases immunogen over a long period and there is the additional advantage that the adjuvant also stimulates the activity of the immune system. However, adjuvants cause local irritation and few (e.g. aluminium hydroxide gel) have been deemed satisfactory for use in humans, although they are used in other animals.

Cost

The production of vaccines for use in humans is subject to legislation by the Food and Drug Administration in the USA and by the Department of Health in the UK. The precautions laid down for the production of a safe vaccine are necessarily stringent. However, this means that the cost of the final product will be correspondingly high. Precautions increase as knowledge is obtained about hitherto undetected viruses in cells used in vaccine manufacture, or other potential hazards, and as they do so the costs increase yet again. Yet another problem to manufacturers in the USA is the cost of defending lawsuits where vaccine-induced damage is alleged; this is a trend now spreading to other countries. Regulations governing the use of veterinary vaccines are not so demanding and the product is correspondingly cheaper. The other major aspect which directly affects the cost is the quantity of vaccine which has to be produced. Live vaccines are cheap to produce compared with killed vaccines, since they only have to initiate an infection, whereas an immunizing dose of killed vaccine has to be provided in a minimum of two injections. Finally, the costs involved with packaging, distribution and administration to the recipient are considerable. These all add up to make vaccination a luxury which poor countries, usually those in the greatest need, cannot afford. Provision of vaccines is one of the ways in which the richer countries can and often do provide valuable aid.

Multiple vaccination

The immune system can respond to more than one antigen at a time; hence it is possible to immunize with a vaccine 'cocktail'. This has advantages in minimizing the number of injections and the inconvenience. In practice, satisfactory immunity results from multiple *killed* vaccines, but problems with multiple *live* vaccines may arise due to mutual interference in multiplication, possibly as a result of the induction of interferon (see later). However, a triple live vaccine of measles, mumps and rubella viruses is in routine use.

INFECTIOUS DISEASES WORLD-WIDE

Table 16.2 puts in context the major virus diseases (poliomyelitis, measles, rotavirus-induced diarrhoea, some respiratory viruses, human immunodeficiency virus (HIV) and hepatitis B virus (HBV) relative to bacterial diseases (tetanus, diphtheria, pertussis, tuberculosis and some sexually transmitted organisms), protozoal diseases (malaria) and the minute snail-borne worm which causes schistosomiasis.

ELIMINATION OF VIRUS DISEASES BY IMMUNIZATION

The World Health Organization (WHO) is responsible for the world-wide control of virus diseases. Table 16.3 shows some of the achievements and current goals, made possible through the development of suitable vaccines.

Table 16.2 The world's leading diseases.

Disease	Comments
Vaccine-preventable diseases (polio, tetanus, measles, diphtheria, pertussis, tuberculosis)	Annually, 46 million infants are not fully immunized; 2.8 million children die and 3 million are disabled annually
Diarrhoeal disease	Annually, at least 750 million children are infected and 4 million die
Acute respiratory infections	4 million children die annually
Tuberculosis	10 000 million people carry the bacterium; annually, there are 10 million new cases and 3 million deaths
Sexually transmitted diseases	1 out of 20 teenagers and young adults contract such diseases each year, and 10 million are infected with HIV
Malaria	There are 100 million cases annually, almost half the world's population lives in malarious areas
Schistosomiasis	There are 200 million cases anually
Chronic hepatitis (hepatitis B virus)	There are 300 million infectious carriers; these have a 200-fold greater risk of developing liver cancer

Table 16.3 Goals, past and future, for the control of some virus diseases.

Virus	Commencement of control programme	Achievement/goal
Variola virus	1966	Eradicated world-wide in 1977
Poliovirus 1, 2 and 3	1988	Eradication world-wide by 2000
Measles virus	1991	Eradication from Europe by 2000
Hepatitis B virus	1992	Universal immunization of infants in Europe by 1997

Eradication of smallpox

Smallpox has been totally eliminated: the last naturally occurring case was recorded in Ethiopia in October 1977. Yet less than 200 years ago mortality from smallpox in England reached 25% of all children born, and in India in 1950 over 41 000 people died. The decline of smallpox in those countries where it was recently endemic has been a triumph for the vaccination programme administered by the WHO. Smallpox is the only disease that has ever been deliberately eliminated and this experience has served to emphasize that successful elimination is possible only if a disease fulfils certain criteria. *There must be overt disease in every instance so that infection can be recognized; the causal virus (variola) must not persist in the body after the initial infection; there must be no animal reservoir from which reinfection of humans can occur; vaccination must provide effective and long-lasting immunity; and the viral antigens must not change.* Apparently, variola virus fulfilled all of these criteria. The elimination of smallpox took several decades and gradually countries could be declared smallpox-free. The final assault on the virus in the Indian subcontinent and Africa was *not* done by mass immunization but by 'ring' immunization. A new case of smallpox was identified, and all people in that location were then immunized to form a ring of immunity. The virus could not break out and so died out. Fortunately, other poxviruses (e.g. whitepox of monkeys, which is antigenically similar to smallpox) are apparently unable to step into the ecological niche vacated by variola and replace smallpox as a pathogen of humans.

A problem arises with laboratory stocks of smallpox virus. As vaccination is no longer required, the population is becoming increasingly susceptible to the danger of an escaped laboratory virus. If we destroy the virus we lose, for ever, 75×10^6 M_r of irreplaceable genetic information. We have no way of knowing the value of the smallpox virus genome, but it encodes proteins which are similar to those found in eukaryotic cells and could conceivably be of use in the future. The problem has been resolved by cloning the entire viral genome in fragments. Viral proteins can now be expressed in the absence of infectious virus. The remaining stocks of infectious virus in the USA and Russia are to be destroyed.

Control of measles

Prior to the development of a vaccine, 130 million cases of measles occurred annually world-wide and 3 million deaths. Even today, in Third World countries, due to malnourishment, mortality can be as high as 80%. Vaccination reached nearly all countries by 1982, but the coverage is still very patchy and in half the African countries reaches less than 60% of children. However, since vaccination commenced, the ocurrence of measles has fallen steadily and in 1990 there were 29 million cases world-wide with 0.9 million deaths.

Some way ahead of the world scene, the Department of Health, Education and Welfare announced in 1978 their intention of elim-inating measles virus from continental USA. Death results from respiratory and neurological causes in about 1 of every 1000 cases and encephalitis also occurs in 1 of every 1000 cases, survivors of the latter often having permanent brain damage. There are also rare cases of subacute sclerosing panencephalitis (SSPE) caused by measles virus (p. 221). Are the criteria for successful elimination met by measles virus? It multiplies only in humans; there is a good live vaccine (95% effective) and only one serological type of virus is known. Usually, measles virus causes an acute infection. In SSPE, virus is present in the brain only and is not transmitted. The measles eradication programme was initially very successful and, by 1985, measles was a rare disease in the USA (Fig. 16.3).

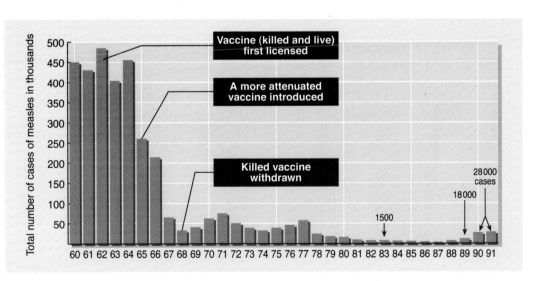

Fig. 16.3 Fall in the number of reported cases of measles in the USA, 1960–1984, following immunization.

However, to achieve this it was necessary to immunize a high proportion of children, so legislation was passed prohibiting children from attending school unless they had been immunized or had had a natural infection. (Other countries will need to debate the ethics of refusing people the right to decide on their own medical treatment.) Despite these measures, there has recently been a disturbing increase of over 10-fold in the number of measles cases (Fig. 16.3), which is thought to reflect partly the problems of maintaining a high level of immunization, particularly in socially deprived inner-city areas, but also a failure of the vaccine to protect some children. However, even if measles were totally eliminated from the USA, there would be a continuing need for immunization to prevent the epidemic spread of the virus, following its inevitable introduction by an infected visitor, to an otherwise totally susceptible population. Measles in non-immune adults is much more severe than in children.

The WHO plans to eradicate measles from Europe by 2000, but does not yet see the way clear to eradicating the disease worldwide.

Eradication of poliovirus

In 1985, the Pan-American health authorities announced a plan to eliminate poliovirus from North, Central and South America by 1995. In 1991, no poliomyelitis was reported, apart from a few cases in two South American countries. However, this does not mean that the *virus* has been eliminated, since 99% of infections are subclinical. This successful policy has now been extended world-wide by WHO. It consists of using oral live vaccine (to types 1, 2 and 3), setting up laboratories in all countries, encouraging good public relations, with events such as national immunization days, and intensive immunization whenever there is an outbreak. Most of Western Europe, Australia and New Zealand had no cases of polio in 1991. Elsewhere, there are encouraging signs of only low numbers of cases of poliomyelitis. In countries where the disease has been virtually abolished, the main danger is of complacency, because of the continued presence of the virus or of the introduction of virus from elsewhere by infected travellers. High levels of immunity still need to be maintained.

Control of other virus diseases

Elimination of other virus diseases world-wide is unlikely to be easy, but WHO is studying the feasibility of eliminating rabies virus. The latter presents a new dimension as it is endemic in wild animals: foxes in Europe, racoons in the USA, vampire-bats in South America, wolves in Iran and jackals and mongooses in India. These transmit virus directly to humans or via domestic animals, notably dogs, but these infections are incidental to the normal life-cycle of rabies virus. Recently, WHO has been experimenting with immunizing wild animals in inaccessible mountainous terrain by dropping from aircraft pieces of meat doctored with live vaccine. It has had notable success with foxes in Belgium, using vaccinia virus genetically engineered to express the rabies virus envelope protein (see p. 248 for an account of this type of vaccine).

The elimination of HBV is being considered by the WHO as part of its cancer prevention programme (see p. 244).

CLINICAL COMPLICATIONS WITH VACCINES

Listed below are some of the circumstances when the normal immunization procedure is ineffective, inadvisable or even dangerous. In all these situations only a killed vaccine should be used.
1 Very young children sometimes develop a more severe infection than occurs in an older child.
2 Maternal antibody is transmitted to offspring via the placenta and can prevent stimulation of the immune response, so immunization is carried out after this has waned, at about 2−6 months of age.
3 Certain clinical conditions can make an infection more severe than normal, e.g. people with eczema are prone to a generalized infection with vaccinia virus instead of the local infection at the site of introduction of the vaccine.
4 Since fetal development can be deranged by virus infection, e.g. by rubella virus, immunization should be avoided during pregnancy.
5 A prior virus infection can sometimes interfere with the multiplication of a live vaccine, with the result that immunity is not established.

Under certain circumstances, it has been found that the acquisition of immunity as a result of administration of a killed vaccine potentiated the disease when the wild virus was contracted. This ocurred with respiratory syncytial virus (Paramyxoviridae, class V), which causes a lower respiratory tract infection in very young children. This virus has two surface proteins, F and G, but antibodies to F are better at providing protection. It is believed that the inactivation process destroyed the immunogenicity of F and that the resulting antibodies to G formed immune complexes in the respiratory tract. These in turn stimulated a hypersensitivity response, and an influx of fluid and immune cells to that site thereby impeding breathing. The danger of this situation is that it is not revealed until the vaccine is tested in humans. In the example cited, the disease was potentiated only in infants, the group that was in greatest need of protection. This demonstrates the insoluble difficulty that possible untoward effects of a vaccine will not be revealed until it is administered to people, and the need to test a vaccine in all age-groups.

It is too late to immunize when signs and symptoms of an

infection are already apparent. Thus it might appear that the very effective post-exposure immunization against rabies is an exception, but this is done at the time of the suspected *inoculation* of virus by the bite of a rabid animal. It works because of the slow transport of virus along peripheral nerves to the spinal cord and brain, and the disease does not start until it gets there. (Incidentally, the modern rabies vaccine is produced in tissue culture and has none of the problems of the infamous infected rabbit brain suspension devised by Louis Pasteur, which required multiple injections into the abdomen.)

One treatment of an already established disease is to administer immunoglobulin with activity against the infecting virus. This is really chemotherapy with antibody, since no immune response is stimulated and this *passive immunity* lasts only as long as the immunoglobulin survives in the body. A further complication is that one must be prepared to treat an anaphylactic response if successive doses of the same 'foreign' immunoglobulin are given. One alternative may be to construct hybrid molecules from the variable region of mouse monoclonal antibody and the constant regions of human antibody, so that the latter replaces most of the immunogenic parts of the foreign mouse molecule (p. 329) or, better still, to clone and express genuine human antibody (p. 331). Other emergency measures not involving the immune system also involve chemotherapy, including treatment with interferon (pp. 249–253).

NEW APPROACHES TO VACCINES

From the foregoing pages, the reader will appreciate that the manufacture of vaccines is an immensely difficult process. The central problem is that of safety — making sure that a killed vaccine contains no residual virulent particles or that a live attenuated vaccine does not revert to virulence. Neither can be guaranteed absolutely. Safety and the attendant problems of expense and efficacy have encouraged research into alternative ways of producing vaccines.

Understanding antigenic determinants

Most killed virus vaccines (Table 16.1) consist of whole particles, but it is known that only a minor part of the surface structure of the virus stimulates protective immunity. The rest is either non-immunogenic or stimulates an immune response which is non-protective. Thus research has pursued a reductive approach to determine if a vaccine can be made that consists only of those vital immunogenic regions of the virus particle.

X-ray crystallographic analysis of the influenza virus hae-magglutinin (HA) protein and of natural or laboratory-derived anti-

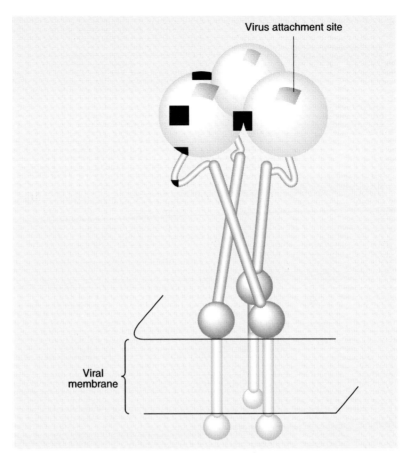

Fig. 16.4 Schematic diagram of an influenza virus haemagglutinin spike deduced from X-ray crystallographic analysis. The spike is composed of three identical monomers. Only when antibody binds to the shaded areas does it neutralize. Antibody binding elsewhere does not neutralize. The attachment site that binds to sialic acid on the host cell is marked. (Adapted from Wilson *et al.*, 1981, *Nature* **289**, 368.)

genic mutants has shown that only five regions of the molecule are involved in the binding of neutralizing antibody (Fig. 16.4). Antibody which binds to other regions of the HA does not neutralize, so, in theory, immunization with one or more oligopeptides corresponding to the antigenic sites would be sufficient to induce protecting antibody. However, life is not so simple and the theory would be borne out only if we were dealing with a continuous antigenic determinant (Fig. 16.5A). On the other hand, if the structure of the determinant were dependent upon its conformation, the oligopeptide might only achieve this if more distal parts of the protein were not required to maintain the shape (Fig. 16.5B) and the correct conformation would not be achieved if residues from two or more polypeptides were needed in conjunction (Fig. 16.5C). In practice, oligopeptides do not stimulate neutralizing antibodies to influenza virus.

From the discussion so far, the reader might conclude that a vaccine could consist solely of oligopeptides in the correct conformation. However, there is more to the story and we should now take a step back to consider how the immune system is stimulated to synthesize antibodies to the antigenic determinants of the HA.

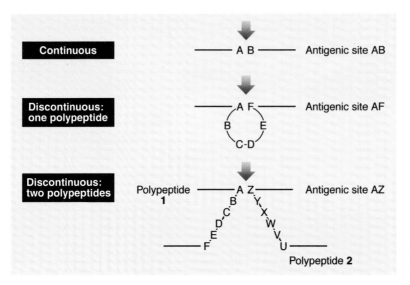

Continuous ———— A B ———— Antigenic site AB

Discontinuous: one polypeptide ————A F———— Antigenic site AF
B   E
C-D

Discontinuous: two polypeptides Polypeptide ————A Z———— Antigenic site AZ
1   B Y
C   X
D   W
E   V
————F   U————
Polypeptide 2

Fig. 16.5 Antigenic determinants represented by a two-letter code and viewed by the immune system in the direction arrowed.

Detailed analysis of protein immunogens, including influenza virus HA has shown that helper T cells are needed to assist B lymphocytes to become plasma cells, and that the helper T cells recognize antigenic determinants that are different from those that bind neutralizing antibodies (p. 207). Thus a peptide vaccine has to have two antigenic determinants, one to stimulate helper T cells and the other to stimulate B lymphocytes.

Prediction of antigenic determinants

Although antigenic determinants have been located for influenza virus HA, as described above, such work is laborious and expensive. One short-cut is based on the understanding that antigenic determinants are likely to be parts of the molecule which are rich in hydrophilic amino acids. These will extend into the aqueous environment and be easily 'seen' by cells of the immune system. Stretches of predominantly hydrophobic amino acids are folded into the internal regions of the molecule, where they are hidden from sight. All one needs to have to determine likely antigenic sites is the amino acid sequence and a computer program to search for hydrophilic regions, such as that devised by Kyte and Doolittle (Fig. 16.6). Hydrophilic oligopeptides are then synthesized chemically, usually but not always covalently linked to a protein carrier molecule and used to raise antibody. Testing the antiserum with the native protein determines if the computer has guessed right. The procedure has been used successfully with hepatitis B surface antigen (HBSag). Recent refinements have suggested that antigenic determinants are likely to be those hydrophilic regions with the highest mobility. Apparently, lymphocytes are like fish and are attracted by the most lively bait.

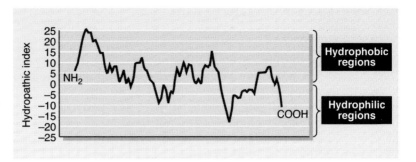

Fig. 16.6 Identification of hydrophilic regions of a protein by calculating the hydropathic value of each amino acid residue across seven residues, i.e. the amino acid in question and the three residues on either side.

Hepatitis B virus vaccine

There are estimated to be more than 300 million carriers of hepatitis B virus (HBV) world-wide. The virus infects people of all ethnic groups, but those of Chinese origin are particularly susceptible. The virus causes acute liver disease, which is extremely debilitating and can progress to a chronic lifelong carrier state which may prove fatal. In addition, the chronic infection may lead to one of the commonest types of cancer, primary hepatic carcinoma (Chapter 17). The virus is transmitted through contact with infected blood and saliva and, when the mother is a carrier, to children early in life. These children have the highest risk of becoming carriers themselves (Fig. 16.7).

The problem with a vaccine is that HBV cannot be grown in culture and infects only humans and higher primates, such as chimpanzees. This has been solved by cloning HBSag and the expression in yeast of a protein which has proved to be a very effective recombinant vaccine. Very importantly, children born to infected mothers can be immunized at birth and protected. Today, this is the standard commercially available vaccine. Three injections are required at a cost of around £56.

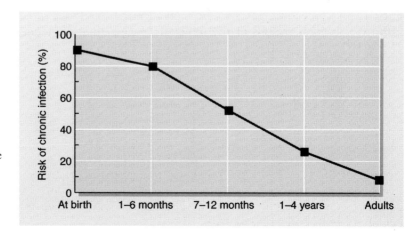

Fig. 16.7 Risk of becoming a carrier, one who is chronically infected with HBV, strongly depends upon the age at which a person is infected.

GENETICALLY ENGINEERED VACCINES

With killed virus vaccines, it is a major problem to produce sufficient material cheaply and ensure that no infectious virus survives the inactivation procedure. There are three solutions, now that deoxyribonucleic acid (DNA) technology can be used to identify the part of the viral genome that encodes the particular virus protein against which protective immunity (usually antibody) is directed. These are: (i) expression of the entire protein; (ii) expression of a fragment of the protein containing the antigenic site; or (iii), chemical synthesis of a peptide which contains the antigenic site. For (i) and (ii), viral DNA, or a DNA copy if the virus had a ribonucleic acid (RNA) genome, can be excised and inserted into an appropriate expression vector, together with control (promoter, stop and polyadenylation) signals. Thus, we have a small part of the viral genome, by definition non-infectious, which by insertion into a host cell growing on an industrial scale will produce very large amounts of protein very cheaply. The ultimate is solution (iii), which has no requirement whatsoever for living cells. Lastly, with any of these approaches, you do not even have to be able to grow the virus in culture, a great advantage for viruses like HBV.

Does it work? Table 16.4 summarizes the state of what is still very much an experimental development. Genetic engineers would very much like to use bacterial expression systems which are well understood from the viewpoints of both their genetics and large-scale industrial production. However, human and veterinary diseases are caused by viruses, whose newly synthesized polypeptides undergo eukaryotic-type cotranslational and post-translational modifications, such as glycosylation and proteolytic cleavage, which prokaryotic cells cannot accomplish; therefore, we have to use eukaryotic cells. However, in addition to cultured cells from higher animals (for which technology has been intensively developed of late for monoclonal antibody production), there is several thousand

Table 16.4 Genetically engineered vaccines.

Advantages	Problems
Non-infectious	Identification of neutralization antigen
Large-scale production methods available	Need for proper co- and post-translational modifications of viral polypeptide
Cheap to produce	Need to achieve proper conformation of viral protein to avoid poor immunogenicity
Can use genes from non-cultivable viruses	Separation of viral protein from cell constituents

years' experience in the bulk culture of yeasts, also eukaryotes, for brewing and baking.

The method works well if a linear determinant gives protection immunity, but antibody is usually directed to conformational sites (Fig. 16.5). At the moment, we do now know how to instruct a peptide to fold in a particular way. This success is only achieved by chance. Nevertheless, the first recombinant vaccine, comprising the HBSag, has proved itself in clinical trials and, in 1986, came on to the commercial market. This was a major breathrough in view of the huge number of HBV-infected people in the world (p. 244) and inability to grow the virus in cell culture.

A chemically synthesized experimental vaccine against foot-and-mouth disease vaccine similarly proved very successful, both in raising neutralizing antibody and in protecting cattle, the main species at risk. Good luck and good science came together here, as it was only realized after the event that the peptide used, residues 146–161 of the coat protein VP1: (i) was a linear determinant; and (ii) contained a helper T cell epitope (p. 243).

There are many problems still to be overcome before the 'new' vaccines replace conventional ones. Some have already been attended to (HBV: see above). Others require a more profound understanding of the immune response which is effective in a particular disease situation and how to stimulate it, particularly in overcoming the poor immunogenicity which is intrinsic to subviral entities.

Genetically engineering a virus as a vaccine

Live vaccines evoke the most effective immunity and are the cheapest to produce but in practice are very difficult to make. Hence, the idea of inserting the gene for a foreign neutralization antigen into a pre-existing live vaccine, so that it is expressed naturally as the virus multiplies, is particularly attractive. This has already been achieved experimentally for antigens of numerous viruses, including influenza, rabies, herpes simplex type 1 and hepatitis B viruses, using vaccinia virus as the live vaccine. Vaccinia virus, once used universally as a vaccine, is no longer employed since smallpox was eradicated in 1977. The virus is administered intradermally by scarifying the skin, and the virus multiplies at that site, causing a reaction about the size of a small boil and stimulating both antibody and cell-mediated immunity to viral antigens.

The mechanics of producing a recombinant vaccinia virus vaccine, outlined in Fig. 16.8, starts by inserting a cloned gene into a plasmid under the control of the promoter of the vaccinia, thymidine kinase (TK) gene. Upon transfection into vaccinia virus-infected cells, recombination takes place, because of the homologous TK sequences which both possess. The TK gene is inactivated by insertion of the plasmid, which means that recombinants can be

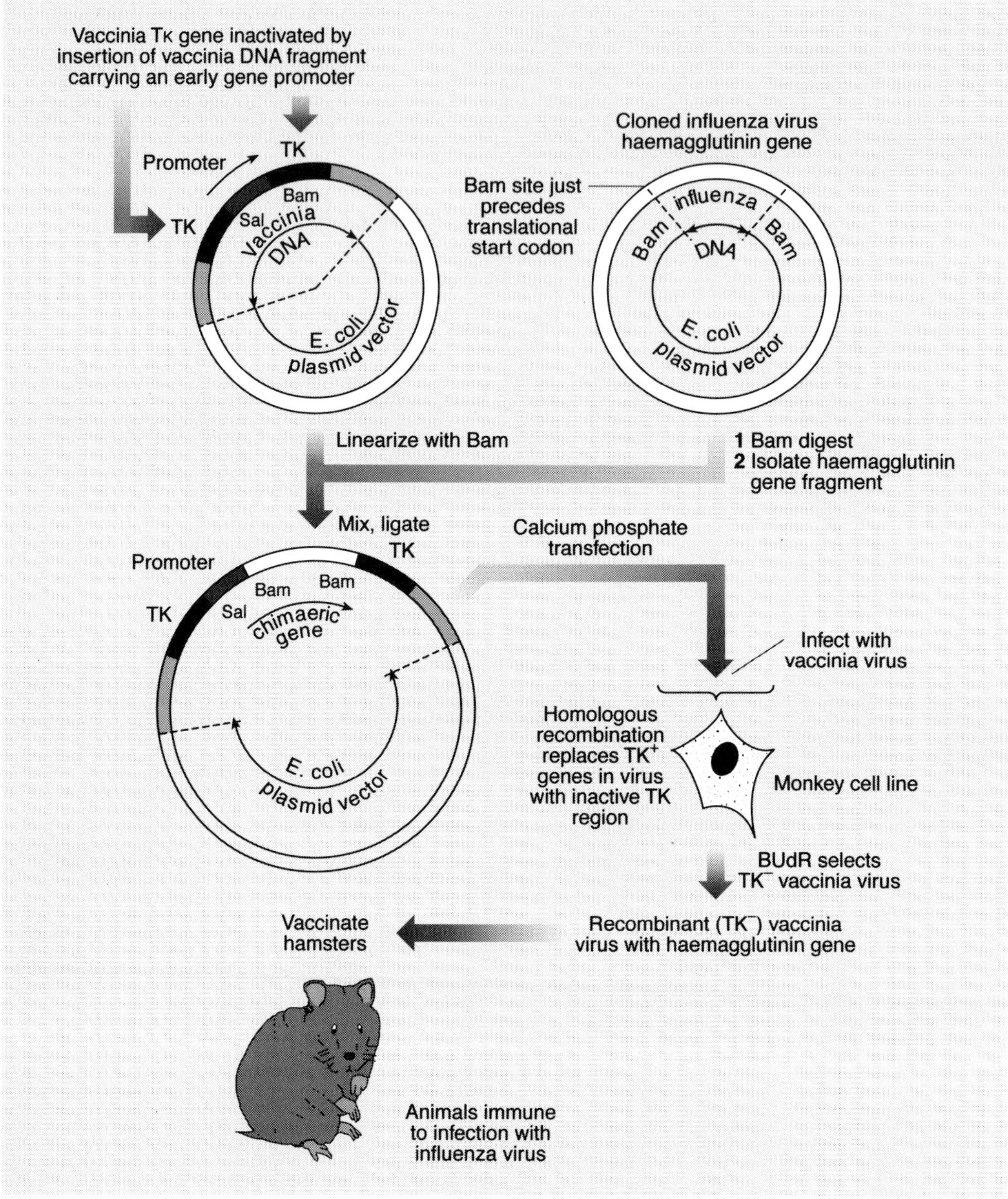

Fig. 16.8 Construction of an infectious vaccinia virus recombinant expressing influenza virus haemagglutinin. BUdR is the DNA synthesis inhibitor bromodeoxoyuridine.

selected, because they are unable to utilize (and hence be inhibited by) the DNA synthesis inhibitor bromodeoxyuridine. Consequently, only recombinant virus produces plaques.

Animals infected with recombinant vaccinia carrying genes for HBSag or influenza virus HA respond with excellent antibody and cell-mediated immune responses, so vaccines of this type are an exciting prospect. The original vaccinia virus vaccine, obtained from fluid which exuded from localized intradermal infections of sheep, would now be produced in tissue culture and would cost about the same as the poliovirus vaccine — about 7 US cents per dose.

One problem is the rare life-threatening situation of generalized vaccinia, in which vaccinia virus infection spreads to the entire body, which occurs at an incidence of about 1 in 25 000 doses. While this was an acceptable risk against the far higher likelihood of death from smallpox, it could not be tolerated, for example, in a vaccine against influenza. However, insertion of a foreign gene, as described above, already reduces the virulence of vaccinia by an order of magnitude, and the co-expression of cytokine genes in vaccinia virus (interleukin-2 and interferon-γ) completely abolish its lethality for althymic (nude) mice. Another problem is that immunity directed against the vaccinia virus antigens themselves may preclude the vaccinia vector from being used on a subsequent occasion. However, the antivaccinial immunity is relatively short-lived and, in the days of smallpox vaccination, was repeated every 3 years. One way to circumvent the difficulty lies in the fact that the vaccinia genome can accommodate around 25 kilobases of DNA without losing infectivity. This means that the recombinant virus could express 10–20 foreign antigens and serve as a one-off polyvalent vaccine which would immunize against a broad spectrum of virus diseases.

CHEMOTHERAPY OF VIRUS INFECTIONS

Ever since the successful introduction of antibiotics to control bacterial infections, there has been the hope that similar treatments for virus infections were just around the corner, but, apart from certain specific circumstances, this hope has not yet been realized (Table 16.5). The reason is that the virus multiplication is tied so intimately to cellular processes that most 'inhibitors' cannot discriminate between them. However, viruses do have unique features, so, in theory, inhibitors specific for these should be able to serve as effective chemotherapeutic agents. An inhibitor would be effective if it blocked or interfered with any of the stages of multi-plication, i.e. with attachment, replication, transcription, trans-lation, assembly or release or progeny particles.

Chemicals are used to treat DNA virus infections of the con-junctiva and cornea of the eye. The agents used are nucleoside analogues, which are incorporated into viral DNA and prevent replication and transcription. These compounds, with the exception of acyclovir (see below), are incorporated into cellular DNA and are toxic for any dividing cell. Hence their use is restricted to the poorly vascularized surface of the eye. In practice, 0.1%

Table 16.5 Chemotherapy of virus infections.

Compound	Mode of action	Susceptible viruses
Nucleoside analogues		
Acyclovir*	No 3′ OH group on deoxyribose, so DNA synthesis is terminated	Herpes simplex viruses 1 and 2; also VZV but is less sensitive
Ganciclovir*	As above	Human cytomegalovirus
AZT*	As above	HIV-1; delays progress of AIDS (see Ch. 19)
Amantadine*	Blocks the M2 proton channel	Type A influenza viruses (see p. 94)
α- and β-interferons*	Up-regulates MHC class I	Selected infections only: chronic hepatitis B and C
	Antiviral state?	Warts caused by human papillomaviruses
Compounds which block attachment/entry		
Soluble CD4	Blocks attachment to CD4$^+$ T cells	HIV-1 (see Ch. 19)
WIN52084	Prevents uncoating	Rhinoviruses
Antibody	Neutralization	Any, but best used prophylactically or during acute infection

* Compounds in routine clinical use.
AZT, 3′azido-2′,3′-dideoxythymidine

5′-iodo-2′-deoxyuridine, an analogue of thymidine, improved healing in 72% infections with herpes simplex type 1, vaccinia and adenoviruses. In extreme circumstances, such as life-threatening cases of encephalitis caused by herpes virus type 1 or vaccinia viruses, analogues of cytidine (cytosine arabinoside) or adenosine (adenine arabinoside) used to be used systemically. These have been superseded by acyclovir (see below). Another antiviral compound is amantadine (p. 94), which has had limited prophylactic and therapeutic use in the treatment of type A influenza virus infections. In one influenza epidemic, the number of cases in the population under study fell from 14.1 to 3.6% with amantadine prophylaxis. As with all types of chemotherapy, a worrying aspect is the frequency with which amantadine-resistant mutants arise.

Acyclovir: a selective antiviral compound

Properly known as 9-(2-hydroxyethoxymethyl)guanine (Fig. 16.9), acyclovir (or Zovirax) has proved to be by far the best antiviral agent so far discovered and, like other pyrimidine analogues, it is particularly effective against acute infection with herpes simplex viruses. It overcomes the problem of other nucleoside analogues, that of being incorporated into cellular DNA, by requiring the

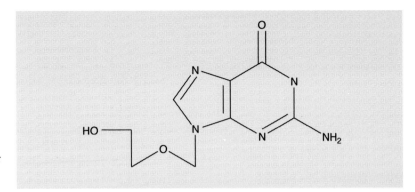

Fig. 16.9 The formula of the chain-terminating nucleoside, acyclovir. Note that most of the cyclic sugar ring is missing—hence the name.

viral thymidine kinase to attach the phosphate groups necessary for acyclovir to be used in DNA synthesis (Fig. 16.10). Thus, with acyclovir, the ultimate aim of creating a compound which is toxic only in infected cells has been achieved. Acyclovir has no activity against latent infections but, when used promptly, has proved effective against corneal eye infections, cold sores and genital infections caused by HSV-1 and HSV-2. It is also invaluable in treating life-threatening generalized infections which can occur in the immunocompromised, particularly those treated with immuno-suppressants in the course of transplant surgery or cancer chemo-therapy. Acyclovir-resistant mutants do arise but do not seem to be a significant problem.

Interferon therapy

Their universal antiviral activity and high specific activity make

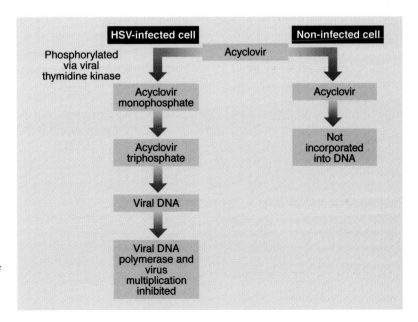

Fig. 16.10 The selective toxicity of acyclovir for herpes simplex virus-infected cells.

α-, β-, γ-interferons ideal therapeutic agents theoretically, but clinical trials have not been encouraging. To achieve success in early experiments at the Common Cold Research Unit in Salisbury, UK, it was necessary to give volunteers 14×10^6 units of α-, β-interferon as a nasal spray four times per day, including the day before they were infected with a rhinovirus (Table 16.6). The problem was then thought to be the difficulty of obtaining interferons in sufficient quantity. Today, cloned interferons can be expressed in high yield and purified with ease but, even with high doses, therapy is only effective if started before the virus is inoculated and is thus not applicable to natural rhinovirus infections.

Experimental treatment of a variety of human virus infections with interferons has been generally disappointing and it is still not clear why this should be. However, interferon has been of particular use in the treatment of persistent infections caused by human papillomaviruses hepatitis B virus (Hepadnaviridae) and hepatitis C virus (Flaviviridae). The (Papovaridae), human papillomaviruses cause benign tumours (warts) on the body, larynx and genitalia, the latter being of particular concern, as they may develop into cancer of the cervix (Chapter 17). Over half the warts injected with interferon regress, but they tend to recur when interferon is stopped. Treatment is best in conjunction with surgery.

After acute HBV infection, a proportion of patients develop a chronic infection (Fig. 16.7), which can lead to immune complex disease, destruction of the liver (cirrhosis) and liver cancer. Virus-specific $CD8^+$ cytotoxic T lymphocytes (CTLs) are present, but these cannot clear the infection, because the virus depresses the expression of major histocompatibility complex (MHC) class I proteins on infected liver cells. Treatment with α-, β-interferon is beneficial, but it acts by enhancing the expression of MHC class I proteins (p. 210) and not through its antiviral activity. HBV-specific peptides are then presented in sufficient amount by MHC class I proteins to allow the $CD8^+$ CTLs to lyse the HBV-infected liver cells. The treatment is about 50% successful and, under optimum circumstances, can clear the infection completely (Fig. 16.11).

Interferons themselves have side-effects, causing fever, local inflammation, muscular pain, fatigue and malaise, effects resulting from their action as cytokines and regulators of the immune system.

There has been hope that interferons might be useful as anti-cancer agents as: (i) they are active in regulating cell division; and (ii) some tumours are, in conjunction with cofactors, caused by

Table 16.6 Treatment of common colds caused by a rhinovirus infection with α-, β-interferon.

	Colds	Virus isolations
Interferon-treated	0/16	3/16
Mock-treated	5/16	13/16

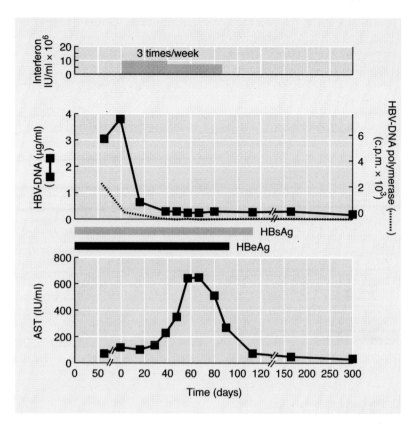

Fig. 16.11 Successful cure of a patient with a chronic HBV infection by treatment with α-interferon. Note the fall in viral DNA, DNA polymerase, and the viral antigens HBe and HBS. These remain undetectable, showing that the infection has been cleared. The transient rise in aspartate aminotransferase (AST), which is a marker of liver damage, indicates the destruction of infected liver cells due to up-regulation of MHC class I antigens on infected liver cells (see text). From Thomas 1990.

viruses (p. 259; Chapter 17). However, these hopes have not been realized, although, as expected, the most promising results were obtained with virus-associated cancers (such as the hairy T cell leukaemia caused by human T cell lymphotropic virus II). Hopefully, some permutation of the 12 different α-interferons with β-interferon or γ-interferon or chemical modification may improve their efficacy, but the major requirement is to gain a much greater understanding of their fundamental properties, including their immunoregulatory functions.

FURTHER READING

Anderson, R. M. (1991) Rabies: immunization in the field. *Nature (London)*, **354**, 502–503.

ASM News (1990) Genetically engineered rotavirus vaccines. *American Society for Microbiology News*, **56**, 258.

ASM News (1991) Overall US health shows some improvement. *American Society for Microbiology News*, **57**, 294–296.

Bach, J. F., Fracchia, G. N. & Chatenoud, L. (1993) Safety and efficacy of therapeutic monoclonal antibodies in clinical therapy. *Immunology Today*, **14**, 421–425.

Behbehani, A. M. (1991) The smallpox story: historical perspective. *American Society for Microbiology News*, **57**, 571–576.

Benjamin, D. C. *et al.* (1984) The antigenic structure of proteins: a reappraisal. *Annual Review of Immunology*, **2**, 67–101.

Bomford, R. (1992) Adjuvants for viral vaccines. *Reviews in Medical Virology*, **2**, 169–174.

Bright, S., Adair, J. & Secher, D. (1991) From laboratory to clinic: the development of an immunological reagent. *Immunology Today*, **2**, 130–134.

Brown, F. (1984) Synthetic viral vaccines. *Annual Review of Microbiology*, **38**, 221–235.

Brown, F. (ed.) (1990) Modern approaches to vaccines. *Seminars in Virology*, **1**, 1.

Brown, F., Schild G. C. & Ada, G. L. (1986) Recombinant vaccinia viruses as vaccines. *Nature (London)*, **319**, 549–550.

Darby, G. (1990) Virus replication and strategies for specific inhibition. In: 'Control of Virus Diseases'. *Society for General Microbiology Symposium*, Vol. 45, pp. 188–212. Dimmock, N. J., Griffiths, P. D. & Madeley, C. R. (eds). Cambridge: Cambridge University Press.

Evans, A. S. (1984) Criteria for control of infectious diseases with poliomyelitis as an example. *Progress in Medical Virology*, **29** 141–165.

Fenner, F., Henderson, D. A., Arita, I., Jezek, Z. & Ladnyi, I. D. (1988) *Smallpox and its eradication*. Geneva: World Health Organization.

Fox, J. L. (1990) Rabies vaccine field test undertaken. *American Society for Microbiology News*, **56**, 579–583.

Hilleman, M. R. (1985) Newer directions in vaccine development and utilization. *Journal of Infectious Diseases*, **151**, 407–419.

Knight, P. (1991) Measles epidemic, vaccine shortfall stirs up controversy. *American Society for Microbiology News*, **57**, 560–561.

Knight, P. (1991) Baculovirus vectors for making proteins in insect cells. *American Society for Microbiology News*, **57**, 567–570.

Kyte, J. & Doolittle, R. F. (1982) A simple method for displaying the hydropathic character of a protein. *Journal of Molecular Biology*, **157**, 105–132.

Lentz, T. L. (1990) The recognition event between virus and host cell receptor: a target for antiviral agents. *Journal of General Virology*, **71**, 751–766.

McKinlay, M. A., Pevear, D. C. & Rossmann, M. G. (1992) Treatment of the picornavirus common cold by inhibitors of viral uncoating and attachment. *Annual Review of Microbiology*, **42**, 635–654.

Marsden, H. S. (ed.) (1992) Antiviral therapies. *Seminars in Virology*, **3**, no. 1.

Mitchell, C. D. & Balfour, H. H., Jr (1985) Measles control: so near and yet so far. *Progress in Medical Virology*, **31**, 1–42.

Mitsuya, H. & Broder, S. (1987) Strategies for antiviral therapy in AIDS. *Nature (London)*, **325**, 773–778.

Moss, B. (1991) Vaccinia virus: a tool for research and vaccine development. *Science*, **252**, 1662–1667.

Nicholson, K. G. (1984) Properties of antiviral agents. *Lancet*, **ii**, 562–564.

Nicholson, K. G. (1984) Antiviral therapy. *Lancet*, **ii**, 617–621.

Oxford, J. S. (1990) Chemotherapy of influenza and respiratory virus infections. In: *'Control of Virus Diseases'. Society for General Microbiology Symposium*, Vol. 45, pp. 213–242. Dimmock, N. J., Griffiths, P. D. & Madeley, C. R. (eds). Cambridge: Cambridge University Press.

Scott, G. M. (1983) The antiviral effects of interferon. In: *'Interferons'. Society for General Microbiology Symposium*, **35**, 277–371.

Stevenson, M., Bukrinsky, M. & Haggerty, S. (1992) HIV-1 replication and potential targets for intervention. *AIDS Research and Human Retroviruses*, **8**, 107–117.

Thomas, H. C. (1990) Management of chronic hepatitis virus infection. In: *'Control of Virus Diseases'. Society for General Microbiology Symposium*, **45**, 243–259. Dimmock, N. J., Griffiths, P. D. & Madeley, C. R. (eds). Cambridge: Cambridge University Press.

Tyrell, D. A. J. (1984) The eradication of virus infections. In: *'The Microbe 1984'. Society for General Microbiology Symposium*, **36**, Pt 1, 269–279.

Tyrrell, D. A. J. & Oxford, J. S. (1985) Antiviral chemotherapy and interferon. *British Medical Bulletin*, **41**, 307–405.

Westhof, E., Altschuh, D., Moras, D., Bloomer, A. C., Mondragon, A., Klug, A. & Van Regenmortel, M. H. V. (1984) Correlation between segmental mobility and the location of antigenic determinants in proteins. *Nature (London)*, **311**, 123–126.

White, D. O. (1984) *Antiviral chemotherapy, interferons and vaccines—a status report. Monographs in Virology*, **16**. Basle: Karger.

Zuckerman, A. J. (1985) New hepatitis B vaccines. *British Medical Journal*, **290**, 492–496.

Also check Chapter 21 for references specific to each family of viruses.

17 Carcinogenesis and tumour viruses

The great majority of viruses of vertebrates are not oncogenic:* that is to say, they do not have the ability to initiate a cancer. However, the normal manifestation of infection by some deoxyribonucleic acid (DNA) viruses is a benign tumour; others can produce malignant tumours when injected into new-born animals, while a few are oncogenic in humans and other animals. Only a small number of ribonucleic acid (RNA) viruses have been unequivocally associated with neoplastic disease, but these too are oncogenic under natural conditions.

As well as causing tumours, both groups of viruses have representatives which are able to transform mammalian cells in culture. Transformed cells are recognized by their changed phenotype (see Table 17.5 below). They may also cause tumours in animals, but tumorigenesis is usually a multistage process, of which transformation is but the first step. In addition, tumour formation involves complex interactions with the immunological system of the host animal; hence transformation in culture is not an infallible indicator of malignancy *in vivo*. Transformation permits the study of early events required for the development of malignant potential by infected cells and the study of the acquisition of new genetic material by mammalian cells.

There is no single mechanism by which viruses induce tumours, and we shall discuss carcinogenesis by members of various virus families along the lines of the summary in Table 17.1. Carcinogenesis is multifactorial and your own experience will tell you that it is a rare event; various factors in addition to virus infection, such as host genotype, diet, environmental carcinogens (other than viruses) and other invading organisms, all contribute to the process. However, it is estimated that viruses are a contributary cause of 20% of all human cancers. To call a virus a 'tumour' virus is misleading, as the name refers to one very infrequent aspect of its normal life cycle; there is no satisfactory explanation of the evolutionary advantage to the virus of causing tumours.

In the study of oncogenesis, two major theories have heen predominant: that there is deletion of genetic material from the cell which is essential for normal functioning or that there is a very specific activation of genetic material, which could be a positive or negative control element, either *de novo* or to an abnormal level. Both mechanisms are relevant in naturally occurring cancers, and both are exploited in virus-mediated transformation and tumorigenesis. As can be seen (Table 17.2), there are two types of *oncogene* that are carried by viruses. The first, which is found in

---

* A glossary of terms is provided at the end of the chapter.

Table 17.1 Mechanisms following infection by which viruses may induce tumours.

Persistence of all/part of the viral genome which
(a) expresses a viral gene (an oncogene) which initiates or maintains transformation
(b) alters the control or expression of cellular genes (insertional mutagenesis)
(c) causes immunosuppression

Table 17.2 Viral and cellular genes involved in oncogenesis.

	Example
Viral genes	
Necessary for replication	Most DNA virus transforming genes, e.g. SV40, polyomavirus large T
Not necessary for replication	All retrovirus oncogenes, e.g. v-*src*, v-*myc*, v-*ras**
Cellular genes	
Homologous to retrovirus oncogenes	c-*myc*, c-*ras*
At specific sites of virus integration	c-*myc*, c-*erb*-B
At specific sites of chromosomal translocation	c-*myc*, c-*abl*
Oncogenic on transfection	c-*ras*, c-B-*lym*
Activated or repressed	Class 1 antigens of the major histocompatibility complex
Whose products interact with viral oncogenes	p53 with SV40 large T, pp60$^{c\text{-}src}$ with polyomavirus middle T
Inactivated in tumours	Tumour suppressor genes, e.g. p53, Rb

* Many more have been identified.
The v- or c- prefix denotes a viral or cellular oncogene.

DNA tumour viruses, comprises genes which have functions essential for virus replication. The second comprises cellular genes which have been transduced by retroviruses, often with the acquisition of subtle mutations which are crucial to their oncogenic activity. (Other oncogenes of this type and their products are listed in Table 17.6) Oncogenes of the first category function by altering the activity of either cellular protooncogenes (see below), cellular tumour suppressor genes or their protein products. Those of the second have a gene dosage effect, as they are not regulated by properly controlled promoters, or have altered biochemical functions due to mutation.

The term 'oncogene' is confusing because so-called cellular oncogenes are in fact normal cellular genes which function in cell division. These normal cellular genes are sometimes called protooncogenes, and were first identified in retroviruses.

Some transformed cells can revert to a normal phenotype. This implies that transformed cells are essentially normal cells which no longer have their normal control over proliferation or differentiation and, secondly, that they require constant positive action for the transformed state to be maintained. Other transformed cells can have their phenotype suppressed by fusion with normal cells, implying that tumour cells lack a normal cell function required to prevent tumorigenic growth. These experiments provide direct evidence for the role of both oncogenes and tumour suppressor genes in transformation.

CARCINOGENESIS

Transformation is one of the earlier stages *en route* to cancer. The full-blown cancer cell is not created at a stroke but is the result of a series of changes, of which one or more may be due to the effect of a tumour virus (Fig. 17.1). Interaction with one or more cofactors (Table 17.3) is all-important and it is this permutation which makes virus-induced cancer a rare event. There is no certainty that the progression will be fulfilled; rather, it is more likely that it will be aborted. The immune system is important, as it provides a check on virus multiplication and the spread of cells bearing viral or non-self tumour antigens.

ONCOGENIC VIRUSES

Various members of the Papovaviridae, Adenoviridae, Herpes-viridae, Retroviridae and Hepadnaviridae listed in Table 17.3 are known to be oncogenic in animals or are suspected of having

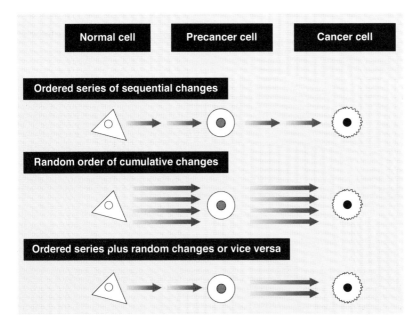

Fig. 17.1 Multistep progression from a normal to a cancer cell. The number of arrows do not indicate the actual number of events involved.

Table 17.3 Viruses suspected of having oncogenic properties in humans, and some non-human tumour viruses.

Virus family	Type	Tumour	Animal	Cofactor
Adeno	2, 5, 12	Sarcoma	Hamster	Age (new-born only), immune competence, species*
Hepadna	Hepatitis B	Primary hepatocellular carcinoma	Humans	Age, aflatoxin, alcohol, smoking
	Woodchuck hepatitis	Hepatocellular carcinoma	Woodchuck	Species
Herpes	Epstein–Barr	Burkitt's lymphoma, nasopharyngeal carcinoma, immunoblastic lymphoma	Humans	Malaria, nitrosamines, HLA genotype, immunodeficiency
	Cytomegalo	Kaposi's sarcoma (?)	Humans	Immunodeficiency (e.g. following HIV-1 infection, HLA genotype
	Herpesvirus saimiri	Experimentally induced lymphomas and leukaemias	Owl monkey	Species; not the natural host
	Herpesvirus ateles	Experimentally induced lymphomas and leukaemias	Marmosets	Species; not the natural host
	Marek's disease virus	T cell lymphoma	Chicken	Species
	Lucké frog virus	Adenocarcinoma of kidneys	Leopard frog	Species
	Herpesvirus sylvilagus	Lymphomas	Cottontail rabbit	Species
Papova	Papilloma (HPV)	Warts, cervical neoplasia	Humans	Smoking, herpes simplex viruses (?)
	Polyoma (BK, JC)§	Neural tumours	Humans	?
	Polyoma	Sarcoma	Mouse	Age, species
	Bovine papilloma	Warts, fibroepithelioma	Cattle, hamster, rabbit	Age, (new-born), consumption of bracken fern
Retro†	Human T cell lymphotropic (HTLV-1)	Adult T cell leukaemia, lymphoma	Humans	?
	(HTLV-2)	Hairy T cell leukaemia	Humans	?
	Rous sarcoma	Sarcoma	Chicken	Genetic
	Avian myeloblastosis	Myeloblastosis	Chicken	Genetic
	Abelson murine leukaemia virus	Leukaemia	Mouse	Genetic
	Bovine leukaemia	Leukaemia, lymphoma, lymphosarcoma	Cow	?
	Feline leukaemia	T cell lymphosarcoma, leukaemia, fibrosarcoma	Cat	?

* Tumours are restricted to the species mentioned.
† And 100+ more (see Weiss *et al.*, 1984/5).
§ BK and JC are the virus strains involved.
HLA, human leucocyte antigen (i.e. human form of MHC).

oncogenic properties in humans. These can all be classified as DNA viruses, provided you grant retroviruses honorary status on account of their DNA intermediate in replication (Chapter 9). There are no 'pure' oncogenic RNA viruses. Thus, representatives of most major families of DNA viruses are associated with cancer. Amongst animal systems, the best understood in molecular terms and the smallest in genome size are the papovaviruses, polyomavirus and simian virus type 40 (SV40).

In humans, good evidence links hepatitis B virus (HBV) with primary hepatocellular carcinoma (PHC), Epstein–Barr virus (EBV) with nasopharyngeal carcinoma and Burkitt's lymphoma and human papillomaviruses (HPV) with cervical cancer. Complete observation of Koch's postulates (p. 214) is out of the question and other, less direct criteria, such as statistical data, are used in addition. The importance of infection at an early age in predisposing to cancer (HBV, HPV) is worth underlining (Fig. 16.7; p. 263; p. 271).

Papovaviridae

One group of papovaviruses causes warts (papillomas) in humans and domestic animals. Polyomavirus infects mice subclinically but, when artificially inoculated into infant rodents, causes a wide variety of tumours: hence its name 'polyoma'. Another, SV40, came to light during the preparation of poliovirus vaccine in monolayers of monkey kidney cells. (SV40 produces vacuoles in infected cells and is sometimes called vacuolating agent. The name of this group, papovavirus, is thus derived from the initial letters of its chief members: PA for papilloma, PO for polyoma and VA for vacuolating agent).

Cells in culture can respond to infection by polyoma or SV40 in the following ways: the virus may undergo a lytic infection, with production of progeny virus, in a permissive cell; an abortive infection, in a non-permissive cell, in which early, but not late, events in multiplication take place; or an intermediate situation, where there is some virus production and some cell survival, in a semi-permissive cell (Table 17.4). Transformation can occur in abortively infected cells but also in permissive cells inoculated with defective virus. In either case, transformation is always a rare $(1 \text{ in } 10^5)$ event.

Table 17.4 How cells of different species respond to infection.

	SV40	Polyomavirus
Monkey	Permissive	Non-permissive
Mouse	Non-permissive	Permissive
Hamster	Semi-permissive	Semi-permissive
Rat	Semi-permissive	Semi-permissive

The properties of transformed cells

Cells that have been transformed are readily detected by their altered growth characteristics (Table 17.5) and by the ease with which they are agglutinated by certain plant proteins, such as wheat germ agglutinin and concanavalin A, which bind strongly to the sugar moiety of glycoproteins and are collectively known as 'lectins'. How viruses with a genome of no more than $5 \times 10^6 M_r$ cause this multiplicity of effects was for long a mystery, but the concept of oncogenes as genetic elements which upset the normal proliferation/differentiation of a cell provides the explanation.

The events leading to transformation by papovaviruses

Integration of viral DNA

Viral DNA is present in transformed cells, since extracted DNA hybridizes specifically with RNA made *in vitro* using purified viral DNA as template. The DNA is known to be integrated, as analysis with restriction endonucleases shows that viral DNA is flanked by covalently linked cellular DNA sequences. The same type of analysis also demonstrated that there was no preferred integration site but that viral DNA could be located at many different sites in the cellular genome. Analysis of transformed cells for 200 generations showed that viral DNA retained the same relationship with flanking host DNA sequences and was therefore stably integrated. As SV40 and polyomavirus genomes are circular, it seems likely that integration is achieved by recombination, in a manner analogous to the way *Escherichia coli* becomes lysogenized by phage. (p. 189). However, there seems to be no homology between host and viral DNA sequences, so integration probably takes place by non-homologous recombination. Cells transformed by SV40 often contain a complete copy of the viral genome, since they yield infectious virus by cocultivation or fusion with permissive cells and the extracted DNA is infectious. SV40-transformed cells may also have fragments of viral genomes integrated, and significantly there is a sixfold excess of the early region which encodes the T antigens. Overall, in both SV40- and polyomavirus-transformed cells it is estimated that there are usually between 0.6

Table 17.5 Some altered properties of transformed cells.

Multiply for ever (immortalization)
Grow to higher saturation density
Have reduced requirement for serum
Grow in suspension in soft agar
Form different cell colony patterns
Grow on top of normal cell monolayers
Readily agglutinated by lectins

(polyomavirus) and 10 copies of the viral genome per diploid cell or, in other terms, viral DNA represents about 0.001% of the cellular genome. It takes only a small amount of virus DNA to subvert the normal functioning of a cell.

Roles of the T antigens

Like the tumour antigens of all other DNA viruses, T antigens of SV40 and polyomavirus are essential for the normal replication of the virus (pp. 160–162). In other words, mutants created by site-specific or conventional mutagenesis of T antigen genes are unable to multiply and have a reduced yield of progeny virus. What are the functions of T antigens and why does polyomavirus have an additional T antigen, middle T? Immunofluorescence shows that large T antigen is found predominantly in the nucleus, and DNA binding studies show that it has a high affinity for both viral and cellular DNA – not surprising for a protein which is responsible for initiating DNA synthesis. Illuminating experiments about the role of large T in transformation have come from the use of conditional lethal mutants of SV40 which are *ts* in that gene. At the permissive temperature, *ts* mutants transform cells, but, when the cells are grown at the higher non-permissive temperature, two types of result are found. Cells transformed by one group of *ts* mutants lose some or all of their transformed properties, indicating that large T is need involved in the transformation process and the maintenance of the transformed state. Cells transformed by a second group of *ts* mutants retain their transformed phenotype, implying that only some domains of SV40 T antigen are required for trans-formation. Polyomavirus large T alone is not sufficient to transform cells completely; middle T antigen is also important and is located mainly on the inner face of the plasma membrane. Viruses which have been manipulated so that only middle T is expressed are transforming, but the resulting cells are still dependent on serum for their growth and therefore are not completely transformed. In the same way, small t antigen has some accessory transforming properties but alone is insufficient to bring about transformation. It seems, therefore, that large T, middle T and small t antigens act synergistically to initiate (and probably to maintain) transformation. There is also evidence from other systems that this may be a common feature of carcinogenesis, and activation of more than one oncogene may be a necessary part of the multistage process of its establishment (Fig. 17.1).

Transformed cells also have tumour-specific transplantation antigens in their plasma membrane. These are recognized by trans-plant rejection tests, in which cell-mediated immunity is predomi-nant. In SV40-transformed cells, the tumour-specific transplantation antigen is a fragment of large T.

So much for the background, but what is known of the mechanism by which T antigens contribute to transformation? Firstly, the site of integration of SV40 or polyomavirus is not crucial, as either virus can integrate at many different sites in the host cell DNA. What is important is an association of T antigens with cellular proteins, as this is an obvious way in which the normal functioning of a cell can be upset. One of these is the product of c-*src*, a cellular tyrosine protein kinase called pp60$^{c\text{-}src}$ (this gene has also been acquired as an oncogene by Rous sarcoma virus (RSV)). pp60$^{c\text{-}src}$ is thought to control regulation of cell division and, by combining with middle T of polyoma virus, to alter the regulation of the cell cycle. As middle T is necessary for virus multiplication, it is likely that pp60$^{c\text{-}src}$ is simply concerned with efficient/rapid synthesis of DNA. The induction of supranormal amounts of pp60$^{c\text{-}src}$ activity on infection is an obvious way in which transformation could come about, and it is known that stable transformation is only achieved by the cells dividing at least once after infection.

Another mechanism involves the very important cell protein p53, which is central to controlling the normal rate of cell proliferation. When inactivated by mutation or by interaction with products of DNA tumour viruses, such as large T of SV40, cell division is no longer restrained. Hence p53 is also known as a tumour suppressor. It is thought that the malfunction of p53 is almost universal in human cancers.

Human papilloma viruses

There is a growing body of evidence that other members of the Papovaviridae, HPVs, are implicated in cervical and penile cancer. There are over 40 HPV serotypes and none can be grown in cell culture. These viruses cause warts (which are benign tumours) of either the skin or the internal epithelial cell layers. Certain serotypes are associated with genital warts, the occurrence of which correlates with early sexual behaviour and a (high) number of sexual partners. Cervical cancer is also strongly correlated with promiscuity, but this does not necessarily implicate HPVs. However, in one study, 93% of women with cervical cancer had HPV antibodies, whereas they were absent from a matching control group. Suspicion falls most heavily on HPV-16 and 18, which are known to integrate with cellular DNA. Three out of four cervical tumour cell lines, including HeLa cells, which have been cultivated since 1953, also contain HPV-18 genomic sequences. The molecular biology of HPVs is different from polyoma and SV40. However, HPV-mediated transformation mechanisms are similar to those of SV40, involving the inactivation of tumour-suppressor proteins in the cell. Again, infection alone is not sufficient for tumour formation, and it is thought that there may be a cofactor, such as genital infection at a relatively early age with herpes simplex virus type 1 or 2, or cigarette smoking, acting in concert with HPV.

Adenoviridae

As with the papovaviruses, each adenovirus-transformed cell contains adenovirus-specific T and transplantation antigens and the equivalent of several molecules of viral DNA sequences integrated into the chromosomes. However, these are incomplete genomes, and infectious virus cannot be recovered from transformed cells. Although only about 0.02% of the DNA from malignant cells will anneal specifically with DNA extracted from purified adenovirus, this viral DNA must be preferentially transcribed, since 2% of the RNA recovered from the polysomes is viral-specific. Hybridization experiments show that this viral RNA is transcribed from less than 10% of the viral genome.

Transformation by adenovirus requires integration of only a specific part of its DNA. All adenovirus-transformed cells contain a region from one end of the genome (known by convention as the left-hand part) which represents 10% of the total genomic DNA. In other experiments, cells could be transformed by fragments of isolated DNA of $1 \times 10^6 \, M_r$ (equivalent to about 5% of the genome) or by a $1.6 \times 10^6 \, M_r$ fragment derived specifically by restriction endonuclease cleavage from the left-hand end of the genome.

The region at the left-hand end of the genome which is responsible for oncogenic transformation by adenovirus type 12 is the early region, E1, so called because of the timing of its expression of the lytic cycle (see Fig. 10.17). E1 is composed of two transcriptional units, E1a and E1b, each of which encodes two polypeptides (Fig. 17.2). Restriction of E1 into its two functional units, E1a and E1b, has allowed the experimental demonstration of the important fact that transformation and oncogenesis are separate, though related, processes.

Baby rat kidney (BRK) cells have a limited lifespan of 2–3 weeks unless immortalized by transformation. When such cells were transfected with E1a DNA alone, foci of cells with increased growth potential emerged. However, these were not immortal (i.e. could not be grown indefinitely in culture). Full transformation of BRK

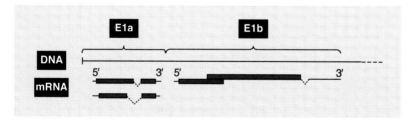

Fig. 17.2 The early region (E1) of adenovirus. E1a polypeptides are made from two different messenger RNAs (mRNA), using the same reading frames, and differ in the size of the spliced-out intron. E1b polypeptides are made from the same mRNA and have different reading frames. The solid blocks indicate the coding regions of mRNAs.

cells required both E1a and E1b genes. Only these fully transformed cells were tumorigenic in congenitally T-cell-immunodeficient mice (known as *nude* mice). Thus, while E1a is essential for transformation, alone it is not sufficient for transformation (Fig. 17.3). The mechanism by which E1a and E1b proteins alter cell growth appears to be analogous to that observed for the Papovaviridae. Firstly, virus-encoded polypeptides interact with proteins encoded by tumour-suppressor genes to overcome the latter's growth-inhibitory effects. Secondly, other mutations accumulate to further alter cell behaviour.

Different adenovirus serotypes show different tumorigenic potential in normal immunocompetent animals, adenovirus type 12 being highly oncogenic while adenovirus type 5 is non-oncogenic. The same distinction is seen when the tumorigenicity of type 5- and type 12-transformed BRK cells is compared in syngeneic immunodeficient rats. (Syngeneic means genetically identical, and such animals are necessary to avoid rejection of the tumour cells through immunological recognition of foreign *rat*

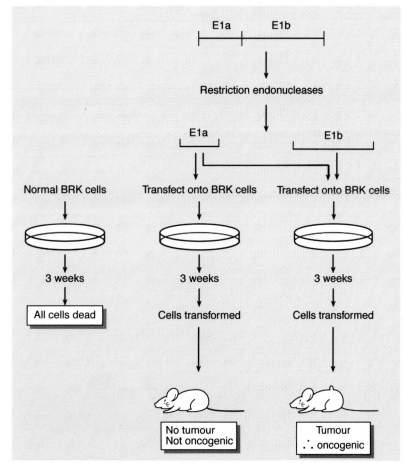

Fig. 17.3 Demonstration of the separate nature of transformation and oncogenesis: while the E1a region alone of adenovirus type 12 is transforming, only cells carrying adenovirus type 12 E1b as well are tumorigenic in syngeneic rats. BRK, baby rat kidney.

antigens.) To determine whether or not this tumorigenic property was associated specifically with an adenovirus type 12 product, BRK cells were transformed with type 5 E1a and E1b and type 12 E1a and E1b in different combinations, and the resulting cells assessed for tumorigenicity. This property clearly correlated with type 12 E1a (Fig. 17.4).

What additional or different function is carried exclusively by adenovirus type 12 E1a? In which of the ways listed in Table 17.1 does adenovirus type 12 affect the physiology of a cell to make it oncogenic? The main clue comes from data which show that, although adenovirus types 12 and 5 express T antigens in transformed cells, only the latter are lysed by cytotoxic T lymphocytes (CTL). CD8$^+$ CTL only function when they recognize a foreign peptide presented by major histocompatibility complex (MHC) class I antigen, so it was a very significant discovery that expression of class I MHC proteins was reduced only in adenovirus type 12-transformed cells. Thus, it seems that tumorigenicity resides in the ability of adenovirus type 12 to 'hide' the cells it has transformed from the immune system, a form of immunosuppression. It has now been shown that it is one of the E1 products of adenovirus type 12 that inhibits the expression of class I MHC proteins at the transcriptional level.

Herpesviridae: Epstein–Barr virus

EBV infects everyone. Childhood infection is subclinical, but infection of young adults can cause prolonged debilitation (glandular fever or infectious mononucleosis, named after the extensive proliferation of host B and T lymphocytes in the blood). Either

Transfected DNA	Transformation	Oncogenesis
E1a$_{12}$ —— E1b$_{12}$	+	+
E1a$_{12}$	+*	−
E1a$_{12}$ —— E1b$_5$	+	+
E1a$_5$ —— E1b$_5$	+	−
E1a$_5$	+*	−
E1a$_5$ —— E1b$_{12}$	+	−

Fig. 17.4 Transfection into BRK cells of E1a and E1b DNA from oncogenic adenovirus type 12 and non-oncogenic adenovirus type 5 in permutation. E1a determines whether or not E1a + E1b − transformed cells can form a tumour in immunocompetent rats. Cells transformed by E1a alone (*) showed an incompletely transformed phenotype, indicating that E1b plays a contributory role in transformation.

way, EBV both causes an acute infection of epithelial cells of the nasopharynx, which may become persistent, and establishes a life long latent infection of circulating B cells.

During latency, the viral DNA circularizes but does not integrate (i.e. it is episomal) and expresses only nine of its 100 or so genes. B cell tumours, such as Burkitt's lymphoma, caused by EBV are rare and the stages by which EBV causes cancer are only now beginning to be unravelled. In the first place, we know that T cells of the immune system are of utmost importance in preventing (the non-oncogenic) proliferation of the latently infected B cells which we all possess. Secondly, we know that B lymphocytes transformed *in vitro* are not oncogenic, showing again the separate nature of transformation and oncogenesis. Thirdly, we suspect that in some way EBV causes a translocation of part of human chromosome 8, usually to chromosome 14 and rarely to chromosome 2 or 22. This is significant, as these contain the highly active transcriptional regions encoding immunoglobulin heavy and light chains respectively. In the translocated fragment of chromosome 8 is a cellular oncogene, c-*myc*, which is concerned with the normal regulation of cell proliferation and/or differentiation. It is believed that as a result of this rearrangement c-*myc* comes under a new set of controlling elements, which are responsible for the expression of the very active immunoglobulin genes. As a result, c-*myc* is dis-regulated and expressed in supranormal amounts. The product of c-*myc* is a DNA-binding protein responsible for aspects of transcriptional control in the cell (Table 17.1b). We know it is expressed in normal cells and helps to regulate the cell cycle. Evidently, excess production of c-*myc* allows unchecked cell division.

One key event in the generation of Burkitt's lymphoma is immunosuppression caused by malaria. This and other cofactors in the expression of viral disease are discussed in Chapter 20.

Retroviridae

Retroviruses are widely distributed throughout vertebrates (humans, monkeys, cats, horses, dogs, mice, rats, birds, snakes, fish) and invertebrates (flies (*Drosophila*), tapeworm). They are enveloped viruses with two identical molecules of a small single-stranded plus-sense RNA of M_r $1-3 \times 10^6$ (see p. 134) and are replicated via a DNA copy, which is inserted into the host genome (Chapter 9). Not all retroviruses are oncogenic (see p. 349); for example, the lentivirus group, which includes human immuno-deficiency virus (HIV) (see Chapter 19), is primarily associated with immunodeficiency diseases. Of the oncogenic retroviruses, most is known about those of murine and avian origin, since these infect cells in culture. Surprisingly, the first human T cell lymphotropic virus type 1 (HTLV-1), was not recognized until 1980.

If fibroblastic, cells are infected with either avian (ALV) or murine leukaemia virus (MLV), very little morphological change occurs.

The cells are neither transformed nor lysed, making a direct plaque assay impossible. However, if mouse cells which are replicating leukaemia virus are plated on to a monolayer of XC rat tumour cells, syncytia are formed. The syncytia can be seen in the layer of XC cells and reflect the number of infected cells. Alternatively, some other parameter of infection is measured, such as reverse transcriptase activity or production of a viral antigen.

ALV and MLV cause leukaemia when injected into the appropriate animal, and transform epithelial, but not fibroblastic, cells *in vitro*. The behaviour of murine sarcoma virus (MSV) is quite the opposite, since it cannot replicate on its own but is capable of transforming fibroblasts (and not epithelial cells) *in vitro*. Foci of transformed cells appear on the Petri dish and can be used to quantify this virus. Such cells produce no virus particles and are known as non-producer cells. MSV can produce progeny particles providing MLV is present to supply the missing functions; this is a form of genetic complementation. If a stock of MSV is allowed to infect mouse cells, two types of foci may be detected: one produces two viruses, an XC plaque-forming virus (i.e. MLV) and a focus-forming virus (i.e. MSV); however there are also foci of transformation which are the result of infection with MSV alone. Note that RSV is the sole example of a non-defective *onc*-containing sarcoma virus (Fig. 10.19).

As a result of intensive molecular study, giant strides have been made in the understanding of carcinogenesis with both viral and non-viral causes. Retroviruses have two fundamental properties in this context: (i) they insert their genomes into host cell DNA, causing mutations; and (ii) their genomes have a great facility for recombination with other homologous viral genomes but also with the genome of the host cell. In other words, like the phages discussed earlier, retroviruses can act as transducing agents. Host−virus recombination almost certainly involves recombination between host DNA and inserted provirus, and precedes transcription of viral RNA. In common with other viruses, there is a limit to the size of genome that a virion can accommodate. Hence, acquisition of cellular sequences usually results in a reciprocal loss of viral genes. Thus, all acutely transforming retroviruses, with the exception of RSV, are defective and require a helper for their replication (Fig. 17.5A).

Retrovirus oncogenesis

The association between the retrovirus *onc* sequence and its transforming/tumorigenic properties was demonstrated by their co-ordinate loss in mutated viruses. However, there was a paradox, as all oncornaviruses are tumorigenic, whether or not they possess a v-*onc* sequence. Clearly, some other feature of the virus was also important. This proved to reside in the long terminal repeat sequences (LTRs) (Fig. 9.2) which define the ends of viral DNA.

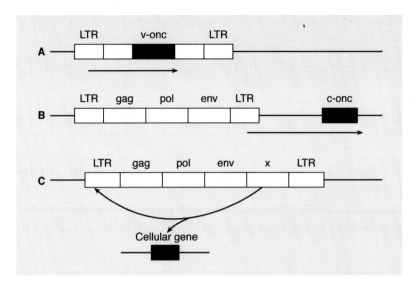

Fig. 17.5 Models of retrovirus oncogenesis involving integration of viral genomes into cellular DNA. (A) v-*onc* in a defective viral genome whose transcription is under the control of strong promoters in the viral LTRs. (B) Integration of a virus without an oncogene, so that its LTR enhances expression of a normal cellular oncogene (c-*onc*). (C) Expression of a transcription-enhancing product from viral gene X which affects both viral and cellular (in this case, not c-*onc*) transcriptional controls. (A) and (B) are *cis*-acting and (C) *trans*-acting on both viral and cellular transcription, as shown by arrows.

During oncogenesis by retroviruses which lack an oncogene, LTRs act by modifying the expression of cellular genes (Fig. 17.5 and see below). In general, viruses with v-*onc* genes transform cells *in vitro* and are rapidly tumorigenic: chickens die of cancer 1–2 weeks after infection with RSV; most viruses without v-*onc* cause a slowly progressive tumorigenesis in only a proportion of infected animals. The latter are by far the commonest in nature. Really, this is an artificial distinction, as later we shall show that retroviral v-*onc* genes and cellular genes activated by LTRs are virtually identical. In addition, gene X of human lymphotropic leukaemia viruses (HTLVs) encodes a product which enhances transcription, not only from its own LTR but from cellular control elements as well (Fig. 17.5C).

Retroviruses as insertional mutagens

Retroviruses do not have precise integration sites. The insertion of retroviral DNA into the host genome may therefore have a variety of effects on cellular gene expression. Insertion within a coding region will generally destroy gene function, and it is self-evident that such ablation must be non-lethal for the cell for this mutation to be oncogenic.

Insertion in the vicinity of a coding region may have positive effects on cellular gene expression, as, within retrovirus LTRs, there are also *enhancer* sequences which increase the efficiency of transcription from promoters within their influence. Random insertion of such control elements can switch on cellular genes that are not being expressed. Such an event might cause tissue-specific or developmentally regulated products to appear out of context causing aberrant cell behaviour. The effect of LTRs are all the more readily realized as they can act when inserted upstream or downstream of the gene in question and in either orientation (Fig. 17.6).

Properties of retrovirus oncogenes

One of the surprises has been the variety of retrovirus oncogenes so far described (Table 17.6), but all have a common property — the potential for modifying cellular behaviour. DNA-binding proteins (like the large T of papovaviruses) can affect gene expression and cell division, leading to immortalization; phosphorylation is well known as a control of enzyme activity (remember glycogen metabolism), and there are plasma membrane receptors for cellular growth-enhancing factors. Despite their names, one viral oncogene or oncogenic function is not enough to be oncogenic *in vivo*. Other functions and possibly other oncogenes or control-modifying events are necessary for the full multistep process to be realized. Great progress has been made, but this is by no means the end of the story and much more remains to be elucidated.

Cellular oncogenes

Cellular oncogenes are involved in oncogenesis by DNA tumour viruses as well as retroviruses. Originally, they were discovered by their homology to retroviral oncogenes, but the discovery that they were oncogenic in their own right established the cellular oncogene theory. Such data were obtained by showing that DNA fragments extracted from human tumours were able to transform

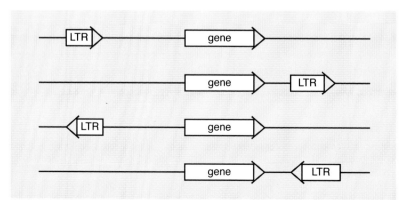

Fig. 17.6 Activation of a cellular gene by insertion near by of a retroviral LTR in any orientation; this is due to the presence of enhancer elements in the LTR.

Table 17.6 Some retrovirus oncogenes and their products.

Tumour	Oncogene	Product and its location
Sarcoma	v-*src*	Tyrosine-specific protein phosphokinases (PM)
B cell lymphoma	v-*abl*	
Carcinoma, sarcoma, and myelocytic leukaemia	v-*myc*	Nuclear DNA binding proteins; transcriptional regulators
Myeloblastic leukaemia	v-*myb*	
Osteosarcoma	v-*fos*	
Erythroleukaemia	v-*erb*-B	EGF receptor (PM)
Erythroleukaemia and sarcoma	v-*ras*	GTP-binding protein (PM)
Sarcoma	v-*mos*	Phosphoprotein in cytosol
Sarcoma	v-*sis*	Cytoplasmic homologue of PDGF

PM, plasma membrane; EGF, epidermal growth factor; GTP, guanosine triphosphate; PDGF, platelet-derived growth factor.

normal cells in culture. However, herein lies a paradox, for how can c-*oncs* be at one and the same time 'normal' genes and oncogenic? Table 17.7 summarizes four possibilities. Enhanced expression, for instance by insertion of viral LTRs, we have met above. Gene amplification is seen in many tumour cell lines in culture, and obviously can also lead to over-expression. Relocation of the cellular oncogene c-*myc* is seen as one of the later stages in the transformation of B lymphocytes by EBV. Such relocation can increase expression of the c-*myc* product, a DNA-binding nuclear protein, by up to 100-fold. Point mutation of a c-*onc*, conferring altered properties on its product, may be an essential for it to act as an oncogene. For example, a nucleotide change in a human c-*ras* gene isolated from a carcinoma of the bladder causes a single amino acid substitution of a valine for a glycine. The effect on this important regulatory protein is to permanently switch on its activity, so leading to altered cell behaviour. To date mutated c-*ras* genes have been identified in 70% of human tumours investigated. Point mutation also brings us full circle to the derivation of v-*onc*. Comparison of amino acid sequence shows that c-*onc* and v-*onc* also differ by only a single amino acid residue.

Hepadnaviridae: hepatitis B virus

Together with unknown cofactors, HBV causes a liver cancer,

Table 17.7 Ways in which cellular oncogenes (c-*onc*) can become oncogenic.

Have their expression enhanced
Be amplified
Be relocated
Acquire point mutations

(primary hepatocellular carcinoma). People who are chronically infected have a 200-fold increased risk of this cancer. With 300 million carriers, this makes HBV the most potent human carcinogen after smoking. All primary hepatocellular carcinoma cells contain integrated HBV sequences, but the exact role of the virus is not known. It is possible that integration of the HBV genome alters the transcription of a cellular gene regulating cell division or that a viral gene product, the HBx antigen, transactivates other viral genes. Indeed insertion of the HBx gene under its own regulatory controls into mice, resulted in pathological changes in the liver, which culminated in the development of liver cancer.

SUMMARY AND PROSPECTS FOR THE CONTROL OF CANCER

Only in the past few years has there been convincing evidence that viruses are agents involved in causing some human cancers. This has been a vindication of the reductionist approach to the analysis of cancer, and we now have a molecular framework of understanding within which to devise new experiments to probe the roles of oncogenes. However, viruses alone do not cause cancer in nature and, as yet, we have a very rudimentary insight into the other nutritional, chemical and genetic aspects. Molecular knowledge of tumour viruses has outstripped understanding of these important cofactors.

Vaccines against cancer

Vaccines which prevent infection with tumour viruses are *ipso facto* vaccines against cancer. This has been amply demonstrated with Marek's disease virus of chickens (Herpesviridae), which causes a cancer of cells of the feather follicles, and with successful experimental vaccines against herpesvirus saimiri and EBV. Immunization against tumour viruses is as easy (or as difficult) as immunization against any non-tumour virus. Any problems are simply those of producing a successful vaccine. An excellent vaccine for humans against HBV is already on the market, but to combat other known human cancer-causing viruses several new vaccines would be needed (to EBV, HTLV-1 and 2 and HPV-16 and 18), a daunting prospect for the manufacturer. One problem is that each tumorigenic virus only rarely causes cancer. Even with a disease as emotive as cancer, would there be sufficient interest to persuade people to avail themselves of immunization? Would the 'man (or woman) in the street' understand that, at best, such vaccines would only protect against virally induced tumours? ... If not, to the common eye, the vaccines would be deemed a failure and acceptance by future generations would be compromised.

Thinking about vaccines must proceed in tandem with the more traditional anti-cancer approaches involving the elimination of cofactors, education about the risk of cofactors, improved early diagnosis and better treatment. On the latter point, we can hope that molecular biology will be able to show how changes brought about by, or in, specific oncogenes can be reversed.

GLOSSARY

Adenocarcinoma A carcinoma developing from cells of a gland.
Benign An adjective used to describe growths which do not infiltrate into surrounding tissues. Opposite of malignant.
Cancer Malignant tumour; a neoplasm; a growth which is not encapsulated and which infiltrates into surrounding tissues, the cells of which it replaces by its own. It is spread by the lymphatic vessels to other parts of the body. Death is caused by destruction of organs to a degree incompatible with life, extreme debility and anaemia or by haemmorrhage.
Carcinogenesis Complex multistage process by which a cancer is formed.
Carcinoma A cancer of epithelial tissue.
Fibroblast A cell derived from connective tissue.
Leukaemia A cancer of white blood cells.
Lymphoma A cancer of lymphoid tissue.
Malignant A term applied to any disease of a progressive and fatal nature. Opposite of benign.
Neoplasm An abnormal new growth, i.e. a cancer.
Oncogenic Tumour causing.
Rhabdomyosarcoma A malignant tumour of striated muscle fibres.
Sarcoma A cancer developing from fibroblasts.
Tumour A swelling, due to abnormal growth of tissue, not resulting from inflammation. May be benign or malignant.

FURTHER READING

Bishop, J. M. (1983) Cellular oncogenes and retroviruses. *Annual Review of Biochemistry*, **52**, 301–354.
Bos, J. L. & Van der Eb, A. J. (1985) Adenovirus region EIA: transcription modulation and transforming agent. *Trends in Biochemical Sciences*, **10**, 310–313.
Cairns, J. & Logan, J. (1983) Step by step with carcinogenesis. *Nature (London)*, **304**, 582–583.
Campo, M. S. (1992) Cell transformation by animal papillomaviruses. *Journal of General Virology*, **73**, 217–222.
Courtneidge, S. & Smith, A. E. (1983) Polyoma virus transforming protein associates with the product of c-*src* cellular genes. *Nature (London)* **303**, 435–439.
Dalgleish, A. G. (1991) Viruses and cancer. *British Medical Bulletin*, **47**, 21–46.

Desrosiers, R. C., Bakker, A., Kamine, J., Falk, L. A., Hunt, R. D. & King, N. W. (1985) A region of the *Herpesvirus saimiri* genome required for oncogenicity. *Science*, **228**, 184–187.

Fanning, E. & Knippers, R. (1992) Structure and function of simian virus 40 large tumor antigen. *Annual Review of Biochemistry*, **61**, 55–85.

Finnegan, D. J. (1983) Retroviruses and transposable elements—which came first? *Nature (London)*, **302**, 105–106.

Gallimore, P. H., Byrd, P. J. & Grand, R. J. A. (1985) Adenovirus genes involved in transformation. What determines the oncogenic phenotype? In: *'Viruses and Cancer'. Society for General Microbiology Symposium*, **37**, 125–172.

Griffin, B. E., Karran, L., King, D. & Chang, S. E. (1985) Immortalizing genes encoded by Epstein–Barr virus. In: *'Viruses and Cancer'. Society for General Microbiology Symposium*, **37**, 93–110.

Hardy, W. D. (1984) A new package for an old oncogene. *Nature (London)*, **308**, 775.

Hunter, T. (1984) The epidermal growth factor receptor gene and its product. *Nature (London)*, **311**, 414–416.

Kim, C.-M., Koike, K., Saito, I., Miyamura, T. & Jay, G. (1991) *HBx* gene of hepatitis B virus induces liver cancer in transgenic mice. *Nature (London)*, **351**, 317–320.

Kingston, R. E., Baldwin, A. S. & Sharp, P. A. (1985) Transcription control by oncogenes. *Cell*, **4**, 3–5.

Klein, G. & Klein, E. (1985) Evolution of tumours and the impact of molecular oncology. *Nature (London)*, **315**, 190–195.

Lane, D. P. (1992) p53, guardian of the genome. *Nature (London)*, **358**, 15–16.

Levine, A. J., Momand, J. & Finlay, C. A. (1991) The p53 tumour suppressor protein. *Nature (London)*, **351**, 453–456.

Melnick, J. L., Ochoa, S. & Oko, J. (eds) (1985) Viruses, oncogenes and cancer. *Progress in Medical Virology*, **32** (a report from a symposium).

Minson, A., Neil, J. & McCrae, M. (1994) Viruses and Cancer. *Society for General Microbiology Symposium*, **51**, 1–309.

Pardoll, D. M. (1993) Cancer vaccines. *Immunology Today*, **14**, 310–316.

Robertson, M. (1984) Message of *myc* in context. *Nature (London)*, **309**, 585–587.

Santos, E., Sukumar, S., Martin-Zanca, D., Zarbl, H. & Barbacid, M. (1985) Transforming *ras* genes. In: *'Viruses and Cancer'. Society for General Microbiology Symposium*, **37**, 291–313.

Teich, N. (ed.) (1991) Viral oncogenes. Part I. *Seminars in Virology*, **2**, no. 5.

Teich, N. (ed.) (1991) Viral oncogenes. Part II. *Seminars in Virology*, **2**, no. 6.

Temin H. (1983) Oncogenes: we still don't understand cancer. *Nature (London)*, **302**, 656.

Tooze, J. (ed.) (1981) *DNA tumor viruses* (2nd edn). Cold Spring Harbor Laboratory. (A comprehensive text but dated in some areas.)

van Beveren, F. & Verma, I. M. (1985) Homology amongst oncogenes. *Current Topics in Microbiology and Immunology*, **123**, 73–98.

Weinberg, R. A. (1984) Cellular oncogenes. *Trends in Biochemical Sciences*, **7**, 131–133.

Weis, R. A. (1984) Viruses and human cancer. In: *'The Microbe 1984.' Society for General Microbiology Symposium*, **36**, Pt 1, 211–240.

Weiss R. A. (1985) Unravelling the complexities of carcinogenesis. In: *'Viruses and Cancer'. Society for General Microbiology Symposium*, **37**, 1–21.

Weiss, R., Teich, N., Varmus, H. & Coffin, J. (1984/5) *RNA tumor viruses* (2nd edn). Cold Spring Harbor Laboratory. (A comprehensive text but dated on oncogenes.)

Also check Chapter 21 for references specific to each family of viruses.

18 The evolution of viruses

Viruses undergo evolutionary change, just like any other living organism. Their genomes are subject to mutations at the same rate as all nucleic acids and, where conditions enable a mutant to multiply at a rate faster than its fellows, that mutant virus will be selected and will succeed the parental type.

In any discussion of evolution, one naturally starts at the earliest time possible, and it is pertinent to ask where viruses first came from (see also p. 335). The absence of any fossil records of viruses and scarcity of other evidence have not, of course, prevented scientists from speculating about their origins! The two prevailing opinions take into account the parasitic existence of viruses today:
1 Viruses have arisen from degenerate cells which have lost the wherewithal for free-living.
2 Viruses have arisen from pieces of cellular nucleic acid which have escaped from the cell.

The molecular biology of bacteriophages and their prokaryotic host cells differs considerably from that of viruses of eukaryotes and their host cells, to the extent that it is not possible to grow bacteriophages in eukaryotic cells or eukaryotic viruses in bacteria. Thus, it appears that phages and viruses of eukaryotes have arisen independently or diverged at a very early stage.

Whatever their origins, viruses have been a great biological success, for no group of organisms has escaped their attentions. In Chapters 12 and 13, we discussed the various ways in which viruses interact with their hosts, and we saw how viruses cause a variety of changes, ranging from the imperceptible to death. Evolution of any successful parasite has to ensure that the host also survives. The various virus—host interactions alluded to earlier can be thought of as ways in which this problem is being solved.

In this chapter we shall discuss, firstly, the evolutionary implications of the distribution of morphologically similar viruses throughout a range of different hosts and, secondly, examples of virus evolution that have occurred in relatively recent times.

SPREAD OF MORPHOLOGICALLY SIMILAR VIRUSES

The most commonly occurring bacteriophage (see Table 3.2) has double-stranded deoxyribonucleic acid (DNA) packaged in a particle of the 'head-and-tail', type which reaches its zenith of complexity in the T-even phages. This type of phage infects a wide variety of bacterial species. There are a range of variations in head size and tail length, and it is easy to construct a gradient of variation from the simple to the complex after electron microscopic examination of the phages (see pp. 358–9). This does not necessarily indicate any evolutionary relationship, but it does mean either that these phages arose many times independently or that the basic pattern has been subjected to evolutionary modification. It is curious that the 'head-and-tail' virus particle has never been found in eukaryotes.

Another example of a widespread group of viruses are the rhabdoviruses of eukaryotes (see Fig. 3.16). Their 'bullet' shape makes them easy to identify. They have a lipid envelope and negative-sense single-stranded ribonucleic acid (RNA). Rhabdoviruses infect both plants and animals, and are found in invertebrates and cold- and warm-blooded vertebrates (but, like all viruses, each rhabdovirus serotype infects a very restricted range of host species). Since these viruses are morphologically indistinguishable, it is tempting to speculate that they have arisen once and subsequently spread in a truly remarkable fashion. The rhabdoviruses also present us with an interesting link between viruses of the animal and plant kingdoms, since at least one representative (e.g. lettuce necrotic yellows virus) multiplies in both the insect vector and the plant it feeds on.

EVOLUTION OF MEASLES VIRUS

Measles virus is a paramyxovirus (class V). In nature it infects only humans, and the infection results in lifelong immunity from the disease. F. L. Black studied the frequency of the disease in island populations (Table 18.1). There is a good correlation between the size of the population and the number of cases of measles recorded on the island throughout the year. A population of at least 500 000 is required to provide sufficient susceptible individuals (i.e. births) to maintain the virus in the population. Below that level, the virus will eventually die out, unless it is reintroduced from an outside source.

Table 18.1 Correlation of the occurrence of measles on islands with the size of the population.

Island group	Population $\times 10^{-3}$	New births per year $\times 10^{-3}$	Months with measles (%) 1949–1964
Hawaii	550	16.7	100
Fiji	346	13.4	64
Solomon	110	4.1	32
Tonga	57	2.0	12
Cook	16	0.7	6
Nauru	3.5	0.17	5
Falkland	2.5	0.04	0

On the geological time-scale, humans have evolved recently and have only existed in populations of over 500 000 for a few thousand years. In answer to the question about where measles virus was in the days of very small population groups, we can conclude that it could not have existed in its present form. It may have had another strategy of infection, such as persistence, which would allow it to infect the occasional susceptible passer-by, but we have no evidence of this. However, F. L. Black has speculated upon the antigenic similarity of measles, canine distemper and rinderpest viruses. The latter infect dogs and cattle, respectively, which have been commensal with humans since their nomadic days. Black suggests that these three viruses have a common ancestor which infected prehistoric dogs or cattle. The ancestral virus evolved to the modern measles virus when changes in the social behaviour of humans gave rise to populations large enough to maintain the infection. This evolutionary event would have occurred within the last 6000 years, when the river valley civilizations of the Tigris and Euphrates were established.

EVOLUTION OF MYXOMAVIRUS

Myxomavirus belongs to the Poxviridae (class I) and causes a benign infection in its natural host, the South American rabbit, producing wart-like outgrowths (benign tumours) as the only visible evidence of virus multiplication. In the European rabbit, a susceptible but foreign host myxomavirus causes myxomatosis. This is a generalized infection, with lesions over the head and body surface, which is usually fatal. In nature, the disease is spread by virus carried on the mouthparts of the mosquitoes that feed on rabbits or by the rabbit flea. However, this virus does not multiply in the vector.

Myxomavirus was released in England and Australia upon a wholly susceptible host population of the European rabbit in an attempt to eradicate the rabbit, a serious agricultural pest. This experiment in ·nature was carefully studied with respect to the changes occurring in the virus and the host populations. As we shall see, it provides an object lesson in biological control.

In the first attempts to spread the disease in Australia, myxomavirus-infected rabbits were released in the wild but, despite the virulence of the virus and the presence of susceptible hosts, the virus died out. It was realized later that this failure was due to the scarcity of mosquito vectors, whose incidence is seasonal. When infected animals were released at the peak of the mosquito season, an epidemic of myxomatosis followed. Over the next 2 years, the virus spread 3000 miles across Australia and even across the sea to Tasmania. However, during this time it became apparent that fewer rabbits were dying from the disease than at the start of the epidemic. The investigators found two significant facts. Firstly, they compared the virulence of the original virus with virus newly isolated from wild rabbits by inoculating isolates into standard laboratory rabbits. They found that: (i) rabbits took longer to die; and (ii) a greater number of rabbits recovered from infection. From this it was inferred that the virus had evolved to a more avirulent form (Table 18.2). The explanation was simple: mutation produced virus variants which did not kill the rabbit as quickly as the parental virus. This meant that the rabbits infected with the mutant virus survived to be bitten by the vectors for a longer period than rabbits infected with the original strain. Hence the mutant would be transmitted to a greater number of rabbits. In other words, there was a strong selection pressure in favour of mutants which survived in the host in a transmissible form for as long as possible.

The second fact concerned the rabbits themselves, and the question was raised whether rabbits which were genetically resistant to myxomatosis were being selected. To test this hypothesis, a breeding programme was set up in the laboratory. Rabbits were infected and survivors were mated and bred. Offspring were then infected and the survivors mated and so on. Part of each litter was tested for its ability to resist infection with a standard strain of myxomavirus. The result confirmed that the survivors of each generation progressively increased in resistance.

This work shows how evolutionary pressures set up a balance between a virus parasite and its host which ensures that both continue to flourish. This fact remains a stumbling-block to the advocates of biological control of pests that attack animals or plants. Today, in the UK and Australia, rabbits remain a serious agriculture problem.

EVOLUTION OF INFLUENZA VIRUS

Background

The three types of influenza virus are distinguished by the antigenicity of their nucleoproteins and are called type A, B and C. Type A causes the world-wide epidemics (pandemics) of influenza,

Table 18.2 Evolution of myxomavirus to avirulence in rabbits after introduction of virulent virus into Australia in 1950.

Mean rabbit survival time (days)	Mortality rate (%)	Year of isolation			
		1950–1951	1952–1953	1955–1956	1963–1964
<13	>99	100*	4	0	0
14–16	95–99		13	3	0
17–28	70–95		74	55	59
29–50	50–70		9	25	31
	<50		0	17	9

* Percentage of rabbits tested.

and both types A and B cause epidemics during the winter. Type C causes minor upper respiratory illness and will not be discussed further. Resistance to infection is determined by whether or not the immune system has been previously exposed to the infecting virus. The viral antigens relevant to protective immunity are the external haemagglutinin (HA) and neuraminidase (NA) glycoprotein spikes (see Fig. 3.16). Earlier in this chapter, we mentioned that infection with measles virus resulted in lifelong immunity to measles. Why, then, is it common experience for people to suffer several attacks of influenza in their lifetime? The answer is that influenza A and B viruses continuously evolve new HAs and NAs against which previously acquired immunity is ineffective. How and why this happens is discussed later. However, immunity acquired to an influenza virus is effective at preventing reinfection by that same strain of virus, and it will be relevant to remember that influenza A virus naturally infect birds (particularly sea and fresh-water birds), pigs, horses and seals, in addition to humans.

Antigenic drift

Influenza A viruses have been isolated from humans since the discovery in 1932 that they could infect ferrets, although embryonated eggs and tissue cultures are used now. Each new isolate was tested serologically with antisera to all other known influenza strains. It soon became apparent that the more recent isolates had slightly different HA and NA antigens from earlier strains. Next it was realized that the 'old' strains were no longer present and only the new strains could be isolated. This phenomenon is aptly called *antigenic drift* (Fig. 18.1). It rests on the assumption that in nature influenza strains carrying new antigenic determinants arise by natural selection of mutants occurring in the current human influ-

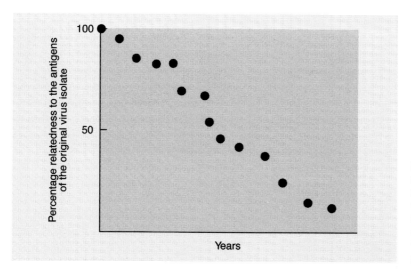

Fig. 18.1 Diagram showing antigenic drift of type A influenza virus in humans. This could represent the HA or NA. Each point is a virus strain isolated in a different year.

enza virus population. Antigenic drift of type A influenza viruses does not occur in other species, presumably because their higher reproductive rates provide a sufficient supply of susceptible individuals. Influenza B viruses (which only infect humans) also undergo antigenic drift, but this is slower and has two main branches; also several different strains cocirculate. The reason for this difference is not known.

Evolution cannot occur at such a fast rate without efficient selection for a favourable mutant. How this might come about was shown by experiments in which virus was grown in cell culture in the presence of critical amounts of neutralizing antibody (Fig. 18.2). After seven passages into new cultures containing antibody, the resulting virus was used to immunize rabbits and compared in reciprocal haemagglutination inhibition tests with the original strain. The HAs were now very different. This 'evolution in a test-tube' is thought to mimic the antigenic drift which occurs in nature in our respiratory tracts.

The antigenicity of another virus which initially causes a respiratory infection, measles virus, is unchanging. Why then, does measles virus not evolve like influenza virus when they both have the same mutation rate? There is no clear answer but it may be relevant that only measles develops into a generalized infection with virus in the blood. This viraemia is open to the full force of the immune system and the resulting immunity can effectively get rid of both parental virus and any variants that arise. Influenza is predominantly a virus of the respiratory tract, which cannot mount such a rigorous immune response as the circulatory system. This apparent defect is a physiological compromise to prevent over-reaction by the immunologically active cells to the vast amounts of foreign antigens which we continuously inhale into our lungs.

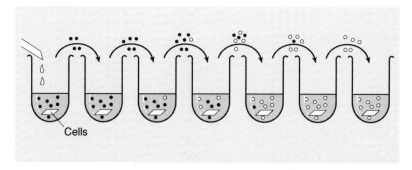

Fig. 18.2 Antigenic drift in a test-tube. Cultures containing antibody insufficient to completely neutralize the virus (●) are inoculated. Progeny virus is transferred to new cultures which also contain antibody. Mutants (○) resistant in some degree to the antibody, arise spontaneously and have a selective advantage. They gradually become dominant in the population.

Antigenic shift

It became clear from examining strains isolated over a period of over 50 years that the evolution of influenza type A virus was not solely antigenic drift but was broken up by the mysterious appearance of new *subtypes*, with an HA or NA or both totally unrelated to those of the previous years (Fig. 18.3). In contrast, influenza type B undergoes only antigenic drift. In addition to the subtypes discovered since human influenza viruses were first isolated, there is serological evidence (explained in the section on 'cycling' below) that humans have in the past been infected by subtypes related to modern H2N2 and H3N2 viruses of humans and H1N1 virus of pigs.

Only since 1977 have *two* subtypes cocirculated. In 1994 they continue to do so and to drift. The significance of these new subtypes is apparent in the close correlation of their appearance with the occurrence of major outbreaks of influenza (Table 18.3). This is not too surprising, since immunity developed to the previous subtype will provide no protection against a virus carrying new antigen(s). More surprising is the fact that antigenic drift is responsible for a greater total amount of influenzal illness than the rarer pandemics caused by antigenic shift. Various explanations have been advanced to explain antigenic shift and some of them are considered below.

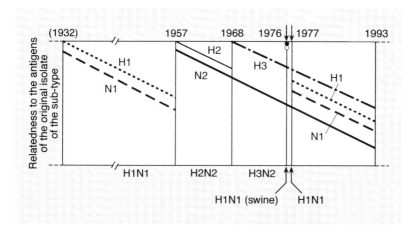

Fig. 18.3 Diagram showing antigenic shift of type A influenza virus. In each of the years written in full, a haemagglutinin (•) or neuraminidase (○) virus was isolated which was antigenically unrelated to the strains of the preceding year. The haemagglutinin of the first human subtype isolated in 1932 is designated H1. The current view is that this virus arose by antigenic drift from the H1 strain which caused the pandemic of 1918. The subtypes H2 and H3 are the result of antigenic shift, while the 1976 and 1977 strains resulted from recurrence of previous strains (see text).

Table 18.3 Correlation between the occurrence of major outbreaks of influenza with the appearance of new subtypes.

Year of epidemic or pandemic	Designation subtype
1889	H2N2*
1900	H3N8*
1918 ('Spanish flu')	H1N1*
1929	H1N1*
1946	H1N1†
1957 ('Asian flu')	H2N2†
1968 ('Hong Kong flu')	H3N2†
1977 ('Russian flu')	H1N1†

* Serological evidence.
† Virus isolations; virus could not be grown in the laboratory until 1932.

Critical alterations of antigenic structure

This postulates that a mutation results in the substitution of an amino acid in a critical part of an antigenic site, which causes the protein to adopt an entirely new configuration and hence antigenicity. The theory is an extreme version of antigenic drift. A prediction that follows from this hypothesis is that the bulk of the protein should have a sequence identical with the strain from which it evolved. This is not borne out by sequence analysis of the relevant HA genes which show major differences between, for example, the 1967 H2 and the 1968 H3.

Explanations that involve other hosts

Some of these hypotheses depend upon the ability of human influenza strains to infect animals and, more importantly, upon the fact that many different subtypes of influenza A viruses have birds, pigs and horses as their natural hosts. Influenza type B infects only humans. There is thus a correlation between the non-existence of type B strains which infect animals and the non-occurrence of antigenic shift or pandemics caused by this virus.

Change of host. The simplest explanation of antigenic shift is that a non-human strain acquires the ability to infect humans. This would readily account for the isolation in 1957 of a virus which had an HA and NA totally different from the strain around in the previous year. (A shift involving a single antigen implies a familial relationship with a strain in the preceding year.) In 1976, the same swine influenza virus was isolated from both pigs and pig-farmers in eastern USA, demonstrating that exchange of influenza viruses between species does take place. Surprisingly, no epidemic resulted, even though the population had no immunity to the virus (H1N1).

It is surmised that the virus lacked the ability of being transmitted, (see p. 226).

Cycling. This theory is based upon the presence of influenza antibodies in sera obtained from people who were alive long before 1932, when techniques for isolating viruses became available. Sera taken before 1957, when the H2N2 subtype influenza was first isolated, were kept frozen and then tested for antibodies to the modern H2N2 and H3N2 viruses. People who were alive in 1889, but not 1888, had H2N2 antibodies, suggesting that they had been infected with an H2N2 virus in 1889. Similarly, it was inferred that an H3N8 virus was around in 1900.

The theory suggests that strains 'go into hiding', perhaps in another host, where they remain until people with immunity have died and there is a substantial population of susceptibles. The virus then emerges and can commence infection in humans. The cycle for the H2N2 subtype turned completely in 69 years, which was about the average life expectancy (Fig. 18.4). However, the appearance in 1977 of an H1N1 strain identical by serology and by nucleotide sequence analysis with the 1950 H1N1 suggests that sufficient susceptible people may accumulate in only 27 years. As yet, there is no clue as to where the virus 'hibernates'; the problem is compounded, as influenza is not known to cause persistent

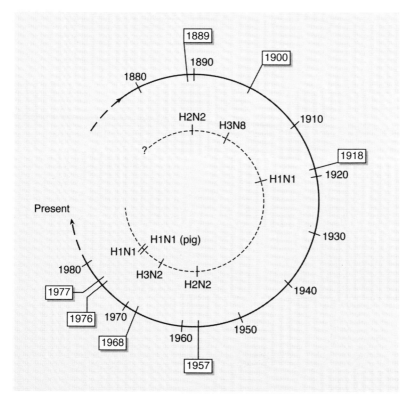

Fig. **18.4** Summary of the evidence that suggests cycling of human influenza type A viruses. The time-scale in the outer circle is marked with the year that a new subtype emerged: e.g. 1889.

infections and, even if it did, it would have to have been in a non-replicating form to emerge after more than a quarter-century with no mutations.

Genetic assortment. The genome of influenza A (and B) viruses consists of eight single-stranded RNA segments, each of which encodes one or two proteins. When a cell is infected simultaneously with more than one strain, newly synthesized RNA segments assort virtually at random to the progeny (Fig. 18.5). Such hybrid strains are genetically stable. They have been produced experimentally in cell culture and in whole animals, and occur in natural infections of humans. Wild sea-birds are the biggest reservoir of influenza virus. There are 14 HA and nine NA antigen subtypes and all permutations have been isolated. Since 1932, only three HA and three NA subtypes have been found in humans (Fig. 18.4). Hybrid strains also have the ability to spread naturally from infected

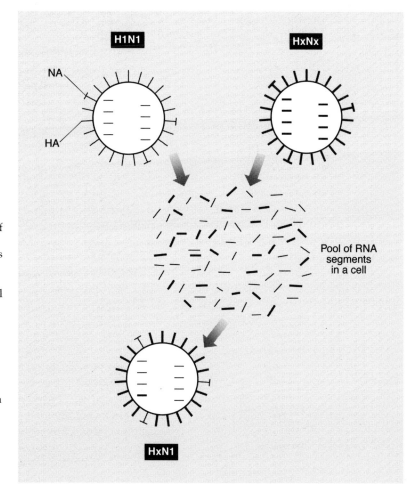

Fig. 18.5 Explanation of antigenic shift in influenza type A viruses resulting from simultaneous infection of a cell by the parental viruses. The eight genome segments from each parent assort independently to progeny virions. In the example shown, only the segment carrying the genetic information for the HA assorted to the progeny. A total of 2^8 (256) genetically different types of progeny virus are possible.

to susceptible animals, an essential property in the evolution of a
new strain. Viruses which could have arisen by genetic assortment
between human and animal strains (e.g. having the HA of an
animal strain and the NA of a human strain) or animal and animal
strains have been isolated in nature. One of these was found in the
harbour seal off the coast of Maine, USA, an animal not previously
known to be infected by influenza virus. The virus caused 20%
mortality, a chilling reminder of the H1N1 human pandemic of
1918, in which 20 million people died world-wide.

Summary

Type A influenza viruses of humans evolve rapidly by antigenic
drift and antigenic shift. Antigenic shift occurs by genetic reassortment, re-emergence of a previously known strain, and infection of
virus from a different host. Type B influenza viruses do not undergo
antigenic shift, although they can undergo genetic assortment.
The absence of non-human type B strains is the missing ingredient.
Type A and B strains do not form hybrids.

  In the laboratory, viruses of stable antigenicity give rise to antigenic variants under the selective pressure of antibody. Clearly,
the capacity for antigenic evolution is there, even if conditions do
not normally permit it.

POSTSCRIPT ON THE WORLD HEALTH ORGANIZATION (WHO)

The WHO co-ordinates viral epidemiology world-wide. There is a
network of national and international reference centres which
study a variety of viruses and, in particular, are monitoring the
appearance of new, potentially pandemic, strains of influenza.
Valuable information is gained on virus evolution, and vaccine
manufacturers can be alerted to the emergence of new epidemic
and pandemic strains.

  Another function of the WHO is the control/eradication of viruses
causing serious disease in humans (such as smallpox, measles,
rabies, poliomyelitis and viral hepatitis; pp. 237–241).

FURTHER READING

Black, F. L. (1966) Measles endemicity in insular populations: critical
    community size and its evolutionary implication. *Journal of Theor-
    etical Biology*, **11**, 207–211.
Fenner, F. & Ratcliffe, F. N. (1965) *Myxomatosis*. London and New
    York: Cambridge University Press.
Holland, J. J. (1992) Genetic diversity of RNA viruses. *Current Topics
    in Microbiology and Immunology*, **176**, 1–226.

Holland, J. J., Spindler, K., Horodyski, F., Grabau, E., Nicol, S. & Vandepol, S. (1982) Rapid evolution of RNA genomes. *Science*, **215**, 1577–1585.

Joklik, W. K. (1974) '*Evolution in Viruses*'. *Society for General Microbiology Symposium*, **24**, 293–320. (A general introduction dealing also with ideas on the origin of viruses.)

Koonin, E. V. (ed.) (1992) Evolution of viral genomes. *Seminars in Virology*, **3**, no. 5.

Reanny, D. (1984) The molecular evolution of viruses. In: '*The Microbe 1984*'. *Society for General Microbiology Symposium*, **36**, Pt 1, 175–196.

Reanney, D. & Ackermann, W. W. (1982) Comparative biology and evolution of bacteriophages. *Advances in Virus Research*, **27**, 205–280.

Steinhauer, D. A. & Holland, J. J. (1987) Rapid evolution of RNA viruses. *Annual Review of Microbiology*, **41**, 409–433.

Webster, R. G., Laver, W. G., Air, G. M. & Schild, G. C. (1982) Molecular mechanisms of variation in influenza viruses. *Nature (London)*, **296**, 115–121.

Webster, R. G., Bean, W. J., Gorman, O. T., Chambers, T. M. & Kawaoka, Y. (1992) Evolution and ecology of influenza A viruses. *Microbiological Reviews*, **56**, 152–179.

Also check Chapter 21 for references specific to each family of viruses.

19 HIV and AIDS

During the past 20 years human immunodeficiency virus (HIV) type 1 has become a common infection of mankind. There is a second, less common but closely related, virus, HIV-2. Both predominantly infect helper T lymphocytes and monocytes, using as a receptor the CD4 protein which they carry on their surface. After a long, essentially symptomless, incubation period of, on average, 8 years, $CD4^+$ lymphocytes decline to such a low level that the immune system can no longer function efficiently, thus resulting in the immunodeficiency which gives the virus its name. The result is that the affected person can no longer restrain certain normally harmless passenger micro-organisms (viruses, bacteria or fungi), which then cause clinically overt disease. A collection of diseases such as this, unrelated except for a common underlying cause, is called a syndrome — hence the name 'acquired immune deficiency syndrome' (AIDS). It is almost always lethal and this makes HIV a very unusual virus (see p. 218).

The HIV/AIDS scenario is tragically grim. The World Health Organization (WHO) estimates that HIV has already infected about 10 million people world-wide (Table 19.1). The projected number of *new* infections each year is shown in Fig. 19.1 . According to this, the total number of infections in Africa peaked in 1992, but there will continue to be one million *new* infections every year until 1995. Asian infections show a staggering linear increase which shows no decline. There are no cure and no vaccine; treatment (with 3'-azido-2',3'-dideoxythymidine (AZT)) (see below) merely delays disease and is prohibitively expensive for all but the richest countries. WHO estimates that, by the year 2000, there will be around 40 000 000 infected people world-wide.

This chapter discusses the immense progress that has been made in understanding HIV and the reasons why modern science has not, at the time of writing (December 1993), come up with an effective countermeasure. The reader should appreciate that this field is moving with immense rapidity and should consult the contemporary literature for an update.

THE BIOLOGY OF HIV-1 INFECTION

What sort of virus is HIV-1?

HIV-1 is a typical member of the lentivirus subgroup of the Retroviridae (p. 349). The name is derived from the latin *lente*, which refers to the *slow* onset of the disease. However, the rate of viral multiplication is normal. There are several well-characterized lentiviruses (Table 19.2), infecting a number of different vertebrate

288

Table 19.1 Estimates of the total number of HIV-1 infected people present in some geographical areas in 1992.

UK	18 000
Far East	1 000 000
New York city, USA	>200 000
USA	2 000 000
Sub-Saharan Africa	7 000 000
World-wide	10 000 000*

* Predicted by WHO to rise to about 40 000 000 by the year 2000.

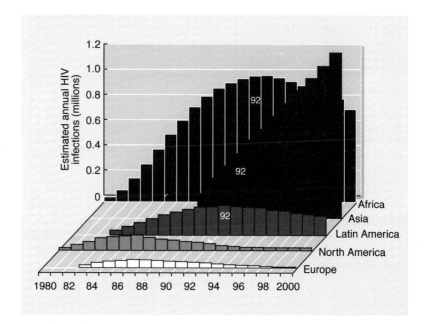

Fig. 19.1 WHO projection of annual new infections with HIV.

species. Common features of lentivirus infections are listed in Table 19.3. HIV-1 causes over 99% of human infections, but the serologically distinct HIV-2 is also on the increase. HIV-2 is more closely related to simian immunodeficiency virus (SIV) than to HIV-1 (Fig. 19.2).

Historically, the disease was recognized before the virus. In 1981, astute clinicians noted that a very rare cancer, Kaposi's sarcoma, was appearing in young homosexual men in New York and that this was associated with immune deficiency. An association with a variety of common infections, leading to death, was then recognized, and AIDS was defined. Various causes were considered and it was not until 1984 that it was thought likely that AIDS was caused by infection with a virus, and in 1985 a virus

Table 19.2 Some typical members of the lentivirus subfamily and their hosts.

Virus	Host	Main target cell	Clinical outcome
Visna–maedi	Sheep	Macrophage	Pneumonia (= maedi) or chronic demyelinating paralysis (= visna), little immunosuppression*
Visna–maedi	Goats	Mammary macrophage	Arthritis, rarely encephalitis, little immunosuppression
Equine infectious anaemia	Horses	Macrophage	Recurrent fever, anaemia, weight loss, little immunosuppression
Bovine immunodeficiency	Cattle	?	Weakness, poor health
Feline immunodeficiency	Cats	CD4$^+$ T cell	AIDS
Simian immunodeficiency	Monkeys	CD4$^+$ T cell	AIDS†
Human immunodeficiency type 1	Humans	CD4$^+$ T cell	AIDS
Human immunodeficiency type 2	Humans	CD4$^+$ T cell	AIDS

* The same virus can cause two different diseases in sheep.
† Different SIV strains are named after the species of monkey from which they were first isolated, e.g. SIV$_{man}$ from the mandrill, SIV$_{agm}$ from the African green monkey, SIV$_{sm}$ from the sooty mangabey. These are all African monkeys, and SIV strains cause no disease in their natural host. However, although it does not naturally infect Asian macaque monkeys, SIV$_{sm}$ does do so under experimental conditions and causes AIDS. SIV$_{mac}$ is thought to be SIV$_{sm}$ which accidently infected a laboratory macaque.

Table 19.3 Common features of lentivirus infections.

Prolonged subclinical infection
Weak neutralizing antibody responses
Persistent viraemia
Continuous virus mutation and antigenic drift
Neuropathology
Infection of bone-marrow-derived cells

new to science, HIV-1 (then also known as human T cell lympho-tropic virus type 3 (HTLV-3) or lymphocyte-associated virus (LAV)) was isolated (Table 19.4). Once viral antigens were available as diagnostic reagents, stored sera could be tested retrospectively for the presence of antibody to HIV and it was demonstrated that there were few HIV infections in humans prior to 1970. However, since we now know that the incubation period averages 8 years, it is clear that HIV had spread in an explosive but silent pandemic throughout that decade.

Where did it all start?

The history of HIV is surprisingly sketchy, despite much effort to

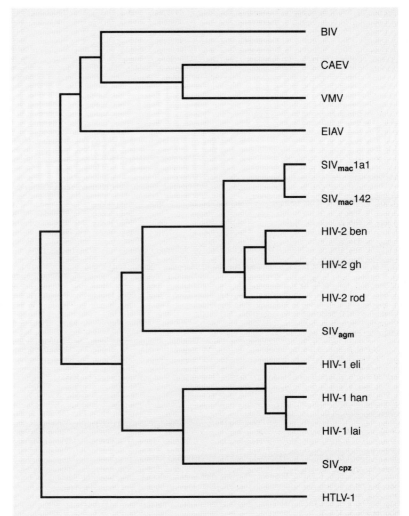

Fig. 19.2 Dendrogram showing the relatedness of different lentiviruses. BIV: bovine immunodeficiency virus; CAEV: caprine arthritis-encephalitis virus; VMV: visna maedi virus; EIAV: equine infectious anaemia virus; SIV: various strains of simian immunodeficiency virus; HIV: various strains of human immunodeficiency virus type 1 and 2; HTLV-1: human T cell lymphotropic virus type 1. (Courtesy of Georg Weiller.)

Table 19.4 Milestones for AIDS and HIV.

1970s	Silent pandemic
1981	AIDS is recognized for the first time
1983	AIDS linked to infection with a 'new' virus
1984	Initial characterization of HIV-1

trace its origins. The earliest HIV-1 infections can be traced by serology to central parts of sub-Saharan Africa. Either it is a long-established but rare infection or a new infection of humans by a virus which has jumped from another species. The latter fits with the generalization that only new infections, not yet in balance with their hosts, cause serious disease. However, no such close relative is known, the nearest being an SIV strain isolated from a chimpanzee in the wild (Fig. 19.2). In any event, the factors which caused the virus to spread, firstly into urban areas in Africa and from there to the USA and to the rest of the world, are not known. There are many gaps in the story.

Molecular biology of HIV-1

Like all retroviruses, HIV-1 is diploid and contains two molecules of single-stranded plus-sense ribonucleic acid (RNA). Its genome conforms to the basic oncovirus plan, described in Chapter 9, but is complicated by containing a number of additional regulatory genes (Fig. 19.3). Both have the basic gene order 5′−gag−pol−env−3′ and are flanked by the characteristic long terminal repeats (LTRs). Both synthesize a single deoxyribonucleic acid (DNA) molecule, which is integrated into the host genome. From this, the simpler oncornavirus synthesizes only two RNAs, the viral genome, which doubles as messenger RNA (mRNA) for the gag (group *antigen*) and polymerase (reverse transcriptase) proteins, and an mRNA with a single splice, which encodes the envelope protein. (Rous sarcoma virus is a unique oncovirus in having an additional *src* oncogene and in being non-defective. It has an additional *src* mRNA (p. 166)). Thus, this very simple virus has proteins which have both structural and enzymatic functions. Gene expression is controlled by *cis*-acting (i.e. self) DNA or RNA sequences and by host-coded factors acting in *trans*. HIV-1 imposes upon this basic plan a number of proteins (tat, rev, vif, vpr, vpu and nef) with positive or negative regulatory activity. Cell transcription factors result in the synthesis in a small amount of full-length viral RNA from integrated HIV-1 DNA, which is then multiply spliced to form small mRNAs encoding tat and rev. Tat and rev are RNA-binding proteins. Tat binds to the LTR and up-regulates transcription. This increases rev to a critical level that down-regulates the synthesis of multiply spliced RNAs and favours the production of non-spliced or singly

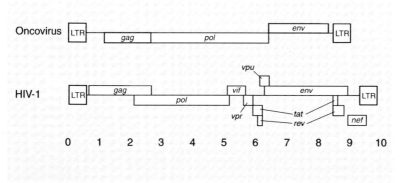

Fig. **19.3** Comparison of the structure of the genomes of a non *src*-containing oncovirus and HIV-1. The scale indicates the genome size in kilobases. Genes not aligned require a frame shift for expression; *tat* and *rev* are expressed from spliced RNAs.

spliced mRNAs, which encode structural proteins (Fig. 19.4). Nef is a negative regulator and reduces production of virus, while vpr and vpu are concerned in morphogenesis of virions and their release from the cell; vif is necessary for virus infectivity. There is much more to be learned about all of these.

Production of the structural proteins from the *gag, pol* and *env* genes and their location in the virion are shown in Fig. 19.5. All are derived from polyprotein precursors cleaved by the virion protease, itself a product of *gag*. The distal part of the envelope or spike protein, glycoprotein 120 (gp120), is not covalently linked to the gp41 anchor and is easily detached. The spike is a dimer or a tetramer and there are 72 spikes per virion. (It is interesting to compare this with influenza virus, which has 10-fold more spikes on a similar surface area.) On gp120 is the virus attachment site, a conformational site that recognizes the CD4 surface protein of T cells and initiates infection.

Infection and disease

CD4 is a protein present on the surface of helper T cells which normally recognizes and interacts with major histocompatibility complex (MHC) class II antigen on 'target' cells. It is the main (but not sole) receptor through which both HIV-1 and HIV-2 initiate infection. There are three ways in which people can contract an HIV infection.

Sexual transmission

Either infected CD4$^+$ T lymphocytes or free virus is transmitted

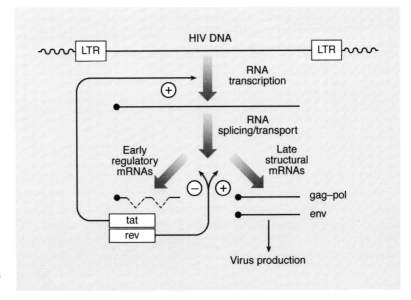

Fig. 19.4 Regulation of HIV-1 gene expression from DNA integrated into the cellular genome. This occurs in two phases: regulatory genes are expressed early and structural genes are expressed late. From Cullen (1991).

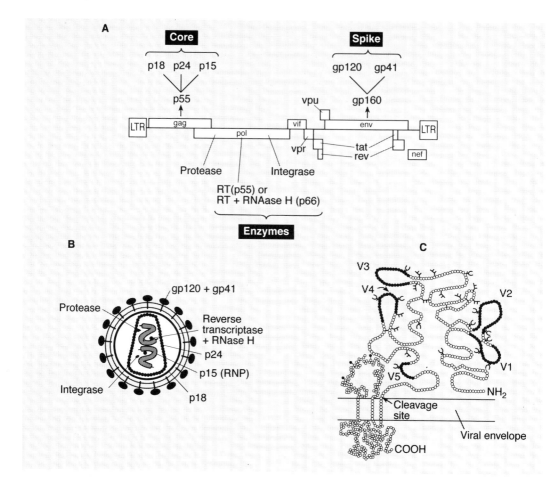

Fig. 19.5 (A) Production of the structural proteins of HIV-1. The nomenclature 'p18' indicates a polypeptide having an M_r of 18 000, etc. The reverse transcriptase holoenzyme is a heterodimer of p51 and p66, the latter having an extra COOH-terminal polypeptide (p15), which is ribonuclease (RNase) H. (B) Structure of the virion. (C) One domain of the tetrameric envelope spike protein adapted from the structures suggested by Modrow *et al.* (1987) and Leonard *et al.* (1990). gp160 is cleaved to form gp41, the COOH-terminal membrane anchor and gp120. gp41 crosses the viral envelope three times. gp120 contains five antigenically variable (V) regions and intervening conserved regions. gp120 is heavily glycosylated and the glycosylation sites are indicated by (Y) or by (↗) in gp120 and potential sites in gp41 by (♈). Intramolecular disulphide binds in gp120 are also shown.

during sexual activity. This is the main route of infection. HIV is spread equally well by heterosexual and by male homosexual activity, but is better spread by infected males than infected females. In Africa, HIV is spread predominantly by heterosexual contact; in Western countries, homosexual contacts have been responsible for the majority of infections, but now the greatest percentage increase in new infections is due to heterosexual activity. Bisexual activity provides a conduit between hetero- and homosexual people.

Transmission through infected blood

Injection with hypodermic needles contaminated with HIV provides a high risk of infection. The virus spreads quickly between injecting drug abusers who share unsterile equipment, and there is a similar risk with hospitals which do not have effective sterilization. Before the virus was recognized, many haemophiliacs were accidentally infected by being given clotting factor VIII prepared from blood contaminated with HIV. Heat sterilization and screening of blood donors have now eliminated this risk.

Vertical transmission

Babies born to HIV$^+$ mothers may be infected. Fortunately, there is only about a 20% chance of infection, but at this age AIDS develops with a shorter incubation period, around 2 years.

The course of infection

This is shown in outline in Fig. 19.6. Remember that an *average* incubation period to AIDS of 8 years means that some will develop disease earlier and some will remain healthy for much longer. Persistent generalized lymphadenopathy (PGL) and AIDS-related complex (ARC) are indicators that immune suppression has started

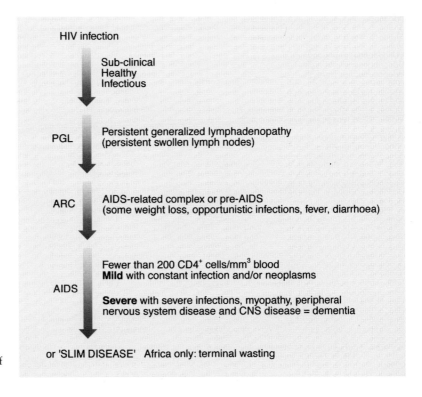

Fig. 19.6 The course of HIV infection.

HIV infection

Sub-clinical
Healthy
Infectious

PGL — Persistent generalized lymphadenopathy (persistent swollen lymph nodes)

ARC — AIDS-related complex or pre-AIDS (some weight loss, opportunistic infections, fever, diarrhoea)

Fewer than 200 CD4$^+$ cells/mm^3 blood
Mild with constant infection and/or neoplasms

AIDS

Severe with severe infections, myopathy, peripheral nervous system disease and CNS disease = dementia

or 'SLIM DISEASE' Africa only: terminal wasting

and is progressing. AIDS is now formally defined by the number of CD4$^+$ T cells in the circulation: in the early stage of infection, people are HIV$^+$ and have >500 CD4$^+$ T cells per mm^3 of blood (the normal range being 800–1200 per mm^3); in the intermediate stage, they have 200–500 cells per mm^3; and, in the late stage, they have <200 cells per mm^3. If an AIDS patient does not die from infection with adventitious micro-organisms (Fig. 19.6), HIV progresses to infect CD4$^-$ cells and causes disease in the muscles and in the peripheral and central nervous systems. 'AIDS dementia' or madness is the final stage. In Africa, AIDS frequently presents as a severe wasting condition, called euphemistically 'slim' disease. This is also a fatal condition. The time taken to progress to AIDS and the symptoms involved vary greatly between individuals. An example of one case-study and the progressive decline in CD4$^+$ T cells is shown in Fig. 19.7.

What cells are infected?

Although CD4$^+$ T cells are the main target, some other cells express the CD4 protein, albeit at a lower level, and are infectable. Yet others have no CD4 protein but are infected, which suggests that here the virus uses a different (and unknown) receptor molecule. In the central nervous system, it is thought that HIV infects microglial cells, which belong to the same cell lineage as the macrophage. Cells of the gut are important in infection resulting from anal intercourse.

During the early and intermediate stages of infection, there is little virus and few infected cells in the circulation. However, there are many more infected cells in lymphoid tissue. About 30%

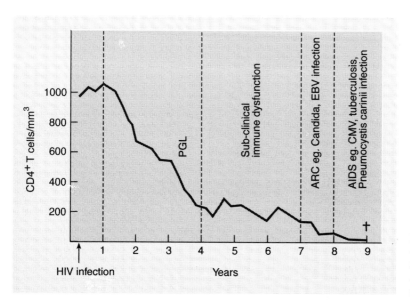

Fig. 19.7 A case-study of AIDS, which ended in death in year 9. Acronyms are explained in Fig. 19.6.

of the lymphoid CD4$^+$ cells are infected, but, as these contain viral DNA but not viral RNA, the infection at this time is largely latent. However, as 98% of all CD4$^+$ T cells are contained in lymphoid tissue, this means that an enormous number of cells are already infected and that eventual activation of the viral genome, with no further need of infection, could account for the observed loss of CD4$^+$ cells in the late stage of disease. In addition, virus is entrapped on the surface of follicular dendritic cells in lymphoid tissue. Destruction of the lymphoid organ accompanies or may be responsible for the late stage of disease and the increase of virus and infected cells in the circulation.

Immunological abnormalities

While loss of CD4$^+$ T cells is the major result of HIV infection, there are many other alterations to immune cells and functions, some of which are listed in Table 19.5. Those in the first category are always found in AIDS patients and are therefore diagnostic. Others are found regularly though not always, and those in the last category are only present occasionally.

What causes the immunological abnormalities?

Some or all of the following may happen (Table 19.6):
1 The virus is directly cytopathic.
2 Envelope protein in the plasma membrane of an infected cell attaches to CD4 molecules on non-infected CD4$^+$ T cells and causes the cell membranes to fuse and a syncytium to form. This is a lethal process and fused cells die.

Table 19.5 Immunological abnormalities in AIDS.

Diagnostic abnormalities
CD4$^+$ T cell deficiency
Reduction in levels of all lymphocytes
Lowered cutaneous DTH
Elevated immunoglobulin concentration in serum
Increased immunoglobulin secretion by individual cells

Regular abnormalities
Decreased proliferative responses
Decreased cytotoxic responses
Decreased response to new immunogens
Decreased monocyte function

Occasional abnormalities
Various other complications*

---

* Students should be aware that the effects of HIV on the immune system extend far beyond the CD4$^+$ T cell and vary greatly between individual patients.
DTH, delayed-type hypersensitivity, a T-cell-mediated reaction.

Table 19.6 Possible ways in which CD4$^+$ cells are eliminated during HIV infection.

Cytopathic action of HIV
Formation of syncytia between infected and non-infected CD4$^+$ cells
Recognition of infected cells by antibody and attack by complement and/or
   phagocytic cells
Attack on infected cells by cytotoxic T cells
Immune attack on *non-infected* CD4$^+$ cells which have bound envelope
   protein

---

3 Infected cells suffer no direct cytopathology but are recognized by antibody to envelope protein expressed in the plasma membrane, and are then attacked by complement or phagocytic cells bearing Fc receptors (Chapter 14).
4 Infected cells proteolytically process HIV proteins and present the resulting peptides on their external surface in conjunction with class I or II MHC antigens. These cells are then attacked by CD8$^+$ or CD4$^+$ cytotoxic T cells (Chapter 14).
5 Any non-infectious virus or free envelope protein which is released from infected cells can attach to non-infected CD4$^+$ cells and render those cells liable to immune attack, either by antibody or, if the antigen is processed and presented by MHC proteins, by cytotoxic T cells, as described above.

Why is the incubation period of AIDS so long?

The onset of AIDS is defined by CD4$^+$ T cells dropping below a concentration of 200 cells per mm^3 of blood. Thus the question can be rephrased in terms of the time taken — 8 years on average — to reduce CD4$^+$ cells to this level. It is clear that this is *not* due to viral latency, as an HIV$^+$ person is probably infectious at all times. Figure 19.8 shows a simplified version of the relationship between the virus concentration in blood and the immune response to HIV. After the initial infection, the immune response increases and virus is present at a low level. At the onset of AIDS, there is a dramatic reversal, coincident with a decline in the HIV-specific immune response. There is a general immune deficiency as well (Table 19.5).

*Hypothesis 1: activation of virus and death of CD4$^+$ cells
is limited by the rate at which CD4$^+$ cell clones
are stimulated by cognate antigen*

Important here is the fact that HIV only multiplies in dividing CD4$^+$ T cells. Now, in nature, T cells only divide when they are stimulated by the antigen which their unique T cell receptor recognizes (termed the cognate antigen), and many T cells probably never divide as, in our lifetime, we simply do not ever meet the

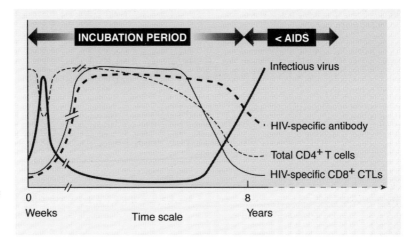

Fig. 19.8 The relationship between infectious virus and the immune response to HIV. CTLs, cytotoxic T lymphocytes.

relevant cognate antigens. For example, HIV will be latent in a T cell specific for antigen *x*, until the cell is stimulated to divide by contact with antigen *x*. Antigen *x*-specific cells will then produce a burst of virus and then die by one of the mechanisms already discussed. Many more CD4$^+$ cells will be infected by the new progeny HIV and the cycle repeats itself. Thus, the time taken to reduce the number of CD4$^+$ cells to the level required for the definition of AIDS depends on the time it takes to come into contact with enough antigens to react with all those T cells. Taking this hypothesis to logical absurdity, if an HIV$^+$ person met no CD4$^+$ cell-specific antigens, he/she would not develop AIDS.

Hypothesis 2: autoimmunity

The envelope protein of HIV-1 has a sequence homology with the MHC class II protein (and class I as well). Normally, the immune system recognizes MHC antigens as self and mounts no immune response to them or to the epitope which is mimicked by HIV. The idea here is that the long incubation period reflects the number of years it takes for the self-tolerance mechanism to be broken and the immune system to respond to the MHC class II-like epitope on the HIV-1 envelope. Once that response is in place, the immune response will then react with the cross-reactive MHC class II epitope, and cells bearing this antigen are then killed. CD4$^+$ T cells, CD8$^+$ T cells and B cells all express MHC class II proteins constitutively, while other cells only express them after exposure to γ-interferon (Chapter 14). This hypothesis assumes that, for some unknown reason, CD4$^+$ cells are the most vulnerable.

*Hypothesis 3: evolution of the envelope protein
in the infected individual*

Sequencing of the gene encoding the envelope protein from isolates of HIV from different infected individuals shows that it can be divided into variable and constant regions (Fig. 19.5C). These isolates also differ antigenically. Of particular importance is that the envelope protein evolves in infected *individuals*, and this is probably driven by the immune response, in a manner similar to antigenic drift in influenza virus (p. 280). This phenomenon of evolution during the lifetime of an infected individual is also found with other lentiviruses. One view is that the long incubation period is determined by the interplay between the immune system and the envelope protein, and that the virus eventually gains the upper hand when it evolves a variant to which the immune system cannot respond. (This evolutionary plasticity of the envelope protein also gives rise to problems in regard to vaccines, which will be discussed later.)

Death and AIDS

The key failure in immune responsiveness is the loss of the helper function of the CD4$^+$ lymphocyte. This cell is the pivotal part of the immune system, as most antibody responses are helper T cell-dependent and helper cells also assist in the maturation of T cell effectors, such as cytotoxic T cells, and can have cytotoxic activity in their own right. One of the functions of T cells is to control a variety of micro-organisms (viruses, bacteria, fungi, protozoa; Table 19.7), a selection of which is carried by all individuals. The immune system is normally unable to evict these micro-organisms but ensures that they remain subclinical and harmless. However, when the T cells are compromised by HIV infection, the brakes are released and the micro-organisms multiply at full-speed. Infections are much more severe and prolonged than normal because of the immunosuppression. Different micro-organisms are activated in AIDS patients in different parts of the world. Tuberculosis is a common complication in Third World countries. This period of chronic adventitious infection may last for several years but eventually becomes overwhelming and death ensues.

PREVENTION AND CONTROL

No virus has been subjected to such intense scrutiny as HIV but,

Table 19.7 Common infections associated with AIDS.

Viral	CMV, EBV	Generalized infection
Bacterial	*Mycobacterium tuberculosis*	Tuberculosis
Fungal	*Candida albicans*	Thrush
	Pneumocystis carinii (a yeast-like organism)	Pneumonia

CMV, cytomegalovirus; EBV, Epstein–Barr virus.

as yet, there is little sign of any effective vaccine. At first sight, this is surprising, since the smallpox vaccine has been in use for nearly 200 years and in the last 30 years several very effective vaccines have been devised and used. The problem lies in the fact that a micro-organism is only sensitive to one particular arm of the immune system (for example, it might be the $CD8^+$ cytotoxic T cell), and other parts of the immune system are ineffective. A different micro-organism might be sensitive to a different part of the immune system. The rules governing immunity are not understood. Thus, all vaccines are made empirically. By definition, a successful vaccine stimulates the required immune response, along with many other irrelevant responses. However, we do not know which element of the immune response is required for protection in any disease or how to stimulate it to order. Further, there are a number of poor vaccines which have defied attempts to improve them (e.g. against cholera and typhoid). HIV is simply an extreme case. We need to know much more about the immune system in general before the rational design of vaccines becomes a reality. That day cannot come soon enough for HIV. The following sections deal initially with the vaccine problem and then discuss chemotherapy.

Prevention of HIV infection

In the absence of any vaccine, avoiding contact with HIV is paramount (Table 19.8). The main risk, already discussed, is through sexual contact. Education programmes have been promulgated world-wide, with advice to have as few sexual partners as possible, to practise safe sex, by using condoms as a barrier to infection, and to avoid anal intercourse, as rectal tissues are more vulnerable to physical damage and infection than those of the vagina. However, there are immense social problems in dealing with any sexually transmitted disease. Historical parallels are not encouraging, as it is well documented that many were undeterred by the risk of contracting syphilis at a time when no treatment was available. The involvement of male and female prostitutes is a particular problem, as, in some areas of the world, a high proportion are infected with HIV.

Table 19.8 Control measures for HIV and AIDS.

Measure	Comment
Avoidance of infection	Available: but how well is it practised?
Diagnosis and detection	Good
Prevention by vaccine	Not available
Chemotherapy	AZT: retards the onset and progress of AIDS; *not* a cure

Diagnosis and detection

Control of any disease needs quick, cheap and reliable methods for detecting signs of infection. These are in place. Possession of *antibody* to the HIV internal antigen, p24, is the main diagnostic criterion, since this provokes the most reliable and strongest antibody response. This test decides if a person is 'HIV-positive'. It is used as a 'dip-stick' in an enzyme-linked immunosorbent assay (ELISA) format (Fig. 19.9). The test is not carried out until 2 months after the suspected infection, to allow antibody to form and avoid a falsely negative diagnosis. With babies born to HIV$^+$ mothers, a period of 3−6 months has to elapse to allow the decay of maternal HIV antibody in the infant's circulation. Only 20% of such babies are infected and make their own antibody.

Virus can be isolated on lines of CD4$^+$ cells grown in the laboratory. This is a slow and labour-intensive process and has the intrinsic problem of all such techniques, that of selecting variants from the infected person which happen to grow well in culture but are unrepresentative of the original virus population. This can be circumvented by use of the polymerase chain reaction (PCR) directly on viral genomes obtained from a patient. However, PCR amplifies efficiently only a few hundred nucleotides, so regions to be amplified have to be selected with great care. Amplified DNA can be sequenced directly to give accurate information on the virus present in the body. However, HIV produces many defective genomes, which are an evolutionary dead-end. PCR does not distinguish between these

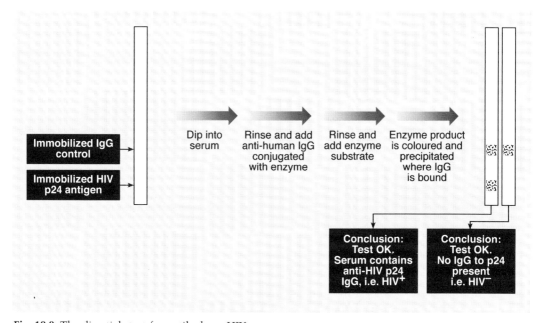

Fig. 19.9 The dip-stick test for antibody to HIV.

and infectious genomes, and the significance of sequence variation, in terms of the evolution of infectious virus, is difficult to evaluate.

Vaccines

Animal model systems

Since it is ethically not possible to test the efficacy of a vaccine or treatment by infecting people with a potentially lethal agent, it is necessary to have a model system. These are limited, as only primates have a CD4 protein sufficiently closely related to the human CD4 protein to permit infection with HIV, but no primate infected with HIV-1 progresses to AIDS (Table 19.9). The use of higher primates, such as chimpanzees, is fraught with many problems, and, aside from ethics, they are very expensive to maintain and there will never be enough to do statistically meaningful experiments.

A major breakthrough came when immunologists decided to reconstitute mice which were congenitally deficient in both B and T cells (known as severe combined immunodeficient (SCID) mice) with human lymphocytes. By conventional immunological wisdom, this experiment was a waste of time, as the transplanted cells should have mounted an immune response against mouse antigens in a classic 'graft-versus-host' reaction. For some reason, not understood, this did not happen and it is now possible to study infection of human lymphocytes and their reaction to experimental vaccines. However, human T cells do not repopulate the mice in their normal proportions and the mice do not develop AIDS.

The alternative approach is to study other non-human immunodeficiency lentivirus–host systems (Table 19.9) and hope to apply lessons learned from these to HIV infections.

Table 19.9 Some animal models for HIV infection and AIDS, and their limitations.

Virus	Animal	Comment
HIV-1, 2	Chimpanzee	Infection but no AIDS; used to test vaccines
HIV-2	Rhesus monkey	Infection and lymphadenopathy – similar to early stages of AIDS
SIV	Macaque monkey	Infection and AIDS
FIV	Cats	Infection and AIDS
HIV-1	SCID mouse + human lymphocytes*	Infection, no AIDS

* SCID, severe combined immunodeficiency: mice which, in their homozygous form, lack both B and T cell responses. These can be reconstituted in part with human lymphocytes (see text).
FIV, feline immunodeficiency virus; SIV, simian immunodeficiency virus.

Immunity to HIV

Infected people and infected animals mount both antibody and T
cell responses which are antiviral, and the appropriate laboratory
assays can demonstrate the presence of anti-HIV neutralizing antibody and cytotoxic T cells. Why, then, do HIV$^+$ people: (i) not
clear the virus; and (ii) go on to develop AIDS? The first question
has many precedents, e.g. herpes simplex and hepatitis B viruses
(already discussed, pp. 219–20), and these viruses have a variety of
ways of evading the immune response. The principle to remember
is that antibody acts only on antigens and T cells act only on
antigen fragments exposed on the outside of cells. However,
despite the presence of antibody or T cells, HIV spreads from cell
to cell. As a virus particle buds from the cell surface, there is a race
between the rate of entry into the safe haven of another CD4$^+$ cell
and the rate of neutralization by antibody. It will depend on a
number of factors, such as a very close association between
infected and non-infected CD4$^+$ cells which would exclude
antibody, and the concentration and affinity of antibody. The
affinity of the envelope protein for the CD4 antigen may be higher
than its affinity for anti-envelope antibody.

   All this discussion refers to people who are already infected. The
aim of immunization is to put in place those immune responses
which will prevent HIV from initiating infection. Regrettably, this
has not so far proved to be possible.

Failure of experimental vaccines to protect against HIV infection

Production of a vaccine is one of the major goals of the world-wide
campaign to prevent AIDS, and the amount of scientific time and
money spent on achieving this goal cannot be criticized. Work has
concentrated on the envelope protein and the best way of presenting
it to the immune system. While is has not been considered practi-
cable, on safety grounds, to develop a live vaccine (how could you
test that it no longer caused AIDS in humans?), many of the other
types of vaccine discussed in Chapter 16 have been investigated:
inactivated whole virus, purified viral protein, recombinant protein,
recombinant protein expressed in a vaccinia virus vector and peptides. All stimulate some part of the immune system, but none
appear to stimulate an immunity — or enough of it — which gives
protection. It is putting it mildly to say that the scientific world is
surprised by this *universal* lack of success — it was expected that at
least one of the preparations would have been protective. One
glimmer of hope comes from work on passive immunity. In these
experiments, chimpanzees were injected with neutralizing monoclonal antibody specific for the gp120 of HIV-1 before being inoculated with infectious virus. They did not develop an infection and
remained free of antiviral immune responses, viral antigens and

viral genomes. The half-life of immunoglobulin G (IgG) (2–30 days, depending on isotype) makes this an unrealistic preventive measure, even using humanized antibodies (p. 329), but it was an important step forward, as it shows that, in principle, antibodies are able to provide protection.

Today, work proceeds on two fronts. The first is empirical and investigates the almost infinite permutation of immunization protocols in terms of type of vaccine, amount inoculated, site(s) of inoculation, adjuvant, number of doses and the interval between them. None of the animal models is ideal, and, while the chimpanzee-HIV-1 model is closest to the human situation, there are not enough animals to carry out preliminary testing of vaccines; however, the SCID mouse reconstituted with human lymphocytes is proving to be a valuable resource. The second front is directed at achieving a fundamental understanding of the immune system. This is a huge endeavour and inevitably slow, but will in the end provide the information necessary to devise an anti-HIV vaccine.

Chemotherapy

Chemotherapy of HIV infections has three aims: (i) to clear the infection; (ii) to prevent the onset of clinical disease (pre-AIDS); and (iii) to prevent the progression of disease to full-blown AIDS by facilitating the recovery of the $CD4^+$ T cell population. The first and second objectives have not been achieved and there has been only limited success with the third. The best treatment, to date, at best only retards the inevitable slide towards death.

The main drug in use is AZT, known commercially as Zidovudine or Retrovir (Fig. 19.10). Its antiretrovirus properties were discovered before the HIV emergency, but at that time there was no known human retrovirus infection. It is an analogue of thymidine and is incorporated by reverse transcriptase into viral DNA. Because it lacks a 3' OH group, no further nucleotide can be added and DNA synthesis ceases. It is not an ideal drug, as it is toxic and expensive and does not prevent death from AIDS. There have been a number of double-blind clinical trials, both in Europe and in the USA, in which HIV^+ people were given either AZT or a placebo. All trials showed some benefit from AZT treatment and were terminated prematurely on ethical grounds, when it became apparent that those receiving AZT were developing AIDS more slowly than those being given the placebo. (The latter were then offered AZT.) However, because the trials were not finished, the *long-term* benefit of AZT treatment is not known. Another trial evaluated the efficacy of different doses of AZT. This was particularly important, as the dose of AZT which was previously deemed necessary (1.5 g per day) produced so many toxic side-effects in some individuals that treatment could not be continued. Table 19.10 shows both the success and the failure of this treatment: limited success, because

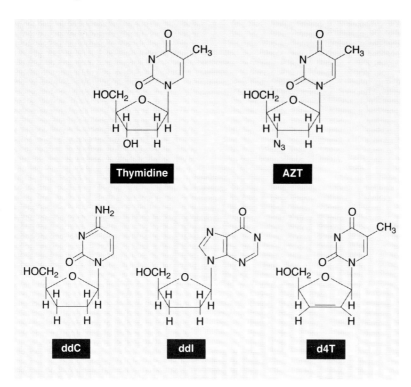

Fig. 19.10 The formulae of AZT (Retrovir or Zidovudine or 3'-azido-2',3'-dideoxythymidine), thymidine and other inhibitors of HIV DNA synthesis. All require phosphorylation by cellular enzymes before being incorporated into DNA.

Table 19.10 Results of a clinical trial to evaluate different doses of AZT*.

	Placebo	AZT	
		High dose (1500 mg/day)	Low dose (500 mg/day)
No. of AIDS cases when the trial was terminated†	38	19	17

* Approximately 430 people per group; all HIV⁺; none with AIDS.
† The trial was terminated prematurely when the benefit of AZT was apparent and patients on placebo were then offered AZT.

the low-dose treatment was apparently effective in halving the rate of progression to AIDS; failure, because it did not protect everyone. The lower dose had advantages in reducing both toxicity and cost. There was a further advantage in treating pre-AIDS patients rather than waiting until disease appears, as they tolerate AZT better. It was recommended that treatment with AZT should start when the concentration of CD4⁺ T cells in blood falls significantly. Now, however, the very large Anglo-French 'Concorde' trial, which used groups of 870 HIV⁺ pre-AIDS people, reported in 1993 that there was no difference in the numbers or rate of progression to clinically apparent AIDS over a period of 3 years, despite AZT increasing the numbers of circulating CD4⁺ T cells.

In summary, the main conclusions from clinical trials are that:

(i) AZT does not influence progression to clinical AIDS; (ii) AZT is of benefit only to patients who have developed AIDS (it slows progression of disease, prolongs life and ameliorates neurological disease); and (iii) AZT does not prevent death. The finding that CD4$^+$ T cell counts improve without affecting progression to AIDS means that this is not always an accurate indicator of disease. This may be because the majority (98%) of CD4$^+$ T cells are in lymphoid tissue and not in circulation. There is hope that use of two or more nucleoside analogues in combination may be more effective, and these are currently being tested in clinical trials. All these compounds have problems of toxicity. Finally, there are financial problems over AZT treatment, as it is intrinsically expensive and long-term (see below), and problems over the emergence of AZT-resistant mutant strains of HIV.

The future of chemotherapy

There is an intensive search to find other antiviral compounds. New nucleoside analogues are being synthesized and tested, and candidate drugs are emerging. Notable amongst these are ddC (2′,3′-dideoxycytidine), ddI (2′,3′-dideoxyinosine) and d4T (2′,3′-dideoxy-2′,3′−dehydrothymidine) (Fig. 19.10). Like AZT, all these lack a 3′ OH group and prevent elongation of DNA. While they seem no more effective than AZT on its own, the hope is that combinations of drugs will be more effective. Also, a second antiviral is needed should AZT-resistant strains, which do indeed emerge after treatment, prove to be a problem. Dual drug therapy will reduce the rate at which resistant strains appear, as the frequency with which one emerges is the *product* of the individual rates of mutation.

Reverse transcriptase itself is an obvious target, but it has taken until 1993 to obtain crystals and an X-ray analysis of its structure. There is a crystal structure of the HIV-encoded protease, and attempts are being made to design inhibitors based on this knowledge. Attempts to crystallize the envelope protein have so far failed. There is great interest, too, in the regulatory elements discussed earlier (pp. 292−3), with hopes that it might be possible, for example, to increase the effect of nef and shut down virus multiplication.

Another approach is to use soluble CD4 to either block attachment of virus to CD4$^+$ cells (Fig. 19.11) or as a 'magic bullet' (Fig. 19.12). The membrane-anchoring sequence is removed from cloned CD4 DNA, which is then inserted into an appropriate expression vector. The expressed soluble CD4 protein binds to the viral attachment site on envelope proteins (there are about 70 per virion) and prevents attachment of the virus to CD4$^+$ cells. This works well in cell culture and, surprisingly, in view of its normal role in binding to MHC class II molecules, soluble CD4 is not toxic when injected into people. However, there is no activity *in vivo*.

A second use of soluble CD4 is to deliver a toxin specifically to

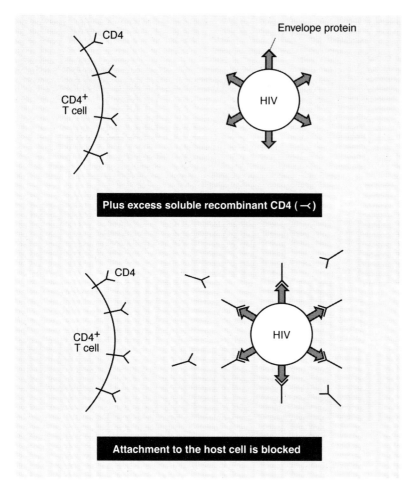

Fig. 19.11 Soluble CD4 blocks attachment of HIV to the CD4 receptor on susceptible cells, but is not effective *in vivo*.

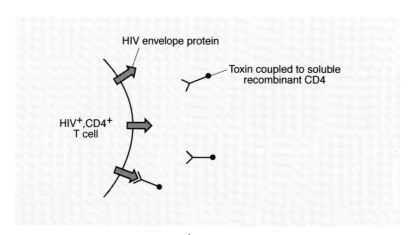

Fig. 19.12 'Magic bullet' treatment of HIV-infected CD4+ cells which express HIV envelope protein. The soluble CD4–toxin complex enters the cell and kills it.

cells expressing HIV proteins in their plasma membrane (Fig. 19.12). Ricin, obtained from the castor oil plant, is an enzyme which cleaves ribosomal RNA and prevents protein synthesis; it is the ultimate toxin, as it takes only one molecule to kill a cell. Although, surprisingly, soluble CD4−ricin does not appear toxic for cells bearing its natural ligand, the MHC class II protein, it would be risky to inject such a powerful toxin into people. Instead, white cells would be obtained from the blood of a patient and treated with soluble CD4−ricin, before being washed and returned, as is done in the treatment of leukaemia.

THE COST OF HIV INFECTION

Every country has a finite amount to spend on health care, and AIDS is making substantial inroads into this budget at a time when health care costs are generally rising. Cost is a major issue in any chronic disease where treatment has to be provided over a number of years. In the USA, even low-dose therapy with AZT costs around $4000 per year and this, together with associated medical and hospital costs, is a major strain on the health budget. Compare this figure with the $5 per person per year, which is the *total* amount allocated for spending on health in many African countries. In Africa there are estimated to be five million people who are HIV$^+$ and they are not being treated.

The cost in human terms of any AIDS case is incalculable. Because of the high incidence of infection, the immense loss of life in Central and East Africa will be devastating to the countries of those regions. Demographic changes are predicted in a manner not seen since the time of the Black Death in Europe in the fourteenth century. There is every indication that HIV is now advancing through Asia on a scale similar to that in Africa.

FURTHER READING

Aldrovandi, G. M., Feuer, G., Gao, L., Jamieson, B., Kristeva, M., Chen, I. S. Y. & Zack, J. A. (1993) The SCID mouse as a model for HIV-1 infection. *Nature (London)*, **363**, 732−736.

Berzofsky, J. & Cease, K. B. (1994) Vaccine strategies with reference to HIV. *Annual Review of Immunology*, **12**.

Cullen, B. R. (1991) Human immunodeficiency virus as a prototypic retrovirus. *Journal of Virology*, **65**, 1053−1056.

Culliton, B. J. (1991) AIDS against the rest of the world. *Nature (London)*, **352**, 15.

Gallagher, R. B. (ed.) (1991) *HIV and the immune system*. Cambridge: Elsevier Science Publishers.

Garcia-Blanco, M. A. & Cullen, B. R. (1991) Molecular basis of latency in pathogenic human viruses. *Science*, **254**, 815−820.

Gardner, M. B. (1991) Simian and feline immunodeficiency viruses: animal lentivirus models for evaluation of AIDS vaccines and antiviral agents. *Antiviral Research*, **15**, 267–286.

Greene, W. C. (1990) Regulation of HIV-1 gene expression. *Annual Review of Immunology*, **8**, 453–475.

Kaneshima, H., Baum, C., Chen, B., Namikawa, R., Outzen, H., Rabin, L., Tsukamoto, A. & McCune, J. M. (1990) Today's SCID-hu mouse. *Nature (London)*, **348**, 561–562.

Koff, W. C. (1991) Advances in AIDS vaccine development. *American Society for Microbiology News*, **57**, 73–76.

Leonard, C. K., Spellman, M. W., Riddle, L., Harris, R. J., Thomas, J. N. & Gregory, T. J. (1990) Assignment of intrachain disulfide bonds and characterization of potential glycosylation sites of the type 1 recombinant human immunodeficiency virus envelope protein (gp120) expressed in Chinese hamster ovary cells. *Journal of Biological Chemistry*, **265**, 10373–10382.

Levy, J.A. (1993) Pathogenesis of human immunodeficiency virus infection. *Microbiological Reviews*, **57**, 183–289.

McCune, J. M. (1991) HIV-1: the infective process *in vivo*. *Cell*, **64**, 351–363.

McLean, A. R. (1993) The balance of power between HIV and the immune system. *Trends in Microbiology*, **1**, 9–13.

Mitsuya, H. & Broder, S. (1987) Strategies for antiviral therapy in AIDS. *Nature (London)*, **325**, 773–778.

Modrow, S., Hahn, B. H., Shaw, G. M., Gallo, R. C., Wong-Staal, F. & Wolf, H. (1987) Computer assisted analysis of envelope protein sequences of seven human immunodeficiency virus isolates: prediction of antigenic epitopes in conserved and variable regions. *Journal of Virology*, **61**, 570–578.

Narayan, O. (ed.) (1992) The lentiviruses. *Seminars in Virology*, **3**, no. 3.

Sheppard, H. W. & Ascher, M. S. (1992) The natural history and pathogenesis of HIV infection. *Annual Review of Microbiology*, **42**, 533–564.

Stevenson, M., Bukrinsky, M. & Haggerty, S. (1992) HIV-1 replication and potential targets for intervention. *AIDS Research and Human Retroviruses*, **8**, 107–117.

Temin, H. M. & Bolognesi, D. P. (1993) AIDS: where has HIV been hiding? *Nature (London)*, **362**, 292–293.

Wain-Hobson, S. (1992) Human immunodeficiency virus type 1 quasispecies *in vivo* and *in vitro*. *Current Topics in Microbiology and Immunology*, **176**, 181–193.

Yarchoan, R., Mitsuya, H. & Broder, S. (1993) Challenges in the therapy of HIV infection. *Immunology Today*, **14**, 303–309.

20 Trends in virology

What will be the trends in virology in the near future? Clearly, the prevention of virus disease in humans, domestic animals and crops still remains the major aim of virologists. For animal diseases, the priorities are improvements in vaccines and the frustrating search for antiviral compounds. Our impotence to combat the human immunodeficiency virus (HIV) pandemic — apart from avoidance of infection — underlines the urgency of these tasks. Plant viruses are no less important, for there are serious virus diseases of virtually all crops, whether grown for fruit, vegetable, grain, beverage or fibre. Such diseases often do not kill the affected host but they can reduce its vigour, leading to reduced yield. There are interesting developments involving the insertion of disease 'resistance' genes through deoxyribonucleic acid (DNA) technology which extend the traditional breeding programmes to obtain virus-resistant cultivars. Meanwhile, the provision of virus-free seeds and plants through elimination of persistent infection remains an important weapon in the armoury of the grower. Recombinant DNA technology and monoclonal antibodies very rapidly fulfilled expectations and, as soon as these were in place, the polymerase chain reaction (PCR) came on stream. All continue to reach out into both applied and academic aspects of virology. We are continually surprised by the speed and precision with which these technologies are able to tackle a variety of fundamental and fascinating problems, particularly in allowing investigation of non-cultivable viruses. Other trends will see the application of modern methods to the analysis of virulence and this, no doubt, will feed back to help in the preparation of better preventive measures. There are grounds for concern about the pollution of our environment, particularly our water resources, with viruses, as water is drawn from major rivers and discharged back therein much enriched with human pathogenic viruses. Further, there is concern now that the appearance of new virus diseases is an inevitable consequence of evolution, replacing the mood of self-congratulation for the successful conquest of smallpox and containment of measles and poliomyelitis. The immense problem of HIV has already been discussed in the preceding chapter. Just as serious is the ever-present possibility of a new, possibly lethal pandemic of influenza.

GENETIC MANIPULATION OF VIRUSES

Every facet of virology is now wide open to investigation by the new recombinant DNA technology and we have already discussed data ranging from the identification of virus isolates to production of specific virus antigens for vaccines to the analysis of virulence.

Methods analogous to those employed for cloning DNA in *Escherichia coli* have been developed to propagate foreign DNA in eukaryotic cells. These involve simian virus type 40 (SV40) and a variety of retrovirus vectors for cloning in animal cells and cauli-flower mosaic virus in plants. Because the initiation site of SV40 is the only *cis* required function for DNA replication, a segment of viral DNA containing the origin of replication can replicate in the presence of a helper SV40 DNA molecule, which provides the necessary *trans* functions. Any foreign DNA that is covalently joined to the viral segment containing the origin should also be replicated in susceptible (monkey) cells. The drawback of SV40 is that the helper virus causes a lytic infection. No helper is needed for retrovirus vectors, as they integrate into the host genome. However, they maintain only a low copy number (about 10) and expression of the cloned gene is correspondingly low.

Major advances continue to be made in systems for expressing cloned genes. Here, it is necessary to provide the signals for tran-scription, initiation and termination to allow production of mess-enger ribonucleic acid (mRNA) (Fig. 20.1). For instance, the SV40 vector system has been modified by insertion of the promoter of the thymidine kinase (TK) gene of herpes simplex virus and a polyadenylation sequence. Termination of translation is ensured by using a synthetic oligodeoxynucleotide sequence which will control all three reading frames. Amongst retrovirus vector systems is the bovine papillomavirus (BPV). This is particularly useful as it is non-lytic and transforming. BPV DNA is maintained largely as an extrachromosomal plasmid in mouse cells. Shuttle vectors, so called because they can be propagated in both bacteria and eukaryotic cells, have been prepared between the bacterial plasmid pBR322 and BPV. These have been maintained in high copy number (up to 200 per cell). The advantage of the BPV vector is that it establishes a *permanent* cell line which efficiently expresses the cloned gene.

Factors controlling the efficiency of expression are poorly under-stood, and it came as a suprise when it was unexpectedly found that insect cells infected with insect baculoviruses p. 347 provided a very superior expression system. Recombinant baculoviruses con-taining an inserted influenza virus type A haemagglutinin (HA) gene have produced up to 5 mg HA/litre and with an inserted β-galactosidase gene 400 mg enzyme/litre. Unfortunately, baculo-viruses are still lytic, so only batch production is possible. None the less, baculoviruses currently outperform any other expression system, whether bacterial, yeast or higher animal cell, and the commercial potential is attracting industrial interest and finance.

Recombinant DNA technology also features highly in disease prevention and control. We have already discussed the new gener-ation of recombinant vaccines centred on vaccinia virus, in which relevant foreign antigens are expressed as if they were proteins

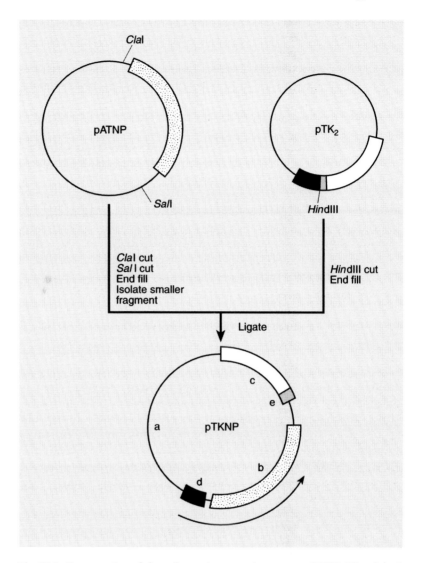

Fig. 20.1 Construction of the eukaryotic expression vector pTKNP. The cloned gene to be expressed (stippled box) was cloned in pBR322 to form pATNP. It was excised from pATNP using ClaI and SalI and end-filled using the Klenow fragment of *E. coli* DNA polymerase 1. This was then blunt-end ligated into the end-filled *Hind* III site of the eukaryotic expression vector pTK$_2$. The resulting expression vector pTKNP contains (a) a fragment of pBR322 (thin line) which carries the origin for DNA replication and the gene conferring resistance to ampicillin, (b) the cloned gene, (c) a fragment of the SV40 genome (open box) containing the polyadenylation signal for the SV40 early transcripts, (d) a fragment of the herpes simplex virus thymidine kinase (TK) gene (solid box), which contains the transcriptional promoter region but lacks the translation initiation codon, (e) a synthetic oligonucleotide containing translation stop codons in all three reading frames. The direction of transcription is indicated by the arrow.

native to the vaccinia virus (Fig. 16.8). There is considerable interest in veterinary circles, too, as pox viruses have representatives native to many animal species and offer the same advantages as a cheap multiple one-dose live vaccine as vaccinia virus does for man.

Genetic engineering of virus virulence is widely talked about and is aimed at either decreasing virulence to provide conventional live vaccines or increasing virulence in viruses for use as biological control agents, such as the insect baculoviruses (p. 349). Such procedures are entirely empirical at present, as we understand so little of the genetic basis of virus virulence. Virulence is usually a polygenic characteristic—not surprisingly, as virulence is the sum of the interactions of all properties of a virus with its host. For our own safety, we are fortunate that genetic manipulation usually decreases virulence, since it alters the delicate balance of various functions. For instance, insertion of foreign antigens into vaccinia virus decreases virulence (measured as plaque-forming units/lethal dose in the mouse eightfold, although the virus still multiplies successfully. None the less, it is a chilling thought that virus virulence could possibly be increased, and utmost care is needed to ensure that no 'Frankenstein's monster' virus is inadvertently created.

The deliberate genetic modification of RNA viruses has been greatly facilitated by the finding that infectious poliovirus could be obtained from recombinant DNA. In an imaginative leap, Vincent Racaniello and David Baltimore transfected monkey cells in culture with a bacterial plasmid containing a complete copy of the genome of poliovirus type 1. Because there were only prokaryotic transcription control signals, the plasmid was not expected to synthesize RNA in the eukaryotic cells. However, after 2 days, there were the typical cytopathic effects of poliovirus and subsequently authentic poliovirus particles were identified. To explain these unlikely events, it is surmised that there is random transcription of the plasmid and that, by a rare chance event, a complete poliovirus RNA is produced. If this were a normal mRNA, it would go undetected, but poliovirus RNA is infectious and is therefore able to replicate itself. Virus particles thus produced are self-sustaining.

From these beginnings, it is now possible to investigate the structure–function relationships in RNA viruses. For example, chimeric polioviruses can be created by cloning and exchanging segments of viral DNA by standard recombinant technology. By transferring under conditions of a plaque assay, viable recombinant viruses can be collected from the plaques that appear. Such a powerful selection procedure allows a very inefficient process to operate successfully (Fig. 20.2).

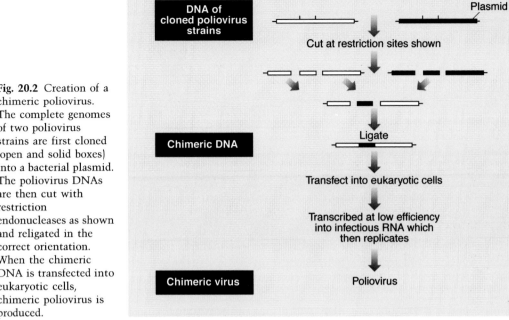

Fig. 20.2 Creation of a chimeric poliovirus. The complete genomes of two poliovirus strains are first cloned (open and solid boxes) into a bacterial plasmid. The poliovirus DNAs are then cut with restriction endonucleases as shown and religated in the correct orientation. When the chimeric DNA is transfected into eukaryotic cells, chimeric poliovirus is produced.

VIRULENCE – THE MAJOR UNSOLVED PROBLEM

There is greater ignorance of virulence than of any other area in the virology of eukaryotes. While we can describe what happens during a virus disease, we have very little notion of the molecular events that distinguish otherwise identical virulent and avirulent strains. There can be no doubt that, whenever their nucleic acids are compared by sequence analysis, differences will be found (see poliovirus below), but as there is more than one nucleotide change the essential problem is to correlate which of these controls virulence, and then to understand in molecular terms how the virulent virus strain interacts with its host to produce a violent end-result while the avirulent strain does not. The analysis is complicated since virulence is the end-product of the very complex reactions of the animal or plant host to the infectious agent and this, by its essence, makes experiments done in cultured cells irrelevant. This is a monumental problem but progress is slowly being made.

Reversion of poliovirus vaccine to virulence

The avirulent type 3 component of poliovirus vaccine is known to revert to virulence through mutation, with the result that there is a very, very low incidence of vaccine-associated paralytic poliomyelitis (one case per 10^8 doses of vaccine). Viruses isolated from such patients are neurovirulent when inoculated into monkeys in the test normally applied to ensure the safety of new batches of vaccine. The ability to clone and rapidly sequence the 7431 nucleotides of poliovirus has enabled the research groups headed by Jeffrey Almond at Leicester and Philip Minor in London to determine what changes in sequence correlate with the reacquisition of virulence.

In all, there are just seven nucleotide changes, one in each of the terminal non-coding regions and five in the coding regions; of the latter, one is silent (at position 6034) and four cause amino acid changes in structural proteins (Fig. 20.3A). Are all or only a few of these mutations responsible for reversion to virulence? Two approaches to answering this problem were used. Firstly, the wild-type strain of type 3 poliovirus 'Leon', from which the vaccine strain was derived by Albert Sabin, was cloned and sequenced (Fig. 20.3B). This gave the interesting information that only ten mutations had occurred in the attenuation process (which, incidentally, was achieved empirically by adaptation of the virus to growth in cell culture at 31°C). There were two mutations in the 5' non-coding region, one in the 3' non-coding region, three leading to amino acid changes and four silent mutations in the coding regions. However, the major point was that there was only a single nucleo-

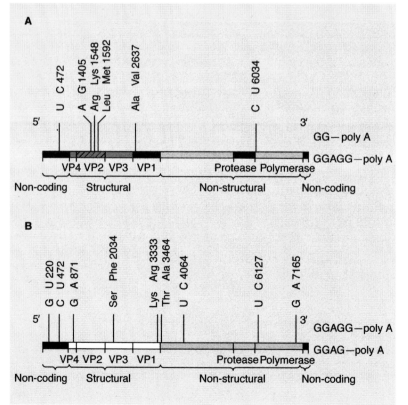

Fig. 20.3 Sequence changes occurring (A) when type 3 poliovirus vaccine strain 'Sabin' reverts to the neurovirulent 119 strain and (B) when the original wild-type neurovirulent Leon strain was attenuated to form the current vaccine.

tide change common to the acquisition of avirulence and subsequent reversion to virulence, a change at residue 472 in the 5' non-coding region from cytosine to uridine and back to cytosine. This was strongly backed up by the finding that the change at residue 472 occurred in *all* vaccine strains that had reverted to virulence.

This is exciting stuff, but the final story has not yet been told, as, firstly, residue 472 is in the non-coding region. Structural considerations indicate that, the change from cytosine to uridine at position 472 greatly alters the secondary structure of the viral RNA. How this affects virulence is not known. Secondly, the mutation at 472 increases but does not fully restore neurovirulence to that of the original Leon strain. It seems that other residue(s) contribute to virulence.

SUBTLE AND INSIDIOUS VIRUS–HOST INTERACTIONS

When it infects an animal, a virus is faced with a variety of differentiated cells which display a great range of structures, physiological functions and biochemical equipment. However, viruses attack and kill only certain cells and it is the specificity of this interaction which can be unexpected and interesting.

Destruction of specialized cells following infection

In Chapter 13, viruses are defined as lytic if they kill the cells which they have infected. However, *in vivo*, viruses only infect and kill cells of their target organ or target tissue. This can be exquisitely specific, as demonstrated by those viruses that have an affinity for the β cells of the pancreas, which produce insulin, the hormone responsible for the storage of glucose as glycogen. If the β cells are destroyed, the animal becomes diabetic. In humans, diabetes is common (occurring at the rate of $1-2/1000$) and has a number of different causes. However, the condition of 'juvenile diabetes' occurs suddenly in previously normal children, and it is tempting to speculate that virus infection of the β pancreatic cells is responsible. Unfortunately, several different viruses are suspected of being able, on occasion, to cause such effects (e.g. in animal models, representatives of the picornaviruses and reoviruses), so the prospects of reducing the incidence of diabetes by immunization are remote. In practice, it is found that repeated infections have an additive effect on the destruction of β cells and are necessary to produce clinically apparent diabetes.

Loss of cellular 'luxury' functions following infection

In Chapter 13, viruses were classified as lytic or non-lytic, with the tacit assumption that the latter exist peacefully with their host cell. Recent data show this to be an over-simplification, as viruses can alter the production of specialized cellular products (termed 'luxury' functions) while leaving the everyday 'house-keeping' functions of the cell unharmed. In one experimental system, Michael Oldstone and colleagues from La Jolla, California, demonstrated the ability of the arenavirus, lymphocytic chorio-meningitis virus (LCMV), to stunt the growth of new-born mice. Fluorescent antibody staining located viral antigen in many tissues, but, significantly, it was present in the pituitary, where growth hormone is synthesized. The pituitary showed no sign of pathology when examined microscopically, but a link between LCMV infec-tion and growth retardation was established when virus was found only in those pituitary cells that make growth hormone. Growth hormone has major effects on growth and glucose metabolism, and infected mice die prematurely of severe hypoglycaemia within 3 weeks after infection. Transplantation into infected mice of growth hormone-producing cells ensures both their survival and their normal development (Fig. 20.4). While it is not known how the virus causes the reduction of growth hormone, it seems more likely to be affecting its synthesis than its structure or function.

Armed with this preknowledge, it seems profitable to re-examine human conditions characterized by hormone, lymphokine or neurotransmitter deficiencies. Such cells would appear normal in all respects save for the loss of a 'luxury' function—luxury for the cell, maybe, but not for the unfortunate person, of course. Michael Oldstone has found that the LCMV can also switch off insulin without damaging the β cells of the pancreas.

Viruses and behavioural changes

There are obvious examples of viruses causing changes in the normal behaviour of infected animals and none more striking than

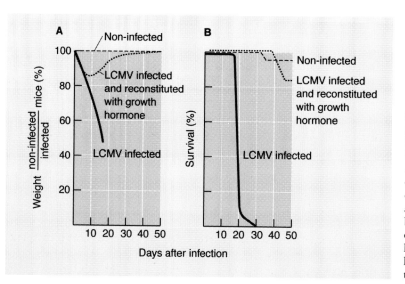

Fig. 20.4 Stunting of growth (A) and death (B) in new-born mice infected by the Armstrong 1371 strain of lymphocytic choriomeningitis virus, and reversal of the trend by transplantation of cells secreting growth hormone (B). Without hormone replacement, the mice die.

rabies virus. This rhabdovirus is usually transmitted by bites, which transfer infected saliva, but also by inhalation of virus in airborne particles. After infection, the virus slowly ascends the peripheral nerves on its way to the brain. The time between infection and the appearance of signs of disease can be up to several months and is proportional to the distance that the virus has to travel. The initial stages of disease in feral animals is characterized by their loss of fear of humans and this progresses through violent rabid behaviour, during which infectious virus can be transferred by biting, to death.

Much more subtle are the effects of Borna disease virus, an unclassified agent which produces slowly progressive disease in a number of experimental vertebrate model systems. Infection of the tree shrew (*Tupaia glis*), a primitive primate, cause a variety of changes in their normal behaviour, including hyperactivity and spatial and temporal disorientation. In breeding pairs, the normal division between the active aggressive male and more passive female behaviour becomes blurred. The normal pattern of socio-sexual activity which is a necessary preamble to successful mating is so disturbed that breeding no longer occurs, although the animals remain in good physical health. Such changes are all consistent with alterations to neurotransmitters, already referred to above. Animals are permanently affected and are thus rendered sterile as a result of infection with Borna disease virus.

The examples given illustrate some of the subtle interactions that occur between host and virus and leave wide open for investigation the possibility that viruses may be the prime causative agent in a variety of human diseases of unknown aetiology. Suspicion that progressive pathological conditions affecting the central nervous system (CNS), such as multiple sclerosis (MS) (p. 328) and Alzheimer's disease, involve virus infections has too often raised false hopes, but still there is no reason to deny the possibility.

COFACTORS IN VIRUS DISEASES

We have learned something of the complexity of virus−host interactions in Chapters 15 and 17, but there are an increasing number of situations where infections or diseases involve a third party.

Immunosuppression by malaria potentiates viral carcinogenesis

It has long been suspected that viruses require the presence of a cofactor to cause a particular disease, but no evidence has come to hand until recently. Epstein−Barr virus (EBV) plays a crucial role in Burkitt's lymphoma by immortalizing B lymphocytes. These form a tumour of the lymphoid issue of the jaw, which is found in high incidence in children, particularly boys, aged 6−9 in tropical Africa and New Guinea. However, EBV occurs throughout the world and causes infectious mononucleosis (glandular fever) in young people and, as evidenced by antibody, by the age of 20 most people have been infected. Therefore, there was *de facto* evidence that a cofactor was responsible for the lymphoma, and epidemiological evidence from Dennis Burkitt himself suggested that it was malaria. How do malaria and EBV interact? EBV, like all herpesviruses, causes a lifelong persistent/latent infection which is kept in check by the immune system, including neutralizing antibody and cell-mediated immunity. People who are immunocompromised can develop a fatal proliferation of EBV-immortalized cells. It has been demonstrated that malaria drastically lowers the control of EBV-infected cells by EBV-specific T lymphocytes, by decreasing the proportion of helper T cells in relation to suppressor T cells. One attack by the malarial parasite, *Plasmodium falciparum*, is not likely to be sufficient, but repeated attacks reduce the immune response to the level needed to establish a tumour. Exactly how malaria achieves this immunosuppression is not known, and an added complication is a likely genetic predisposition of the unfortunate victim; boys, as already mentioned, have a higher incidence of Burkitt's lymphoma than girls.

Enhanced transmission of viruses by mosquitoes infected with nematodes

Rift valley fever virus (Bunyaviridae) is an important pathogen of cattle and sheep, which is transmitted by the mosquito *Aëdes taeniorhynchus*. The virus is ingested with blood from a viraemic animal but is not passed on unless it gets to the insect's salivary gland. From there, it is injected with salivary anticoagulant into the next meal. Normally, this occurs in only 5% of mosquitoes but, if they are fed on animals which, in addition to Rift valley fever virus, are infected with a microfilarian nematode (a microscopic parasitic 'worm'), transmission of virus is increased to 30%. It appears that the microfilariae burrow out of the gut and enhance the passage of virus to the salivary gland. In addition, the presence of microfilariae allows mosquitoes exposed to very low titres of virus to transmit infection.

As many other bunya-, alpha- and flaviviruses are transmitted by biting arthropods, it will be important to discover whether microfilariae promote infection in other situations or if there are other examples of synergism between viral and other parasites.

VIRUS INFECTION CAN BE AN EVOLUTIONARY ADVANTAGE

Certain parasitic wasps (of the Ichneumonidae and Braconidae) are infected with viruses of the Polydnaviridae, which replicate in the ovaries. The female wasp deposits eggs by injection into caterpillars, which are simultaneously infected by virus carried along with the eggs. Normally, the eggs hatch out and the wasp larvae feed on the living caterpillar until it dies, whereupon the wasp larvae pupate and emerge in due course as new adult wasps. If eggs from the female wasp are separated from virus by centrifugation and injected artificially into caterpillars they fail to develop. On examination, the eggs are found to have been overcome by the caterpillar's immune system. However, if purified virus is injected together with eggs, normal development takes place (Fig. 20.5).

Thus, virus is needed to suppress (by some unknown mechanism) the immune responses of the caterpillar, and we have the situation where a virus has become essential to the successful life cycle of its eukaryotic host. We do not know if other viruses bestow evolutionary advantages upon their hosts, but this example stimulates us to turn the normal virus—host relationship on its head and look for positive aspects of infection. Who knows, it may be proved one day that some virus infections are good for you!

THE ERADICATION AND CONTROL OF VIRAL DISEASES

The euphoria that followed the successful elimination of variola (smallpox) virus has been replaced by a realization that other

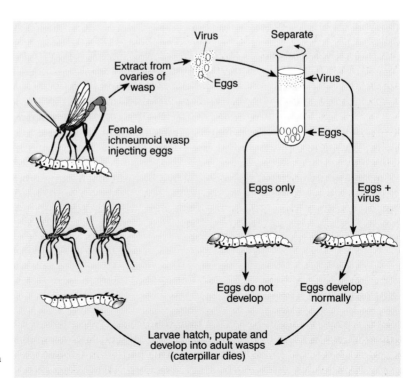

Fig. 20.5 Eggs of the ichneumoid wasp only develop in their host caterpillar when injected together with a polydnavirus.

viruses may not yield so easily (pp. 237–241). While eradication remains the ultimate goal, containment by the better application of existing resources is perhaps more realistic, at least in the short term. However, even in the wealthy countries, there are still enormous problems, ranging from having to devise new vaccines when none currently exists (e.g. against HIV), to improving existing vaccines (e.g. against influenza) and, even where there is an effective vaccine, to persuading people to use it. For instance, in the UK in recent years the take-up rates in children for measles vaccine have been 50–60%, rubella vaccine 80–90% and poliomyelitis vaccine 70–80%. In The Third World, problems of finance, communication, education and priorities all limit virus immunization campaigns, but fortunately there is help from the World Health Organization, the World Bank, governments and relief organizations. Nor should we lose sight of the fact that in some countries more serious diseases, such as malaria and schistosomiasis, have priority for already limited resources. So far, there is no sign of any chemotherapeutic agent which will be of major benefit. Acyclovir is a welcome sign that such agents do exist, but the problem has been long and intractable and will, without lucky breaks, continue to be so.

'NEW' VIRAL DISEASES

It is difficult to pin-point the reasons for the appearance of so-called 'new' viral diseases. It is likely that they are caused either by mutation of a currently existing human virus and/or mutation (in the behavioural sense) of the host which affects the way in which such a virus is transmitted or the location in which it multiplies. Of even greater potential for infection of humans is the vast reservoir of viruses that are specific to every other animal species. It is well known that when a virus crosses a species barrier it can be infinitely more pathogenic than in its normal host. The normally subclinical infections of Lassa fever virus in *Mastomys natalensis*, a peridomestic rodent, and of herpes B virus of monkeys can be lethal when contracted by humans. Presumably, this reflects an upset in the normally favourable host–parasite balance, which has evolved over a long period of time in the natural host species. HIV and acquired immune deficiency syndrome (AIDS) are the subject of Chapter 19. Discussion of other examples of new diseases follows below.

Seal influenza

In 1979, wildlife officials reported that seals on the eastern coast of the USA in the state of Massachusetts were dying. Such sporadic epidemics were not unknown but from this particular episode was isolated influenza virus, a virus not previously known to infect seals. Molecular analysis of the RNA of A/Seal/Mass/1/80 virus gave some suggestions about its possible origins, as each of the eight segments proved to be closely related to that of eight different and previously known avian (particularly sea-bird) strains of type A influenza virus. The inference is that, by a complex series of reassortment events (p. 285), a virus was created with the ability to infect a new species, mammal not bird, and cause lethal disease. Its 20% mortality rate is close to that of better-known diseases, such as smallpox, and it had become neurovirulent.

There are several hundred strains of type A influenza virus known, all with the potential of reassortment to form 'new' viruses, and some of the avian strains (H5 and H7), cause a 100% lethal generalized infection, quite unlike human influenza. Fortunately, most reassortants are not viable in nature, but nature is not short of time to achieve a virulent permutation. Perhaps this was the way that the HINI strain arose in 1918, less than 80 years ago, which killed 20 000 000 people world-wide. Apart from high mortality, the disease was unusual in that it killed people in the 18–25-year-old group rather than the normal risk group, the over 65s and those with chronic cardiorespiratory problems. There is no reason why such a situation should not recur, and it is doubtful whether resources (vaccine, amantadine) are adequate or can be mobilized rapidly enough to be effective.

Postviral fatigue syndrome (PVFS)

PVFS, also known as myalgic encephalitis (ME), is a distressing condition in which patients suffer extreme fatigue of muscle after moderate exertion, and this may be prolonged for a year or more, although sometimes with partial remissions. Described as a syndrome (meaning a collection of different diseases with a common expression), it is defined by exclusion of other causes of chronic fatigue. PVFS has been recognized for over 30 years, although until very recently there has been considerable uncertainty about its cause—whether it was due to a persistent or latent infection or was psychosomatic or plain malingering. Now, new sensitive methods of detection have shown the presence of viral genomes in muscle biopsies of some, but not all, people with PVFS.

The breakthrough came through the use of cloned viral DNA. A biopsy needle was used to remove a small core of muscle, from which nucleic acid was extracted. This was then tested, as described in Table 20.1. A highly significant proportion of people with PVFS had enterovirus RNA in their muscle; a smaller proportion had EBV DNA instead. None had both. An important question is how enterovirus RNA persists, when the virus normally causes an acute infection. Data from Len Archard's group in London show that viral RNA synthesis is abnormal: lytically infected cells contain 99% RNA$^+$ (mRNA and virion RNA) with very little RNA$^-$ (template), but in PVFS biopsies there are similar amounts of RNA$^+$

Table 20.1 Evidence for the presence of enterovirus (Picornaviridae) and Epstein–Barr virus (EBV: Herpesviridae) nucleic acid in muscle biopsy samples from patients with postviral fatigue syndrome (PVFS).

Technique	Primer/probe	PVFS	Controls
PCR	Enterovirus sequence from the 5′ non-translated region	32/60 (53%)	6/41 (15%)
Hybridization	Common cDNA from the enterovirus polymerase gene	34/140 (23%)*	0/152 (0%)
Hybridization	cDNA from EBV nuclear antigen 1 gene	8/89 (9%)*	0/48 (0%)

* Same set of biopsies tested; none was positive for both viruses.
cDNA, complementary DNA.

and RNA⁻. It will be of interest to sequence full-length viral RNA to determine if it is mutated in regions controlling RNA synthesis.

These data are in good agreement with earlier serological evidence of enterovirus infection. Here, 51% of PVFS patients were found to have virion protein 1 (VP1)–antibody complexes circulating in serum and 20% excreted virus–antibody complexes in faeces. Finally, Drs Gow and Behan report from Glasgow that electron microscopy (EM) shows the presence of abnormal mitochondria in muscle of PVFS patients. Obviously, any defect in energy production by the mitochondria would make muscle prone to fatigue. We now have some of the pieces of the PFVS jigsaw puzzle, but they have yet to be fitted together.

Bovine spongiform encephalopathy (BSE), scrapie and the prion controversy

BSE suddenly appeared in dairy cattle in the UK in the 1980s and was formally recognized in 1986. It developed into a large-scale epidemic, with serious economic repercussions, which did not show any sign of abating until late 1993. BSE is a disease of the central nervous system (CNS) which produces changes in mental state, movement and sensation. It lasts for several weeks and is always fatal. BSE appeared to be an infectious disease, but no organism could be isolated. However, the pathology produced was diagnostic of the scrapie family of diseases: the tissue becomes vacuolated, but with no sign of inflammation — no invasion of immune cells; hence the name spongiform encephalopathies used to describe this group of diseases (Table 20.2). Characteristically, all have an incubation period of approximately half a lifetime (e.g. 1 year for a mouse; 4–6 years for a cow; 35 years for humans).

It is almost certain that BSE resulted from the scrapie agent of sheep adapting to cattle, as a result of the practice of giving cattle artificial food concentrates which contain certain tissue residues of slaughtered sheep. (In abattoirs, almost nothing of the animals

Table 20.2 Spongiform encephalopathies related to BSE in the date order that transmissibility was demonstrated.

Disease and occurrence	Host species	Date
Scrapie Common in several countries throughout the world	Sheep, goats	1936
Transmissible mink encephalopathy (TME) Very rare, but adult mortality nearly 100% in some outbreaks	Mink	1965
Kuru Once common among the Fore-speaking people of Papua New Guinea, now rare	Humans	1966
Creutzfeld–Jakob disease (CJD) Uniform world-wide incidence of 1 per million per annum	Humans	1968
Gerstmann–Sträussler–Scheinker (GSS) syndrome A familial form of CJD; less than 0.1 per million per annum	Humans	1981
Chronic wasting disease (CWD) Colorado and Wyoming, USA	Mule-deer Elk	1983 *
Bovine spongiform encephalopathy (BSE)	Cattle	1988
Feline spongiform encephalopathy (FSE)	Domestic cat	1991

* Not demonstrated.

we eat for food is wasted, and other residues go to make fertilizer, machine lubricants and cosmetics.) However, scrapie has been endemic in sheep in the UK for nearly 300 years, so what was the cause of the transmission to cattle? It is likely that this resulted from an abrupt change in the treatment of animal carcass waste products between 1980 and 1983, during which the proportion of meat-and-bone-meal cattle food produced by extraction of tallow with organic solvents decreased by nearly 50%. This treatment had involved processing abbatoir products for 8 hours at 70°C and then the application of superheated steam to remote traces of solvent. Both would have been effective in inactivating scrapie infectivity. Such extreme measures are needed as the infectivity is immensely stable, far more so than with any other micro-organism (see below). It is likely that there was a single origin to the epidemic, made possible by mutation of a scrapie agent, which was then propagated by the recycling of cattle residues from the abattoir in cattle food concentrates, which gave optimum conditions for a cattle-adapted variant to emerge. BSE has now been passed into mice and, indeed, different isolates appear to be homogeneous in regard to incubation period and brain pathology, whereas scrapie strains differ widely in both these properties.

What can be done?

There is no treatment for any of the spongiform encephalopathies. None of their responsible agents express any proteins, so it is impossible to immunize against them. No infectious agent has been purified. Prevention is the only option available. Most of the infectious agent is present in the CNS and lymphoid tissue (e.g. thymus and spleen), and now (since 1988) legislation has changed abattoir practices so that these tissues do not enter the ruminant food chain. At the time of writing (end of 1993), the first sign of a decrease in the numbers of infected cattle (from about 800 per month), occurred, a hopeful indication that the epidemic may be abating. However, we are still waiting to see if there is horizontal cow-to-cow transmission or vertical cow-to-calf transmission, as occurs with scrapie. This is unlikely, as BSE normally appears as solitary cases in herds, and clusters of infected animals, to be expected if there was transmission, are not seen. Further, there is no natural transmission between mink with transmissible mink encephalopathy (TME) or between people with kuru, apart from cannibalism.

Could BSE infect humans?

Historically, scrapie has been endemic in the UK for nearly three centuries, but there is no evidence of transmission to humans, despite the number of sheep consumed. There is little scrapie agent in muscle (meat) and CNS tissue is not widely consumed.

Sheep-to-sheep transmission, as mentioned above, is usually maternal. None the less, there has been considerable proper and natural concern about the risk, not only from the viewpoint of food but because animal products, for example calf serum, are widely used in the pharmaceutical industry, for example for the growth of cultured cells to make virus vaccines. As far as ingestion of the BSE agent is concerned, it is noteworthy that the infection is predominantly in dairy cattle, which are not widely eaten for their meat.

Altogether it seems unlikely, but not impossible, that BSE could infect people. This would, after all, involve another jump across the 'species barrier', and there is no cannibalism which would recycle any mutants and give them a selective advantage, as occurred in cattle. However, BSE has been passed to cats without the advantage of recycling (see below). Only time (30–40 years?) will tell if people were infected during the 1980s.

Necessary precautions have been taken and, since 1989–1990, specified offals (brain, spinal cord and lymphoid tissues) have been legally banned from entering the human food chain. Mice have been fed various tissues from BSE-infected cattle, but only those fed brain developed disease. The appearance of a new spongiform encephalopathy in domestic cats since 1990 underlines the need for care, and it is thought that feline spongiform encephalopathy (FSE) resulted from food, probably offals, containing the BSE agent. A ban on offals in pet food was put in place in 1989.

Scrapie: virino or prion?

Scrapie is a natural infection of sheep, which is usually spread maternally but can also be spread horizontally. The disease has been studied and much is known about it. The name arises from the tendency of diseased sheep to scrape themselves against fence posts and so on, presumably to relieve itching of the skin. The agent has been isolated in mice, and several different strains have been defined by their very precise, but different, incubation periods in mice, which range from around 100 to 300 days, and by the severity and distribution of lesions in the CNS (Fig. 15.5). The lifespan of a mouse is about 730 days. These incubation periods are also host strain-specific and are controlled by a single mouse gene, called *sinc* (scrapie *inc*ubation period). It seems highly likely that *sinc* is the same gene as *PrP*, which encodes the precursor of the protein that forms scrapie-associated fibrils (SAF). These fibrils are a diagnostic feature of the scrapie group of diseases and are identified by EM examination of sections of brain. PrP^c is a normal host glycoprotein found on the surface of neurones in the brain (and cells elsewhere), which in the course of scrapie (or BSE) infection becomes resistant to protease digestion, accumulates extracellularly and results in the loss of surrounding neurones, possibly by apoptosis. Its normal function is unknown, and transgenic mice which have been bred with the PrP^c gene 'knocked out' develop normally.

Scrapie has all the features of an infectious agent. It passes through filters which admit only viruses and can be titrated in mice (often to titres of over 10^7 infectious units/ml). However, it differs from conventional viruses, since it is much more resistant than they are to inactivation by heat, radiation and chemicals, such as formaldehyde. Radiation inactivation kinetics allows the M_r of its genome to be estimated as equivalent to about 250 nucleotides of single-stranded nucleic acid or 125 nucleotides of double-stranded nucleic acid. Thus, it is probably too small to code for its own coat protein or replicase. Both, according to the *virino* hypothesis, would be provided by the host cell. For comparison, the genome of plant pathogenic viroids comprises 359 nucleotides of single-stranded RNA and encodes no protein, while some defective−interfering animal viruses have single-stranded genomes of around 400 nucleotides and also encode no protein.

Another view of the data is that scrapie has no nucleic acid but consists of an infectious protein, or *prion*, which is the modified PrP (or PrP^{sc}). This reversal of the conventional view of the flow of information in modern molecular biology of DNA → RNA → protein has a strong following, led by Stanley Prusiner in the USA. One of its strengths is that PrP^{sc} is the only molecule to have been identified in highly purified and highly infectious extracts of scrapie-infected brain. However, it is not clear how variation in incubation period and pathology of different strains of scrapie can be explained on the basis of a post-translationally modified normal host protein. The thinking is that PrP^{sc} is simply an abnormal conformation of the PrP^{c} protein which is resistant to protease digestion. However, there are inherited forms of human disease (Gerstmann–Straüssler–Schenker syndrome) where there are mutations in the prion gene. There is also an infectious human spongiform encephalopy, called 'Creutzfeld–Jakob disease, and tragically this has been inadvertently transmitted by treating people with contaminated human products (e.g. hormones) obtained *post mortem*.

Conclusive answers are still to come, but transgenic mice genetically engineered to lack the normal *PrP^{c}* gene are proving invaluable. Very recently (end of 1993), it has been reported that these mice cannot be infected with scrapie, which suggests that the normal gene is somehow induced by the exogenous infectious agent. Secondly, mice made transgenic for the human Gerstmann–Straüssler–Schenker mutant *PrP* gene spontaneously develop disease. To complete the circle, we now need to know if extracts of brains of the latter mice contain infectious prions.

Conclusion

BSE is a 'new' disease and represents the ever-present ability of infectious agents to evolve under novel conditions. It seems highly probable that the source of BSE infection has been found and eliminated and that the disease will literally die out, unless, as seems unlikely, there is frequent cow-to-cow transmission. The significance of the prion group of diseases remains to be unravelled, and the nature of the infectious agent has yet to be determined.

Viruses and multiple sclerosis

MS is a disease of the CNS (the brain and spinal cord). About 10% of the cells in the CNS are neurones and the rest are support cells, called glia. About one-third of glial cells are oligodendrocytes, which insulate the electrical activity of nerve cell axons by surrounding them with a cytoplasmic 'Swiss roll' containing myelin. MS is associated with a number of focal areas in the CNS in which oligodendrocytes have degenerated, and this results from attack by myelin-specific CD4^{+} T cells. The disease is chronic, affects adults and involves mainly the limbs, sight and incontinence. It is characterized by periods of remission of a few months to a year or more, when there is at least partial recovery, alternating with reappearance of the MS symptoms. During remission, myelin sheaths regrow around the neuronal axons and neuronal function is restored. The incidence of MS is about 1 per 1000 in the UK and USA, but in other areas of the world this may be higher or lower.

The major question is what initiates the antimyelin auto-immunity. It is known that there are environmental and predisposing genetic factors. It is suspected that viruses are the environmental factor, but there is still no direct evidence. One attractive hypothesis is that MS involves enveloped viruses and is caused not by infection with one virus but by successive infections with several different viruses. It is postulated that each virus enters the CNS and incorporates myelin components into its envelope, which are then presented to the immune system, together with the foreign viral antigens. Myelin components in this context are seen by the immune system as foreign, and autoimmunity results. Continued presence of the virus(es) is not required once the antimyelin CD4$^+$ T cells have been activated.

HUMANIZED AND HUMAN MONOCLONAL ANTIBODIES

Since the initial technology for the preparation of monoclonal antibodies was devised by George Köhler and Cesar Milstein at Cambridge, the face of virology has been transformed by their multifarious uses. We see monoclonal antibodies in the forefront of new analytical, preparative and diagnostic technologies, and nowhere have they had more impact than in the creation of a new sphere of employment in the commercial sector. Now, it is difficult to imagine how virology functioned without them.

There is, however, one important area where the use of mono-clonal antibodies has been frustrated by technical limitations and this is in their therapeutic use. The problem is that monoclonal antibodies are made by immunizing an animal, extracting primed B cells from the spleen and then immortalizing the B cell by fusion with a myeloma (B cancer) cell. While the system works for mice, it is not applicable to humans and no alternative source of B cells (e.g. from blood) has proved effective. The problem of treating people with mouse monoclonal antibody is that their immune system very rapidly reacts to the mouse antigens: at best, this destroys the antibody and, at worst, initiates a potentially lethal anaphylactic response.

The answer may be to invoke recombinant DNA, that other major innovative technology (and job provider). Thus, the new antibody would contain only the complementarity-determining regions (CDRs) of the hypervariable parts of the variable (V) regions of the mouse light (L) and heavy (H) chains, with the rest provided from cloned human antibody (Fig. 20.6). In the native protein, the CDRs are juxtaposed to form the unique antigen-binding site. By minimizing the amount of foreign (mouse) protein, the problem of rejection is overcome.

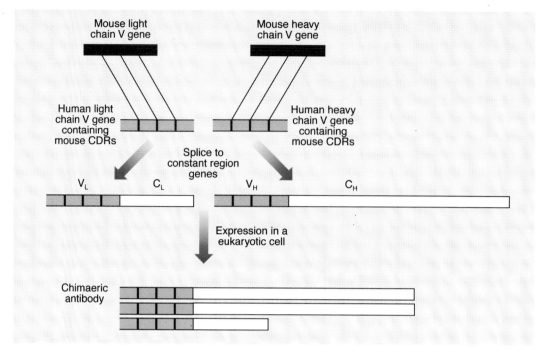

Fig. 20.6 Construction of chimeric genes and synthesis of chimeric antibody. Only the DNA of the complementarity-determining regions (CDRs) are from the mouse monoclonal antibody. The rest is human immunoglobin DNA. Only one set of human genes is required and the DNA of the desired mouse CDRs are then spliced in.

An advantage of this procedure is that the mouse V region can be made into any class of immunoglobulin, though most usefully IgG1, 2, 3 or 4, by recombination with the appropriate human H chain genes. Immunoglobulins have evolved properties which allow them to fulfil specialized defence roles in the body and it is obviously advantageous to use the correct class of immunoglobulin for therapeutic use. Such properties reside in the constant (C) regions of the H chain, so, with a range of 'off-the-peg' human C_H region DNA, the desired result can easily be effected.

All this is extremely labour-intensive, and simpler alternatives are coming forward, using mRNA isolated from antibody-synthesizing cells and the reverse transcriptase PCR (Fig. 20.7). Such cells are withdrawn, using a syringe, from the bone marrow or from blood. In the first reaction, specific primers are used to amplify all V_L regions present and, in the second reaction, all V_H regions are similarly amplified. One V_L and one V_H are then linked together at random, using an oligonucleotide encoding a flexible 15-mer peptide linker $(Gly_4-Ser)_3$, so that, when expressed, the resulting polypeptide folds up to form an antigen-binding site. Any antibody of interest will be present at a very low frequency, but screening has been solved in a novel fashion by expressing the V_L-linker$-V_H$ polypeptide on the surface of a filamentous bacteriophage (fd or M13). These phages encode an attachment protein,

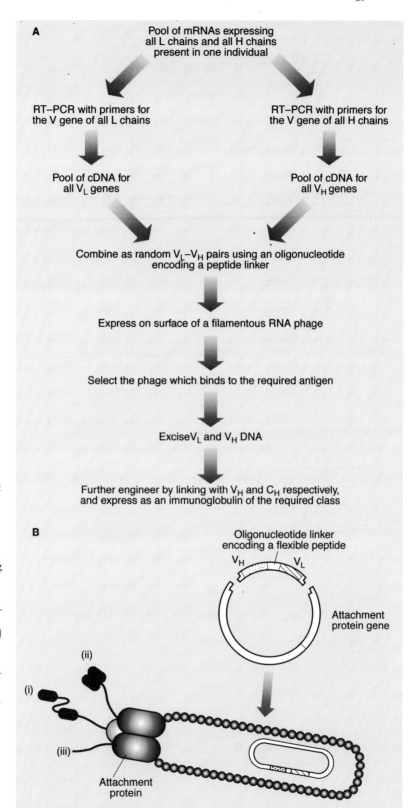

Fig. 20.7 (A) Scheme outlining the random combinatorial method for the cloning of human antibody Fv fragments. Its success lies in the ability to select rare recombinant phages expressing the required antibody activity (about 1 in 10^6 phages). (B) Construction of the fd phagemid containing the fused V_L−linker−V_H gene. This is transformed into *E. coli*. The bacterium is then infected with phage (fd) and progeny phage are produced having a single recombinant V_L−linker−V_H polypeptide (i). This folds to form a functional antibody-binding site (ii). The 'phage antibody' requires at least one molecule of non-modified attachment protein (iii) to be infectious.

Text within the figure:

A

Pool of mRNAs expressing all L chains and all H chains present in one individual

RT−PCR with primers for the V gene of all L chains

RT−PCR with primers for the V gene of all H chains

Pool of cDNA for all V_L genes

Pool of cDNA for all V_H genes

Combine as random V_L−V_H pairs using an oligonucleotide encoding a peptide linker

Express on surface of a filamentous RNA phage

Select the phage which binds to the required antigen

Excise V_L and V_H DNA

Further engineer by linking with V_H and C_H respectively, and express as an immunoglobulin of the required class

B

Oligonucleotide linker encoding a flexible peptide

V_H V_L

Attachment protein gene

(i)

(ii)

(iii)

Attachment protein

expressed in low number—about four copies per virion. This binds to the F pilus of *E. coli*. The V_L–linker–V_H is expressed as a fusion protein with the phage attachment protein. The phage is still infectious and the V_L–V_H is expressed in a form which is able to bind to antigen. This is called a 'phage antibody'. Selection of the required antibody activity is achieved by panning a phage population expressing a myriad of antibody specificities on to the required antigen. Non-bound phage is washed away and bound phage eluted and grown up on a lawn of *E. coli*. Plaques are picked and retested. The V_H–V_L gene can be excised and V_L joined to C_L and V_H joined to C_H and reconstituted as Fabs or as complete antibody molecules. The phage technique allows 1 in 10^6 antibody specificities to be selected. This type of procedure has been used recently to obtain powerful neutralizing antibody from an HIV^+ man. The system is known as the random combinatorial method of making antibody, since it allows the formation of any pair of V_L and V_H genes, even those which may not exist as such in the donor.

Therapy with recombinant mouse–human monoclonal antibodies is not intended to replace the stimulation of immunity with vaccines, but, rather, to provide immunity in circumstances where no other treatment is possible or effective. Examples are the congenitally immunodeficient, or those whose systems are compromised by accident or in the course of transplant surgery or cancer chemotherapy, cases of acute life-threatening disease where there is not time to immunize and in the protection of the newborn from infection by the mother.

ECOLOGY OF VIRUSES

Where a virus causes a serious disease of humans, domestic animals or crops, considerable effort has been devoted towards understanding the ecology of that virus. Obviously, in the absence of suitable antiviral agents to minimize the effects of infection, the only way to control virus diseases is to identify, and where possible to eliminate, vectors and reservoirs of the disease. Thus, virus yellows in British sugar-beet crops can be largely controlled by removing weeds, mangel-wurzel clamps and old beet crops, which serve as overwintering reservoirs of both the virus and its aphid vector. In the past, relatively little attention has been paid to the ecology of viruses not directly affecting humans, but this situation is clearly changing, for reasons best exemplified by the baculoviruses.

Insect viruses

More than 600 viruses have been isolated from dead or moribund insects, although very few have been studied in detail. Since only a small proportion — probably less than 1% — of the total number of insects species have been examined, many more viruses of insects remain to be discovered. Of the eight groups of insect viruses described so far, only the baculoviruses appear to be restricted to insects and, because of their specificity, they have been considered as biological control agents. Most of the insect pests of world-wide importance in agriculture and forestry are members of the Lepidoptera and Hymenoptera and it is fortunate that baculoviruses commonly infect certain species in these groups. Such viruses can cause epizootics in their host population to such a level that very effective natural control is exerted. For example, in recent years, large areas of spruce in Wales were attacked by the spruce saw-fly. Previously, the saw-fly had only been a local pest and the reasons for the widespread population growth are not clear. However, in 1973, a baculovirus was observed to be causing considerable mortality of the saw-fly and, by 1976, effective natural control had been established.

It is significant that, when baculovirus preparations have been applied to insect populations in the field, the degrees of mortality, and therefore control, have been extremely variable. Unfortunately, the reasons for success or failure in such applications are not fully understood and are obviously a hindrance to further development, but the dose of baculovirus administered, its retention of infectivity and the state of the insect host are important factors.

In any introduction into the environment of pathogenic insect viruses, great care must be taken to assess the ecological dangers. The potential for accidental infection of invertebrates or vertebrates, with the possibility of overt disease or, more insidiously, widespread inapparent infections, is both important and difficult to estimate. To this end, there are plans to genetically engineer markers into the baculovirus genome, so that the virus can be easily identified when it is reisolated from the wild. Four baculoviruses are now approved by the US Environmental Protection Agency as insecticides. They are produced commercially and enough virus to treat an acre costs between $2.50 and $10.

Viruses in water

There are other reasons why the ecology of viruses will assume growing importance. For example, there has been a dramatic world-wide tendency toward urbanization, with a concomitant increase in discharge of raw and processed sewage into waterways. These waterways, in turn, are being increasingly used both recreationally and as sources of potable water. Unfortunately, the basic design of sewage treatment plants was elaborated before the nature of viruses was fully appreciated, and one consequence is that effluent from sewage plants contains most of the viruses which entered with the raw sewage. Those viruses that are removed frequently end up in sludge, which is used for landfill or as an agricultural fertilizer. All too frequently, sludge is sprayed on land used for grazing sheep and cattle!

Because of concern about the possible presence of serious viral pathogens in potable and recreational waters, a number of research groups around the world are concentrating on elaborating methods for concentrating viruses from large volumes of water. A number of such devices have been described, one of which can successfully concentrate small numbers of poliovirus particles from 380 litres of water. Undoubtedly, the next decade will see a considerable improvement in the design for such concentrators. Unfortunately, methods for the satisfactory recovery of viruses from solids have yet to be developed and this will hinder much ecological work.

Exotic virus infections of man

In recent years, the public has learned to fear the name 'Lassa fever', which they recognize as a highly lethal infectious disease acquired in tropical Africa. There are a number of other notable exotic viruses ('exotic' meaning introduced from abroad) which at this moment pose a potential rather than an actual threat to the general population. All of these viruses are zoonotic, that is, they are spread to humans from their natural hosts, infected animals, birds or arthropods (Table 20.3). Humans are an accidental host and, as sometimes happens in these circumstances, the resulting disease is more severe than in the natural infection.

The natural host of Lassa fever virus is *Mastomys natalensis*, a peridomestic rodent, which appears to suffer no ill effects from infection. Virus is excreted in the urine and infects humans. However, despite its bad public image, Lassa fever virus causes clinical disease in only 1% of people infected. The other 99%, who are known to have been infected because they have specific antibodies, have a silent infection. Of those affected, about 30% will die, giving an overall mortality rate of about 0.3%. In humans, epidemic disease does not commonly arise, as the virus does not spread efficiently from person to person.

Because exotic viruses are zoonotic, there is no hope for their elimination, and only control measures, affecting, for example, their rodent and arthropod hosts, are contemplated. We understand little about many of these viruses, and there is need of investigation into the natural history as well as the properties of the viruses themselves. However, because of the extreme biohazard, they can only be handled in the highest-containment laboratories and work progresses slowly as a consequence. The entire story of exotic viruses can be summarized as the result of human intrusion into environments in which natural selection has not fitted humans to live.

Table 20.3 Some exotic viruses.

Virus	Group	Comment
Smallpox	Poxviridae	Eradicated. Possible risk from laboratory stocks
Yellow fever	Flaviviridae	Endemic in Central Africa. Has not yet reached Asia. Capable of invading new areas in the wake of its mosquito vector *Aëdes aegypti*. Excellent vaccine available
Venezuelan equine encephalitis	Togaviridae	Epidemic. Serious morbidity and mortality in children
Chikungunya O'Nyong Nyong	Togaviridae	Usually endemic locally and flares into epidemics
Dengue, St Louis, Murray Valley, Japanese, West Nile and various tick-borne encephalitides	Flaviviridae	Usually endemic locally and flares into epidemics
Rift Valley fever	Bunyaviridae	Infects humans and cattle in explosive epidemics, spread by arthropods. High mortality and morbidity rates
Rabies	Rhabdoviridae	Mostly endemic although exotic in UK. Difficult to eradicate when introduced. Spreading in Europe. There is now an effective tissue culture-grown vaccine which can be used as a preventive as well as a therapeutic measure. Few deaths (around five per year) in Europe but many in India and countries of the Middle East
Lassa fever	Arenaviridae	Endemic in certain rodents with direct infection to human via contaminated food or water. Danger from international travel and indistinguishable in its early stages from many other fevers. Only immune serum available as treatment
Marburg and Ebola	Filoviridae	Discovered as a result of infection from preparation of primary monkey cell cultures. Epidemic in 1976 in Sudan involving 600 people with 50–90% mortality. Could re-emerge. Natural history unknown

THE ORIGIN OF VIRUSES

There are two theories concerning the origin of viruses: they are either degenerate bacteria or vagrant genes. Just as fleas are descended from flies by loss of wings, viruses may be derived from micro-organisms which have dispensed with many of their cellular functions. Alternatively, some nucleic acid might have been transferred accidentally to a cell of a different species and, instead of being degraded, as would normally be the case, might have survived and replicated. Although over 40 years have elapsed since these two theories were first proposed, we still do not have any firm indications about the origin of viruses. Now, modern techniques for the rapid sequencing of both viral and cellular genomes are providing the data base for computer analysis. However, while such analyses may identify the progenitors of a virus, they cannot indicate whether it arose by degeneracy or gene escape.

Studies on the potato spindle tuber viroid (PSTV) suggest that it may, in fact, be an RNA copy of host DNA. PSTV hybridizes with cellular DNA of several uninfected host species (family Solanaceae), and infection of such plants with PSTV does not alter the hybridization pattern. Detailed analysis of the kinetics of hybridization indicates that PSTV hybridizes to unique DNA sequences. At least 60% of PSTV is represented by sequences in the DNA of solanaceous host plants but, more interestingly, the DNA from plants not known to be hosts of PSTV also contain sequences related to part of PSTV. Phylogenetically, the more distant a plant species is from the Solanaceae the fewer PSTV-like genes its DNA will contain. Although by no means proof, these results do suggest that viruses are derived from genes in the solanaceous hosts. As PSTV RNA sequences contain no initiator adenine−uracil−guanine (AUG) and cannot be translated into any oligo- or polypeptide, the virion is unlikely to represent the coding sequence of a vagrant gene. However, sequence similarities of PSTV with a cellular RNA, U1, which defines the ends of introns and controls their excision, suggested that viroids might be escaped introns. However, avocado sunblotch viroid has no U1 RNA sequence homology and its origin must be sought elsewhere.

It is unlikely that all the viruses currently known have evolved from a single progenitor. Rather, viruses have probably arisen numerous times in the past by one or both of the mechanisms outlined above. However, once formed, viruses would be subject to evolutionary pressures, just like prokaryotic and eukaryotic organisms. One process which must contribute significantly to virus evolution is recombination between two unrelated viruses. Thus, the *Salmonella* DNA phage P22 can recombine with the morphologically unrelated *Salmonella* phages Fels 1 and Fels 2, as well as coliphage λ, to yield novel hybrid phages. Such hybrids are easily detected by growing P22 on Fels 1 lysogens, for example, and plating the progeny virus on Fels 1 lysogens resistant to P22. The only plaque-forming particles observed are hybrid phages with the appearance of Fels 1 but carrying a considerable portion of the genome of P22, including the immunity region. Such illegitimate recombination events have so far only been detected in bacteriophages. Recombination between RNA molecules has only been detected between the genomes of some animal viruses, such as the polioviruses and foot-and-mouth disease virus, but it seems that these viruses are not subject to such sudden, gross changes by this mechanism. However, new viruses do arise, and one example is a new type 3 poliovirus, isolated from an outbreak of poliomyelitis in Finland, which ran from October 1984 to February 1985. The spontaneous creation of 'new' viruses which are viable in nature, cause disease and are resistant to the immunity provided by vaccine to types 1, 2 and the 'old' type 3 is extremely worrying. What was responsible was probably the rapid evolution of RNA genomes, which, in the absence of a molecular proof-reading mechanism for RNA (unlike DNA), accumulate mutations at a rate of 3×10^{-4} per nucleotide per doubling, compared with DNA at 10^{-9} to 10^{-10}. In other words, an RNA virus can achieve in one generation the degree of genetic variation which would take an equivalent DNA genome 300 000 generations to achieve. Thus the new poliovirus may have arisen through antigenic drift arising from uncorrected mutations, as we have already seen for influenza viruses. However, this has not occurred before in the previous 30 years of known poliovirus history and we do not understand why it has arisen now. Of far greater significance is the potential for genetic change between related viruses with segmented genomes. Here, we are dealing not with mutations, of which many will be non-viable, but with the reassortment of functional genes. The only restriction is the compatibility between the various individual segments making up a fully functional genome. Fortunately, this seems to be a real, though not invincible, barrier to the unlimited creation of new viruses. Remember, seal influenza virus is thought to have arisen in this way (p. 322) and remember antigenic shift in humans (p. 285).

EDUCATING THE GENERAL PUBLIC

Even amongst the quality press and media sources, microbiologists are constantly saddened and annoyed by fundamental misconceptions about viruses. If these professionals make mistakes, what hope is there for the general public? The answer, perhaps, lies in all virologists helping to inform those around them. With greater understanding will come, for example, the willingness to accept vaccines, for, as mentioned earlier, fear and apathy make acceptance rates abysmally low. Every medical treatment carries a finite risk of personal damage, and vaccines are no exception. However, it is to be hoped that the public will appreciate that not taking a vaccine exposes them to far greater risk. By understanding more about infectious agents, sensible precautions against infection can be taken, and the hysteria which has in turn accompanied increases in hepatitis B virus, genital herpes viruses and HIV will be replaced by more objective reactions. Surely, the increase in the number of students taking courses in virology will have an impact on the understanding of the public at large.

FURTHER READING

Behan, P. O., Golberg, D. P. & Mowbray, J. F. (eds) (1991) *Postviral fatigue syndrome.* Edinburgh: Churchill Livingstone.
Berg, G. (ed.) (1983) *Viral pollution of the environment.* Boca Raton, FL: CRC Press.

Büeler, H., Aguzzi, A., Sailer, A., Greiner, R.-A., Autenreid, P., Aguet, M. & Weissman, C. (1993) Mice devoid of PrP are resistant to scrapie. *Cell*, **73**, 1339–1347.

Carr, K. (1993) Prion diseases. *Nature (London)*, **365**, 386.

Chesebro, B. (1992) Spongiform encephalopathies: PrP and the scrapie agent. *Nature (London)*, **356**, 560.

Cooper, J. I. & McCallum, F. O. (1984) *Viruses and the environment.* London: Chapman and Hall.

Epstein, M. A. (1984) Clues to the role of malaria. *Nature (London)*, **312**, 398. (About EBV and Burkitt's lymphoma.)

Evans, D. M. A., Dunn, G., Minor, P. D., Schild, G. C., Cann, A. J., Stanway, G., Almond, J. W., Currey, K. & Maizel, J. V. (1985) Increased neurovirulence associated with a single nucleotide change in a noncoding region of the Sabin type 3 poliovaccine genome. *Nature (London)*, **314**, 548–550.

Fraser, H., Bruce, M. E., Chree, A., McConnell, I. & Wells, G. A. H. (1992) Transmission of bovine spongiform encephalopathy and scrapie to mice. *Journal of General Virology*, **73**, 1891–1897.

Gow, J. W., Behan, W. M. H., Clements, G. B., Woodall, C., Riding, M. & Behan, P. O. (1991) Enteroviral RNA sequences detected by polymerase chain reaction in muscle of patients with postviral fatigue syndrome. *British Medical Journal*, **302**, 692–696.

Hertz, L., McFarlin, D. E. & Waksman, B. H. (1990) Astrocytes: auxillary cells for immune responses in the central nervous system. *Immunology Today*, **11**, 265–268. (About autoimmunity and MS.)

Howard, B. H. (1983) Vectors for introducing genes into cells of higher eukaryotes. *Trends in Biochemical Sciences*, **8**, 209–212.

Hull, R. & Davies, J. W. (1983) Genetic engineering with plant viruses and their potential as vectors. *Advances in Virus Research*, **28**, 1–33.

Kaneshima, H., Baum, C., Chen, B., Namikawa, R., Outzen, H., Rabin, L., Tsukamoto, A. & McCune, J. M. (1990) Today's SCID-hu mouse. *Nature (London)*, **348**, 561–562.

Kimberlin, R. H. (1992) Bovine spongiform encephalopathy. *Review of Science and Technology of the Office International des Epizooties*, **11**, 347–390.

Kinnunen, L., Pöyry, T. & Hovi, T. (1992) Genetic diversity and rapid evolution of poliovirus in human hosts. *Current Topics in Microbiology and Immunology*, **176**, 49–60.

McCafferty, J., Griffiths, A. D., Winter, G. & Chiswell, D. J. (1990) Phage antibody: filamentous phage displaying antibody variable domains. *Nature (London)*, **348**, 552–554.

Neuberger, M. S., Williams, G. T. & Fox, R. O. (1984) Recombinant antibodies possessing novel effector functions. *Nature (London)*, **312**, 604–608.

Notkins, A. L. (1984) Diabetes: on the track of viruses. *Nature (London)*, **311**, 209–220.

Old, R. W. & Primrose, S. B. (1994) *Principles of gene manipulation* (5th edn). Oxford: Blackwell Scientific Publications.

Oldstone, M. B. A. (1989) Viruses can cause disease in the absence of morphological evidence of cell injury: implication for uncovering new diseases in the future. *Journal of Infectious Diseases*, **159**, 384–389.

Oldstone, M. B. A., Rodriguez, M., Daughaday, W. H. & Lampert, P. W. (1984) Viral perturbation of endocrine function to disordered cell function, disturbed homeostasis and disease. *Nature (London)*, **307**, 278–281.

Oldstone, M. B. A., Southern, P., Rodriguez, M. & Lampert, P. (1984) Virus persists in β cells of Islets of Langerhans and is associated with chemical manifestations of diabetes. *Science*, **224**, 1440–1443.

Pattison, J. R., Brown, F. & Brunt, A. A. (1984) New virus disease syndromes. In: 'The Microbe 1984'. *Society for General Microbiology Symposium*, **36**, Pt1, 241–267.

Plückthun, A. (1991) Antibody engineering: advances from the use of *Escherichia coli* expression systems. *Bio/Technology*, **9**, 545–551.

Pollock, R. R., Teillaud, J.-L. & Scharff, M. D. (1984) Monoclonal antibodies: a powerful tool for selecting and analyzing mutations in antigens and antibodies. *Annual Review of Microbiology*, **38**, 389–417.

Prusiner, S. B. (1993) Biology of prion diseases. *Journal of Acquired Immune Deficiency Syndromes*, **6**, 663–665.

Racaniello, V. E. & Baltimore, D. (1981) Cloned poliovirus complementary DNA is infectious in mammalian cells. *Science*, **214**, 916–919.

Sänger, H. L. (1984) Minimal infectious agents: the viroids. In: 'The Microbe 1984'. *Society for General Microbiology Symposium*, **36**, Pt 1, 281–334.

Sprankel, H., Richarz, K., Ludwig, H. & Rott, R. (1978) Behaviour alterations in tree shrews (*Tupaia glis*, Diard 1820) induced by Borna disease virus. *Medical Microbiology and Immunology*, **165**, 1–18.

Stanway, G., Hughes, P. J., Mountford, R. C., Reeve, P., Minor, P. D., Schild, G. C. & Almond, J. W. (1984) Comparison of the complete nucleotide sequences of the genomes of the neurovirulent poliovirus P3/Leon/37 and its attenuated Sabin vaccine derivative P3/Leon 12a,b. *Proceedings of the National Academy of Sciences, USA*, **81**, 1539–1545.

Turell, M. J., Rossignol, P. A., Spielman, A., Rossi, C. A. & Bailey, C. L. (1984) Enhanced arboviral transmission by mosquitoes that concurrently ingested microfiliariae. *Science*, **225**, 1039–1041.

Tyrrell, D. A. J. (1984) The eradication of virus infections. In: 'The Microbe 1984'. *Society for General Microbiology Symposium*, **36**, Pt 1, 269–279.

Weissman, C. (1991) Spongiform encephalopathies: the prion's progress. *Nature (London)*, **349**, 569–571.

Winter, G. & Milstein, C. (1991) Man-made antibodies. *Nature (London)*, **349**, 293–299.

Also check Chapter 21 for references specific to each family of viruses.

21 The classification and nomenclature of viruses

Listed below are the major groups of viruses, together with a brief description and sketches of the particles (*not* to scale). Viral taxonomy is still in progress and many viruses are as yet unclassified. The list is subdivided for convenience into viruses which infect vertebrates and other hosts, vertebrates only, invertebrates, plants, fungi and bacteria, although certain viruses cross such boundaries.

It can be seen that the list does not reflect the distribution of viruses amongst their various hosts but the emphasis is on research into viruses of medical or veterinary importance, plus those for which we have the good fortune to have excellent culture systems. However, work is expanding into the other areas, which will redress the situation and make this list in need of continuous revision.

Key references to each virus family are indicated below. If possible, readers should consult: Francki, R. I. B. *et al.* (1991) Classification and nomenclature of viruses. *Archives of Virology Supplementum*, 2; and for animal viruses: Fields, B. N. & Knipe, D. M. (1990) *Virology*, Vols 1 and 2 (2nd edn). New York: Raven Press.

VIRUSES MULTIPLYING IN VERTEBRATES AND OTHER HOSTS

Note that some or all of the Reoviridae, Bunyaviridae, Rhabdoviridae and Togaviridae multiply in both vertebrates and other hosts. Other families include genera which multiply solely in vertebrates.

Family: **Iridoviridae** (class I)

125–300 nm icosahedral particle consisting of a spherical nucleocapsid surrounded by lipid modified by morphological protein subunits. This may be surrounded by an envelope derived from the plasma membrane, but the envelope is not required for infectivity. Double-stranded deoxyribonucleic acid (DNA) of M_r $100-250 \times 10^6$, which is circularly permuted with direct terminal repeats. Contain several enzymes. Transcription and DNA synthesis are nuclear. Messenger ribonucleic acids (mRNAs) have no poly-adenine (poly-A) tails. Cytoplasmic.

Genera: Iridovirus (small 120 nm blue iridescent viruses of insects)
        Chloridovirus (large 180 nm iridescent viruses of insects)
        Ranavirus (frog virus 3)
        Lymphocystivirus (viruses of fish)

See: Willis, D. B. (ed.) (1985) Iridoviridae. *Current Topics in Microbiology and Immunology*, **116** (8 chapters).

Family: **African swine-fever virus** (class I)

Has properties between Irido- and Poxviridae. 200 nm icosahedral particle containing an 80 nm core surrounded by one or two lipid envelopes. No glycoproteins. Double-stranded DNA of M_r 100×10^6 with inverted terminal repeats. Virion contains all enzymes needed for mRNA synthesis. Cytoplasmic; matures by budding through membranes. Spread by ticks.

See: Vinuela, E. (1985) African swine fever virus. *Current Topics in Microbiology and Immunology*, **116**, 151–170.

Family: **Poxviridae** (class I)

Double-stranded DNA of M_r $85-240 \times 10^6$ with inverted terminal repeats. Largest viruses $170-260 \times 300-450$ nm. Complex structure composed of several layers and includes lipid. Core contains all enzymes required for mRNA synthesis. Cytoplasmic multiplication.

Genera: Orthopoxvirus (vaccinia and
  related viruses)
Avipoxvirus (fowlpox and
  related viruses)
Capripoxvirus (sheep pox and
  related viruses)   } Poxviruses of vertebrates
Leporipoxvirus (myxoma and
  related viruses)
Parapoxvirus (milker's node
  and related viruses)
Entomopoxvirus   Poxviruses of insects

See: Moyer, R. W. & Turner, P. C. (1990) Poxviruses. *Current Topics in Microbiology and Immunology*, **163**, 1–211.

Family: **Parvoviridae** (class II)

Single-stranded DNA of M_r $1.5-2.0 \times 10^6$. Particle is a 18–22 nm icosahedron which contains no enzymes. Multiplication is nuclear.

Genera: Parvovirus — viruses of vertebrates, including humans. Virions mostly DNA$^-$
  Dependovirus — adeno-associated virus. Infects vertebrates. Particles contain either DNA$^+$ or DNA$^-$, which form a double strand upon extraction. Efficient replication is dependent on helper adenovirus or herpesvirus
  Densovirus — viruses of insects. Virions DNA$^-$ or DNA$^+$. Helper not required

See: Berns, K. I. (ed.) (1984) *The parvoviruses*. New York: Plenum Press.
Berns, K. I. & Hauswirth, W. W. (1979) Adeno-associated viruses. *Advances in Virus Research*, **25**, 407–449.
Siegl, G. *et al.* (1985) Characteristics and taxonomy of Parvoviridae. *Intervirology*, **23**, 61–73.

Family: **Reoviridae** (class III)

10–12 segments of double-stranded RNA of total M_r $12–20 \times 10^6$. Particle is a 60–80 nm icosahedron. Has an isometric nucleocapsid with transcriptase activity. Cytoplasmic multiplication.

Genera: Reovirus — of vertebrates
Orbivirus — of vertebrates, but also multiply in insects
Rotavirus — of vertebrates
Cytoplasmic polyhedrosis viruses — of insects
Phytoreovirus — clover wound tumour virus
Fijivirus — Fiji disease of plants

See: Estes, M. K. & Cohen, J. (1989) Rotavirus gene structure and function. *Microbiological Reviews,* **53,** 410–449.
Estes, M. K., Palmer, E. L. & Obijeski, J. F. (1983) Rotaviruses: a review. *Current Topics in Microbiology and Immunology,* **105,** 123–184.
Joklik, W. K. (1985) Recent progress in reovirus research. *Annual Review of Genetics,* **19,** 537–575.
Nuss, D. L. & Dall, D. J. (1990) Structural and functional properties of plant reovirus genomes. *Advances in Virus Research,* **38,** 249–306.
Roy, P. & Gorman, B. M. (1990) Bluetongue viruses. *Current Topics in Microbiology and Immunology,* **162,** 1–200.

Family: **Birnaviridae** (class III)

Two segments of double-stranded RNA of M_r 2.5×10^6 and 2.3×10^6 in one 60 nm particle. Icosahedral with 45 nm core. RNA transcriptase present. Cytoplasmic.

Genus: Birnavirus (pancreatic necrosis virus of fish; infectious bursal disease of chickens; *Drosophila* X virus)

See: Becht, H. (1980) Infectious bursal disease virus. *Current Topics in Microbiology and Immunology,* **90,** 107–121.

Family: **Picornaviridae** (class IV)

Single-stranded RNA of M_r 2.5×10^6. Icosahedral particles of 30 nm. Multiplication is cytoplasmic.

Genera: Enterovirus (acid-resistant, primarily viruses of gastrointestinal tract)
Rhinovirus (acid-labile, mainly viruses of upper respiratory tract)
Aphthovirus (foot-and-mouth disease virus)
Cardiovirus (encephalomyocarditis (EMC) virus of mice)
Hepatitis A virus (of humans)
Also various viruses of insects

See: Macnaughton, M. R. (1982) The structure and replication of rhinoviruses. *Current Topics in Microbiology and Immunology,* **97,** 1–26.

Minor, P. D., Kew, O. & Schild, G. C. (1982) Poliomyelitis — epidemiology, molecular biology and immunology. *Nature (London)*, **299**, 109–110.

Racaniello, V. R. (1990) Picornaviruses. *Current Topics in Microbiology and Immunology*, **161**, 1–192.

Stanway, G. (1990) Structure, function and evolution of picornaviruses. *Journal of General Virology*, **71**, 2483–2501.

Family: **Togaviridae** (class IV)

Single-stranded RNA of M_r 4×10^6. Enveloped particles 60–70 nm diameter contain an icosahedral nucleocapsid. Haemagglutinate. Cytoplasmic, budding from plasma membrane. Have a subgenomic mRNA.

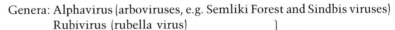

Genera: Alphavirus (arboviruses, e.g. Semliki Forest and Sindbis viruses)
        Rubivirus (rubella virus)
        Arterivirus (equine arteritis virus,   not
        lactate dehydrogenase-elevating virus)   arboviruses
        (about to be reclassified)

See: Garoff, H., Kondor-Koch, C. & Riedel, H. (1982) Structure and assembly of alphaviruses. *Current Topics of Microbiology and Immunology*, **99**, 1–50.

Schlesinger, S. & Schlesinger, M. J. (eds) (1986) *The Togaviridae and Flaviviridae*. New York: Plenum Press.

Family: **Flaviviridae** (class IV)

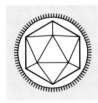

Single-stranded RNA of $M_r \sim 4 \times 10^6$. Enveloped particles 40–60 nm diameter. Differ from Alphaviridae by presence of a matrix protein, the lack of intracellular subgenomic mRNAs and budding from the endoplasmic reticulum. Haemagglutinate. Cytoplasmic.

Genera: Flavivirus (arboviruses, e.g. yellow fever virus)
        Pestivirus
        Hepatitis C virus (no culture system; only humans and chimpanzees)

See: Chambers, T. J. *et al.* (1990) Flavivirus gene organization, gene expression, and replication. *Annual Review of Microbiology*, **44**, 649–688.

Collett, M. S. *et al.* (1989) Recent advances in pestivirus research. *Journal of General Virology*, **70**, 253–266.

Schlesinger, S. & Schlesinger, M. J. (eds) (1986) *The Togaviridae and Flaviviridae*. New York: Plenum Press.

Westaway, E. G. (1987) Flavivirus replication strategy. *Advances in Virus Research*, **33**, 45–90.

Westaway, E. G. *et al.* (1985) Flaviviridae. *Intervirology*, **24**, 183–192.

Family: **Rhabdoviridae** (class V)

Single-stranded RNA of M_r $3.5–4.6 \times 10^6$ complementary to mRNA.

The particle is bullet-shaped or bacilliform ($100-430 \times 70$ nm) and enveloped with 5–10 nm spikes. Inside is a helical nucleocapsid with transcriptase activity. Cytoplasmic, budding from plasma membrane. Some are arboviruses.

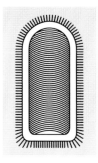

Genera: Vesiculovirus (vesicular stomatitis virus group) viruses of
    vertebrates and insects
    Lyssavirus (viruses of vertebrates, e.g. rabies, and of insects,
    e.g. sigma)
    Plant viruses (e.g. lettuce necrotic yellows, potato yellow
    dwarf and many others)

See: Banerjee, A. & Barik, S. (1992) Gene expression of vesicular
 stomatitis virus RNA. *Virology*, **188**, 417–428.

Bishop, D. H. L. (ed.) (1979/80) *Rhabdoviruses*, Vols 1–3. Boca Raton,
 FL: CRC Press.

Family: **Bunyaviridae** (class V)

Three molecules (large, medium and small) of single-stranded RNA of total M_r 6×10^6. Enveloped 100 nm particles with spikes and three internal ribonucleoprotein filaments 2 nm wide. Cytoplasmic; bud from Golgi. Arthropod-transmitted except hantavirus genus.

Genera: Bunyavirus (Bunyamwera and 150 or so related viruses)
    Phlebovirus (sand-fly fever virus group; 30 or so members)
    Nairovirus (Nairobi sheep disease virus; 30 or so members)
    Hantavirus (Korean haemorrhagic fever or Hantaan virus) *not*
    arboviruses
    Tospovirus (tomato spotted wilt virus group; thrips vector)

See: De Haan, P. *et al.* (1989) Molecular cloning and terminal sequence
 determination of the S and M RNAs of tomato spotted wilt virus.
 Journal of General Virology, **70**, 3469–3473.

Elliott, R. M. (1990) Molecular biology of the Bunyaviridae. *Journal of
 General Virology*, **71**, 501–522.

Kolakofsky, D. (1991) Bunyaviridae. *Current Topics in Microbiology
 and Immunology*, **169**, 1–256.

VIRUSES MULTIPLYING ONLY IN VERTEBRATES

Family: **Herpesviridae** (class I)

Double-stranded DNA of M_r $80-150 \times 10^6$. Particle is a 130 nm icosahedron enclosed in a lipid envelope. Buds from nuclear membrane. Latency for the lifetime of the host is common.

Subfamily: *Alphaherpesvirinae*
    Herpes simplex virus types 1 and 2
    Varicella-zoster virus

Subfamily: *Betaherpesvirinae*
       Human cytomegalovirus
       Mouse cytomegalovirus

Subfamily: *Gammaherpesvirinae* (lymphoproliferative viruses)
       Epstein–Barr virus
       Herpesvirus saimiri
           Unclassified: Marek's disease virus

See: Davison, A. J. (1991) Varicella-zoster virus. *Journal of General Virology*, **72**, 475–486.

Mach, M. *et al.* (1989) Human cytomegalovirus: recent aspects from molecular biology. *Journal of General Virology*, **70**, 3117–3146.

Roizman, B. (1990) Herpesviridae: a brief introduction. In: *Virology* (2nd edn), Vol. 2, pp. 1787–1794. Fields, B. N. & Knipe, D. M. (eds.). New York: Raven Press.

Roizman, B. & Sears, A. E. (1990) Herpes simplex viruses and their replication. In: *Virology* (2nd edn), Vol. 2, pp. 1795–1841. Fields, B. N. & Knipe, D. M. (eds.). New York: Raven Press.

Rouse, B. T. (1992) Herpes simplex virus: pathogenesis, immunobiology and control. *Current Topics in Microbiology and Immunology*, **179**, 1–179.

Stevens, J. G. (1989) Human herpesviruses: a consideration of the latent state. *Microbiological Reviews*, **53**, 318–332.

Family: **Adenoviridae** (class I)

Double-stranded DNA of M_r $20-30 \times 10^6$. Particle is a 70–90 nm icosahedron which replicates and is assembled in the nucleus.

Genera: Mastadenovirus (adenoviruses of mammals)
        Aviadenovirus (adenoviruses of birds)

See: Doefler, W. (ed.) (1983/1984) The molecular biology of adenoviruses. *Current Topics in Microbiology and Immunology*, **109** (1983), **110**, **111** (1984).

Ginsberg, H. S. (ed.) (1984) *The adenoviruses*. New York: Plenum Press.

Horwitz, M. S. (1990) Adenoviridae and their replication. In: *Virology* (2nd edn), Vol. 2, pp. 1679–1722. Fields, B. N. & Knipe, D. M. (eds). New York: Raven Press.

Family: **Papovaviridae** (class I)

Double-stranded circular DNA. Particles have 72 capsomers in a skew arrangement and are assembled in the nucleus. Haemagglutinate. Oncogenic.

Genera: Papillomavirus (producing papillomas in several mammalian species including man) 50–55 nm particle; DNA 5×10^6 M_r
        Polyomavirus (found in rodents, humans and other primates) 40–45 nm particle; DNA 3×10^6 M_r. Includes simian virus type 40 (SV40) and polyomavirus itself

See: Lambert, P. F. (1991) Papillomavirus DNA replication. *Journal of Virology*, **65**, 3417–3420.

Salzman, N. P. (1986) *The Papovaviridae*, Vol. 1, *The polyomaviruses*. New York: Plenum Press.

Salzman, N. P. & Howley, P. M. (1987) *The Papovaviridae*, Vol. 2, *The papillomaviruses*. New York: Plenum Press.

Tooze, J. (1981) *DNA tumor viruses* (2nd edn), Chapters 2–6. Cold Spring Harbor Laboratory, NY.

Family: **Hepadnaviridae** (class I)

One complete DNA minus strand of M_r 1×10^6 with a 5′ terminal protein. DNA is circularized by an incomplete plus strand of variable length (50–100%) which overlaps the 3′ and 5′ termini of DNA minus. There is a 42 nm enveloped particle containing a core with DNA polymerase and protein kinase activities. Includes hepatitis B (HBV) of humans, Pekin duck hepatitis, beechy ground squirrel hepatitis and woodchuck hepatitis viruses. HBV is strongly associated with liver cancer.

See: Ganem, D. & Varmus, H. E. (1987) The molecular biology of the hepatitis B viruses. *Annual Review of Biochemistry*, **56**, 651–693.

Howard, C. R. (1986) The biology of hepadnaviruses. *Journal of General Virology*, **67**, 1215–1235.

Marion, P. L. & Robinson, W. S. (1983) Hepadnaviruses: hepatitis B and related viruses. *Current Topics in Microbiology and Immunology*, **105**, 99–121.

Mason, W. S. & Seeger, C. (1991) Hepadnaviruses: molecular biology and pathogenesis. *Current Topics in Microbiology and Immunology*, **168**, 1–206.

Tiollais, P., Pourcel, C. & Dejean, A. (1985) The hepatitis B virus. *Nature (London)*, **317**, 489–495.

Family: **Coronaviridae** (class IV)

Single-stranded RNA of M_r $2–11 \times 10^6$. Enveloped particles of 60–220 nm with club-shaped sparse spikes. Contains a helical nucleocapsid 9 nm diameter. Cytoplasmic, budding from Golgi and endoplasmic reticulum.

Genera: Coronavirus (avian infectious bronchitis virus and related viruses, including equine arteritis virus)

Torovirus (enveloped biconcave 130 nm particles with spikes. Helical nucleocapsid. Nucleus required for replication. Berne virus of horses).

See: Horzinek, M. C. *et al.* (1987) A new family of vertebrate viruses: Toroviridae. *Intervirology*, **27**, 17–24.

Lai, M. M. C. (1990) Coronaviruses: organization, replication and expression of the genome. *Annual Review of Microbiology*, **44**, 303–333.

Siddell, S., Wege, H. & ter Meulen, V. (1983) The biology of corona-viruses. *Journal of General Virology*, **64**, 761–776.

Spaan, W. J. M. *et al.* (1988) Coronaviruses: structure and genome expression. *Journal of General Virology*, **69**, 2939–2952.

Wege, H., Siddell, S. & ter Meulen, V. (1982) The biology and patho-genesis of coronaviruses. *Current Topics in Microbiology and Immunology*, **99**, 165–200.

Weiss, M. & Horzinek, M. C. (1986) Morphogenesis of Berne virus (proposed family Toroviridae). *Journal of General Virology*, **67**, 1305–1314.

Family: **Calicivirus** (class IV)

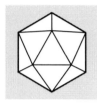

Single-stranded RNA of M_r 2.7×10^6. Icosahedral 37 nm particle with calix-like (cup-shaped) surface depressions.

Genus: Calicivirus (vesicular exanthema of swine virus, Norwalk virus, hepatitis E virus)

See: Cubitt, W. D. (1989) Diagnosis, occurrence and clinical significance of the human 'candidate' caliciviruses. *Progress in Medical Virology*, **36**, 103–119.

Schaffer, F. L. (1979) Caliciviruses. *Comprehensive Virology*, **14**, 249–284.

Family: **Arenaviridae** (class V)

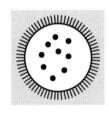

Two (one large and one small) segments of single-stranded RNA of M_r 3×10^6 and 1×10^6. Latter RNA is ambisense. Enveloped 50–300 nm particles with spikes. Contain ribosomes which have no known func-tion. Cytoplasmic multiplication; buds from plasma membrane.

Genus: Arenavirus (lymphocytic choriomeningitis virus and related viruses)

See: Bishop, D. H. L. & Auperin, D. D. (1987) Arenavirus gene structure and organization. *Current Topics in Microbiology and Immunology*, **133**, 5–17.

Compans, R. W. & Bishop, D. H. L. (1985) Biochemistry of arenaviruses. *Current Topics in Microbiology and Immunology*, **114**, 153–175.

Howard, C. R. (1986) Arenaviruses. In: *Perspectives in Medical Virology*, Vol. 2. New York: Elsevier.

Lehmann-Grube, F. (1984) Portraits of viruses: arenaviruses. *Inter-virology*, **22**, 121–145.

Salvato, M. S. (ed.) (1993) *The Arenaviridae*. New York: Plenum Publishing.

Family: **Paramyxoviridae** (class V)

Single-stranded RNA of M_r $5–7 \times 10^6$. Enveloped 150 nm particles have spikes and contain a nucleocapsid 12–17 nm in diameter with

transcriptase activity. Filamentous forms common. Cytoplasmic, budding from plasma membrane. Airborne transmission.

Genera: Paramyxovirus (Newcastle disease virus group). Only this
genus has a neuraminidase activity which is on the same
protein (HN) as the haemagglutination activity
Morbillivirus (measles virus group). Haemagglutinate
Pneumovirus (respiratory syncytial virus group)

See: Kingsbury, D. W. (ed.) (1991) *The paramyxoviruses*. New York:
Plenum Publishing.
Klenk, H.-D. & Rott, R. (1980) Cotranslational and post-translational
processing of viral glycoproteins. *Current Topics in Microbiology
and Immunology*, **90**, 19–48.
Rima, B. K. (1983) The proteins of morbilliviruses. *Journal of General
Virology*, **64**, 1205–1219.
Stott, E. J. & Taylor, G. (1984) Respiratory syncytial virus: brief review.
Archives of Virology, **84**, 1–52.

Family: **Orthomyxoviridae** (class V)

Eight segments of single-stranded RNA of total M_r 4×10^6. Enveloped 100 nm particles have spikes and contain a helical nucleocapsid 9 nm in diameter with transcriptase activity. Only A and B virions have separate haemagglutinin and neuraminidase proteins. Multiplication requires the nucleus. RNA segments in a mixed infection readily assort to form genetically stable hybrids within a virus. Buds from plasma membrane.

Genera: Influenza A and B virus
Influenza C virus. Has seven RNA segments and a receptor-
destroying activity (a sialic acid – O-acetyl esterase) which is
on the haemagglutinin protein

See: Kilbourne, E. D. (1987) *Influenza*. New York: Plenum Medical
Books.
Krug, R. M. (ed.) (1989) *The influenza viruses*. New York: Plenum
Publishing.
Lamb, R. A. & Choppin, P. W. (1983) The gene structure and replication
of influenza virus. *Annual Review of Biochemistry*, **52**, 467–506.
McCauley, J. W. & Mahy, B. W. J. (1983) Structure and function of the
influenza virus genome. *Biochemical Journal*, **211**, 281–294.
Stuart-Harris, C. H. *et al.* (1985) *Influenza: the viruses and the disease*
(2nd edn). London: Edward Arnold.

Family: **Filoviridae** (class V)

Long filamentous particles 800–900 (sometimes 14 000) × 80 nm with helical nucleocapsid of 50 nm diameter. RNA of M_r 4.2×10^6. Buds from plasma membrane. Contains Marburg and Ebola viruses which are highly pathogenic for man. Transmitted by contact.

See: Francki, R. I. B. *et al.* (1991) *Archives of Virology Supplementum*,
2, 247.

Family: **Retroviridae** (class VI)

Two identical molecules of single-stranded RNA$^+$ per virion of M_r $1-3 \times 10^6$. Not infectious. Enveloped 100 nm particles with a core containing a helical nucleoprotein. Contains an RNA-dependent DNA polymerase. The DNA provirus is nuclear. Transmission is horizontal or vertical. Associated with many different diseases. Not all viruses are oncogenic.

Mouse mammary tumour virus group (oncogenic)
Murine leukaemia virus group (oncogenic)
Mason–Pfizer monkey virus group
Avian leucosis virus group (oncogenic)
Foamy virus group (spumaviruses; no disease known)
Human T-cell lymphotropic virus type 1 group (oncogenic)
Lentivirus group (includes human immunodeficiency virus (HIV)-1, HIV-2, visna–maedi virus; associated with immunodeficiency diseases)

See: Levy, J. A. (ed.) (1992) *The Retroviridae*, Vol. 1. New York: Plenum Publishing.
Levy, J. A. (ed.) (1993) *The Retroviridae*, Vol. 2. New York: Plenum Publishing.
Weiss, R. A., Teich, N., Varmus, H. & Coffin, J. (1984/1985) *RNA tumor viruses*, Vols 1 and 2 (2nd edn). Cold Spring Harbor Laboratory, NY.

VIRUSES MULTIPLYING ONLY IN INVERTEBRATES

Viruses occur not only in insects, crustacea and molluscs but probably in all groups of invertebrates. The Poxviridae, Reoviridae, Parvoviridae, Rhabdoviridae and Togaviridae (see earlier) have representatives which multiply in invertebrates. Some plant viruses are transmitted by, but do not multiply in, these vectors.

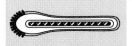

Family: **Baculoviridae** (class I)

Double-stranded circular DNA of M_r $60-110 \times 10^6$. Bacilliform particles $30-60$ nm $\times 250-300$ nm with an outer membrane. May be occluded in a protein inclusion body containing usually one particle (granulosis viruses, upper illustration) or in a polyhedra containing many particles (polyhedrosis viruses, lower illustration).

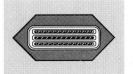

Genus: Baculovirus (*Bombyx mori* nuclear polyhedrosis virus group)

See: Doerfler, W. & Bohm, P. (1986) The molecular biology of baculoviruses. *Current Topics in Microbiology and Immunology*, **131**, 1–168.
Kelly, D. C. (1982) Baculovirus replication. *Journal of General Virology*, **63**, 1–13.

Family: **Polydnaviridae** (class I)

Segmented genome of double-stranded polydisperse circular supercoiled DNA in multiple copies of M_r $1.5-20 \times 10^6$. Group A have particles of 85×330 nm (upper illustration) and group B (lower illustration) are variable in size. Have two membranes as bud from the nucleus and then from the plasma membrane.

See: Stoltz, D. B. & Vinson, S. B. (1979) Viruses and parasitism in insects. *Advances in Virus Research*, **24**, 125–171.

Stoltz, D. B., Krell, P. J., Summers, M. D. & Vinson, S. B. (1984) Polydnaviridae — a proposed family of insect viruses with segmented, double-stranded, circular DNA genomes. *Intervirology*, **21**, 1–4.

Family: **Tetraviridae** (class IV)

Single-stranded RNA of M_r 1.8×10^6 in a 35 nm particle. T = 4 (whereas in Picornaviridae T = 1). All isolated from Lepidoptera. Nudaurelia β virus. No infection of cultured cells.

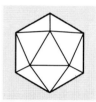

See: Moore, N. F., Reavy, B. & King, L. A. (1985) General characteristics, gene organization and expression of small RNA viruses of insects. *Journal of General Virology*, **66**, 647–659.

Family: **Nodaviridae** (class IV)

Two molecules of single-stranded RNA of M_r 1.17 or 0.48×10^6 in one 29 nm particle. No 3′ poly-A tract. Both RNAs required for infectivity. Cytoplasmic multiplication. Insect viruses but some grow unnaturally in suckling mice or vertebrate cells.

Genus: Nodavirus (*Nodamura* and black beetle viruses)

See: Moore, N. F., Reavy, B. & King, L. A. (1985) General characteristics, gene organization and expression of small RNA viruses of insects. *Journal of General Virology*, **66**, 647–659.

VIRUSES MULTIPLYING ONLY IN PLANTS

Knowledge of virus multiplication is relatively rudimentary since the cell culture systems are less manageable than the animal cell cultures. Work has concentrated on physical properties and disease characteristics. Designation into families or genera has not yet been decided. Remember that the Reoviridae and Rhabdoviridae have members which multiply in both plants and invertebrates. The plant viruses listed below are not known to multiply in their invertebrate vector. Differences in virus proteins and translation strategy (which may not be mentioned below) are important criteria in plant virus classification. There are fewer accounts of individual virus groups than with animal viruses, so the following general texts may be useful.

See: Bos, L. (1983) *Introduction to plant virology*. London: Longman.
Davies, J. W. & Hull, R. (1982) Genome expression of plant positive strand RNA viruses. *Journal of General Virology*, **61**, 1–14.
Davies, J. W., Covey, S. N. *et al.* (1985) Structure and function of plant DNA virus genomes. In: *Molecular form and function of the plant genome*. van Vloten-Doting, L., Groot, G. S. P. & Hall, T. C. (eds). New York: Plenum Press.
Francki, R. (1985) *Atlas of plant viruses*, Vols 1 and 2. Boca Raton, FL: CRC Press.
Matthews, R. E. F. (1992) *Plant virology* (3rd edn). New York: Academic Press (for advanced students).
Matthews, R. E. F. (1992) *Fundamentals of plant virology*. New York: Academic Press (a more manageable text).
Robertson, H. D. Howell, S. H., Zaitlin, M. & Malmberg, R. L. (1983) *Plant infectious agents: viroids and satellites*. Cold Spring Harbor Laboratory, NY.
Wilson, T. M. A. & Davies, J. W. (1992) *Genetic engineering of plant viruses*. Boca Raton, FL: CRC Press.

Caulimovirus (class I) — cauliflower mosaic virus group

Open circular DNA with single strand discontinuties like hepadnavirus. M_r 4–5 × 10^6. Isometric 50 nm particles. Aphid vectors.

See: Francki, R. I. B. *et al.* (1991) Classification and nomenclature of viruses. *Archives of Virology Supplementum*, **2**, 153.
Hohn, T., Hohn, B. & Pfeiffer, P. (1985) Reverse transcription in CaMV. *Trends in Biochemical Sciences*, **10**, 205–209.

Geminivirus (class II)

Circular single-stranded DNA of M_r 0.7–0.8 × 10^6. Two incomplete icosahedral particles 18 × 30 nm joined as a pair and usually found in the nucleus. One (subgroups I and II) or two (subgroup III) molecules of DNA per pair of particles. Persistent in whitefly or leafhopper vectors.

See: Howarth, A. J. & Goodman, R. M. (1982) Plant viruses with genomes of single-stranded DNA. *Trends in Biochemical Sciences*, **7**, 180–182.
Stanley, J. (1985) The molecular biology of geminiviruses. *Advances in Virus Research*, **30**, 139–177.

Commelina yellow mottle virus group (class II)

Bacilliform 130 × 30 nm particles containing incomplete circles of double-stranded DNA (like caulimoviruses and Hepadnaviridae). Transmitted by beetle larvae or leafhoppers.

See: Francki, R. I. B. *et al.* (1991) Classification and nomenclature of viruses. *Archives of Virology Supplementum*, **2**, 153.

Cryptovirus (class III)—white clover cryptic virus group

Isometric 30–38 nm particles containing two molecules of linear double-stranded RNA of M_r $1-1.4 \times 10^6$ and $1.2-1.4 \times 10^6$. Contain a transcriptase. Spread by seed or pollen.

See: Francki, R. I. B. *et al.* (1991) Classification and nomenclature of viruses. *Archives of Virology Supplementum*, **2**, 212.

Luteovirus (class IV)—barley yellow dwarf virus group

Single-stranded RNA of M_r 2×10^6. Isometric 25–30 nm particle. Persistent retention by aphid vectors.

See: Francki, R. I. B. *et al.* (1991) Classification and nomenclature of viruses. *Archives of Virology Supplementum*, **2**, 309.

Maize chlorotic dwarf virus group (class IV)

Single-stranded RNA of M_r 3.2×10^6 in 30 nm particle. Only infects grasses and transmitted by leafhoppers.

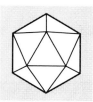

See: Francki, R. I. B. *et al.* (1991) Classification and nomenclature of viruses. *Archives of Virology Supplementum*, **2**, 312.

Necrovirus (class IV)—tobacco necrosis virus group

Single-stranded RNA of M_r 1.5×10^6. Isometric 28 nm particle. Cytoplasmic. Fungal vector.

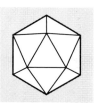

See: Francki, R. I. B. *et al.* (1991) Classification and nomenclature of viruses. *Archives of Virology Supplementum*, **2**, 316.

Sobemovirus (class IV)—southern bean mosaic virus group

Single-stranded RNA of M_r 1.4×10^6 in 30 nm particle. No 3' poly-A tract or transfer RNA (tRNA) structure. Virions present in the nucleus and cytoplasm. Beetle vectors or seed transmission.

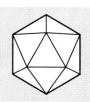

See: Francki, R. I. B. *et al.* (1991) Classification and nomenclature of viruses. *Archives of Virology Supplementum*, **2**, 327.

Tombusvirus (class IV)—tomato bushy stunt virus group

Single-stranded RNA of M_r 1.5×10^6. Not polyadenylated. 30 nm particle. Cytoplasmic and nuclear. Transmitted through the soil.

See: Francki, R. I. B. *et al.* (1991) Classification and nomenclature of viruses. *Archives of Virology Supplementum*, **2**, 332.

Marafivirus (class IV) — maize rayado fine virus group

Single-stranded RNA of M_r 2.2×10^6 in a 31 nm particle. Cytoplasmic. Leafhopper vector.

See: Francki, R. I. B. *et al.* (1991) Classification and nomenclature of viruses. *Archives of Virology Supplementum*, **2**, 314.

Tymovirus (class IV) — turnip yellow mosaic virus group

Single-stranded RNA of M_r 2×10^6. 29 nm particle. Associated with chloroplasts. Beetle vectors.

See: Francki, R. I. B. *et al.* (1991) Classification and nomenclature of viruses. *Archives of Virology Supplementum*, **2**, 336.

Carmovirus group (class IV)

Isometric 33 nm particle. Single-stranded RNA of M_r 1.3×10^6. Atomic structure resolved for turnip crinkle virus. Transmission is mostly mechanical.

See: Francki, R. I. B. *et al.* (1991) Classification and nomenclature of viruses. *Archives of Virology Supplementum*, **2**, 303.

Bromovirus (class IV) — brome mosaic virus group

Four single-stranded RNAs of M_r 1.1, 1.0, 0.8 and 0.3×10^6. Particles 26 nm diameter. Infectivity requires three largest RNAs. Each is in a different particle. Assembly in cytoplasm. Some with a beetle vector.

See: Francki, R. I. B. *et al.* (1991) Classification and nomenclature of viruses. *Archives of Virology Supplementum*, **2**, 382.
van Vloten-Doting, L. & Jaspars, E. M. J. (1977) Plant covirus systems: three component systems. *Comprehensive Virology*, **11**, 1–53.

Cucumovirus (class IV) — cucumber mosaic virus group

Single-stranded RNAs of M_r 1.3, 1.1, 0.8 and 0.3×10^6. 29 nm particles. Infectivity requires three largest RNAs, each of which is in a separate particle. Cytoplasmic. Non-persistent in aphid vector.

See: Francki, R. I. B. *et al.* (1991) Classification and nomenclature of viruses. *Archives of Virology Supplementum*, **2**, 386.
van Vloten-Doting, L. & Jaspars, E. M. J. (1977) Plant covirus systems: three component systems. *Comprehensive Virology*, **11**, 1–53.

Ilarvirus (class IV)—tobacco streak virus group

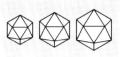

At least three quasi-isometric particles of different sizes from 26 to 35 nm in diameter, with four single-stranded RNAs of M_r 1.1, 0.9, 0.7 and 0.3×10^6 in different particles. All RNAs, or three largest plus coat protein, required for infectivity. Seed- and pollen-borne.

See: Bol, J. F., Cornelissen, B. J. C., Huisman, M. J. & van Vloten-Doting, L. (1985) Structure and function of the tripartite RNA genome of ilarviruses. In: *Molecular form and function of the plant genome.* Davies, J. W. *et al.* (eds). New York: Plenum Press.

Francki, R. I. B. *et al.* (1991) Classification and nomenclature of viruses. *Archives of Virology Supplementum,* **2**, 389.

Alfalfa mosaic virus group (class IV)

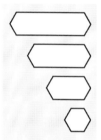

Four bacilliform particles 18×56, 18×43, 18×35 nm and 18×30 nm. Three largest contain single-stranded RNA of M_r 1.1, 0.8, 0.7 and 0.3×10^6, all needed for infectivity. Smallest contains two molecules of the coat protein mRNA. Cytoplasmic. Non-persistent in aphid vector.

See: Francki, R. I. B. *et al.* (1991) Classification and nomenclature of viruses. *Archives of Virology Supplementum,* **2**, 392.

van Vloten-Doting, L. & Jaspars, E. M. J. (1977) Plant covirus systems: three component systems. *Comprehensive Virology,* **11**, 1—53.

Comovirus (class IV)—cowpea mosaic virus group

Two 28 nm particles containing single-stranded RNA of M_r 2.4 or 1.4×10^6. Two coat polypeptides encoded by the smaller RNA. Both RNAs needed for infectivity. Cytoplasmic. Transmitted by beetles or seed.

See: Breuning, G. (1977) Plant covirus systems: two-component systems. *Comprehensive Virology,* **11**, 55—141.

Francki, R. I. B. *et al.* (1991) Classification and nomenclature of viruses. *Archives of Virology Supplementum,* **2**, 360.

Dianthovirus (class IV)—carnation ringspot virus group

Two molecules of single-stranded RNA of M_r 1.5 or 0.5×10^6 in two 32 nm particles; both needed for infectivity. Neither is polyadenylated. The larger RNA codes for a single coat protein. Cytoplasmic. Transmitted through soil.

See: Francki, R. I. B. *et al.* (1991) Classification and nomenclature of viruses. *Archives of Virology Supplementum,* **2**, 364.

Hiruki, C. (1987) The dianthoviruses: a distinct group of isometric plant viruses with bipartite genome. *Advances in Virus Research,* **33**, 257—300.

Nepovirus (class IV) — tobacco ringspot virus group

Two 28 nm particles containing single-stranded RNA of M_r 2.4−2.8 or 1.3−2.4 × 10⁶. Both RNAs needed for infectivity. One coat protein encoded on larger RNA. Some nepoviruses have a third particle containing two molecules of the second RNA. Cytoplasmic. Nematode vectors.

See: Breuning, G. (1977) Plant covirus systems: two-component systems. *Comprehensive Virology*, **11**, 55−141.

Francki R. I. B. *et al.* (1991) Classification and Nomenclature of Viruses. *Archives of Virology Supplementum*, **2**, 368−371.

Fabavirus group (class IV) — broad-bean wilt virus group

Two 30 nm particles containing single-stranded RNA of M_r 2.1 or 1.5 × 10⁶. Both needed for infectivity. Cytoplasmic replication. Non-persistent transmission by aphids.

See: Francki, R. I. B. *et al.* (1991) Classification and nomenclature of viruses. *Archives of Virology Supplementum*, **2**, 366.

Pea enation mosaic virus group (class IV)

Two particles (∼28 nm each) containing single-stranded RNA of M_r 1.7−1.3 × 10⁶. Both needed for infectivity. Nuclear replication. Persistent in aphid vector.

See: Francki, R. I. B. *et al.* (1991) Classification and nomenclature of viruses. *Archives of Virology Supplementum*, **2**, 375.

Tobamovirus (class IV) — tobacco mosaic virus group

Single-stranded RNA of M_r 2 × 10⁶. Rigid cylindrical particles 300 × 18 nm. Transmitted mechanically or by seed.

See: Dawson, W. O. (1992) Tobamovirus−plant interactions. *Virology*, **186**, 359−367.

Francki, R. I. B. *et al.* (1991) Classification and nomenclature of viruses. *Archives of Virology Supplementum*, **2**, 357.

Hirth, L. & Richards, K. E. (1981) Tobacco mosaic virus: model for structure and function of a simple virus. *Advances in Virus Research*, **26**, 145−199.

Tobravirus (class IV) — tobacco rattle virus group

Two straight tubular particles. The larger is about 200 nm long with an RNA of 2.4 × 10⁶ M_r. Shorter particles are 46−114 nm and contain RNA of 0.6−1.4 × 10⁶ which specifies the coat protein. The larger RNA is infectious alone but both RNAs are needed for synthesis of new particles. Cytoplasmic. Nematode vector.

See: Francki, R. I. B. *et al.* (1991) Classification and nomenclature of viruses. *Archives of Virology Supplementum*, **2**, 380.

Harrison, B. D. & Robinson, D. J. (1978) The tobraviruses. *Advances in Virus Research*, **23**, 25–77.

Furovirus (class IV) — soil-borne wheat mosaic virus

Two rods of $92-160 \times 20$ nm and $250-300 \times 20$ nm containing single-stranded RNA of M_r $1.8-2.4 \times 10^6$ or $3.5-4.3 \times 10^6$. Not poly-adenylated. Cytoplasmic. Fungal vector.

See: Brunt, A. A. & Richards, K. E. (1989) Biology and molecular biology of furoviruses. *Advances in Virus Research*, **36**, 1–32.

Francki, R. I. B. *et al.* (1991) Classification and nomenclature of viruses. *Archives of Virology Supplementum*, **2**, 377.

Hordeivirus (class IV) — barley stripe mosaic virus group

Rigid tubular particles with helical symmetry $110-150 \times 20$ nm. Three single-stranded RNAs of M_r $1-1.5 \times 10^6$ (depending on the strain). All required for infectivity. Mostly cytoplasmic. Transmitted mechanically and by seed.

See: Francki, R. I. B. *et al.* (1991) Classification and nomenclature of viruses. *Archives of Virology Supplementum*, **2**, 395.

Potexvirus (class IV) — potato virus X group

Single-stranded RNA of M_r 2.1×10^6. Particle is a flexuous rod with helical symmetry, $470-580 \times 13$ nm. Cytoplasmic but some have nuclear inclusions. Transmitted mechanically.

See: Francki, R. I. B. *et al.* (1991) Classification and nomenclature. *Archives of Virology Supplementum*, **2**, 348.

Potyvirus (class IV) — potato virus Y group

Single-stranded RNA of M_r 3×10^6. Particle is a flexuous rod with helical symmetry, $680-900 \times 11$ nm. Cytoplasmic but some have nuclear inclusions. Non-persistent in aphid vectors.

See: Reichmann, J. L., Laín, S. & García, J. A. (1992) Highlights and prospects of potyvirus molecular biology. *Journal of General Virology*, **73**, 1–16.

Carlavirus (class IV) — carnation latent virus group

Single-stranded RNA of 2.7×10^6 M_r. Particle is a flexuous rod of helical symmetry, about 690×13 nm. Cytoplasmic. Non-persistent in aphid vectors.

See: Francki, R. I. B. *et al.* Classification and nomenclature. *Archives of Virology Supplementum*, **2**, 341.

Capillovirus group (class IV) — apple stem grooving virus group

Flexuous filamentous 640×12 nm particles. RNA of M_r 2.5×10^6. Transmitted through seed.

See: Coffin, R. S. & Coutts, R. H. A. (1993) The closteroviruses, capilloviruses: a short review. *Journal of General Virology*, **74**, 1475–1483.

Yoshikawa, N. & Takahashi, T. (1988) Properties of RNAs and proteins of apple stem grooving and apple chlorotic leaf spot viruses. *Journal of General Virology*, **69**, 241–245.

Closterovirus (class IV) — beet yellows virus group

Long (700–2000 × 12 nm), very flexuous rods with helical symmetry containing single-stranded RNA of M_r $2.5–6.5 \times 10^6$. Aphid vectors.

See: Coffin, R. S. & Coutts, R. H. A. (1993) The closteroviruses, capilloviruses and other similar viruses: a short review. *Journal of General Virology*, **74**, 1475–1483.

Francki, R. I. B. *et al.* (1991) Classification and nomenclature of viruses. *Archives of Virology Supplementum*, **2**, 345.

Tenuivirus group (class V?) — rice stripe virus group

Variable-length filamentous, sometimes branched, narrow particles 8 nm in diameter. Four or five molecules of single-stranded RNA, possibly RNA$^-$. Particles have a transcriptase. Persistent transmission by leafhoppers.

See: Francki, R. I. B. *et al.* (1991) Classification and nomenclature of viruses. *Archives of Virology Supplementum*, **2**, 398.

VIRUSES MULTIPLYING ONLY IN ALGAE, FUNGI AND PROTOZOA

Family: **Phycodnaviridae** (class II)

Large polyhedral 130–200 nm diameter particles containing linear double-stranded DNA of M_r $150–210 \times 10^6$. Contain lipid. Infect *Paramoecium* and *Chlorella* sp.

See: Francki, R. I. B. *et al.* (1991) Classification and nomenclature of viruses. *Archives of Virology Supplementum*, **2**, 137.

Van Etton, J. L. *et al.* (1991) Viruses and virus like particles of eukaryotic algae. *Microbiological Reviews*, **55**, 586–620.

Family: **Totiviridae** (class III)

Isometric 40 nm particles with a genome consisting of a single molecule of double-stranded RNA of M_r 4×10^6. One major capsid protein. Single shell. Transcriptase present. Cytoplasmic.

Genus: Totivirus (*Saccharomyces cerevisiae* virus LA)

See: Adler, J. P., Wood, H. A. & Bozarth, R. F. (1976) Virus-like
particles from killer, neutral and sensitive strains of *Saccharomyces
cerevisiae. Journal of Virology*, **17**, 472–476.

Francki, R. I. B. *et al.* (1991) Classification and nomenclature of viruses.
Archives of Virology Supplementum, **2**, 203.

Family: **Partitiviridae** (class III)

Isometric 32 nm particles with a genome consisting of two molecules
of double-stranded RNA each of M_r $0.9–1.6 \times 10^6$. Each separately
encapsidated and both required for infectivity. Transcriptase present.
Cytoplasmic.

Genus: Partitivirus (*Gaeumannomyces graminis* virus)

See: Buck, K. W., Almond, M. R., McFadden, J. J. P., Romanos, M. A.
& Rawlinson, C. J. (1981) Properties of thirteen viruses and virus
variants obtained from eight isolates of the wheat take-all fungus,
Gaeumannomyces graminis var. *tritici. Journal of General Virology*,
53, 235–245.

Francki, R. I. B. *et al.* (1991) Classification and nomenclature of viruses.
Archives of Virology Supplementum, **2**, 208.

VIRUSES MULTIPLYING ONLY IN BACTERIA

Surprisingly little is known of the comparative biology of bacterial
viruses, as the molecular biology is based upon a detailed study of
only a few representatives. For a cross-family account of various
aspects of bacteriophages

See: Eiserling, F. A. (1979) Bacteriophage structure. In: *Comprehensive virology*, Vol. 13, pp. 543–580. Fraenkel-Conrat, H. &
Wagner, R. R. (eds). New York: Plenum Press.

Fraenkel-Conrat, H. & Wagner, R. R. (eds) (1977) Bacterial DNA
viruses: reproduction. *Comprehensive virology*, Vol. 7. New
York: Plenum Press.

Fraenkel-Conrat, H. & Wagner, R. R. (eds) (1977) Bacterial DNA
viruses: regulation and genetics. *Comprehensive virology*, Vol. 8.
New York: Plenum Press.

Lewin, B. (1977) *Gene expression—3. Plasmids and phages*,
Chs 4–9. New York: Wiley.

Reanney, D. C. & Ackerman, H. W. (1982) Comparative biology of
bacteriophages. *Advances in Virus Research*, **27**, 205–280.

Family: **Myoviridae** (class I) (coliphage T-even group)

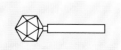

Linear double-stranded DNA of M_r 120×10^6. Head isometric or
elongated, 110×80 nm; complex contractile tail 113 nm long. Includes
T2, T4, T6, PBS1, SP8, SP50, P1, P2, 21, and 34, Mu.

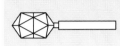

See: Francki, R. I. B. *et al.* (1991) Classification and nomenclature of viruses. *Archives of Virology Supplementum*, **2**, 161.

Lewin, B. (1977) Phage T4. In: *Gene expression—3*, pp. 536—681. New York: Wiley.

Matthews, C., Kutter, E. M., Mosig, G. & Berget, P. B. (1983) *Bacteriophage T4*. Washington DC: American Society for Microbiology.

Family: **Siphoviridae** (class I) (coliphage λ group)

Linear double-stranded DNA of M_r 33×10^6. Head 60 nm diameter; long non-contractile tail up to 570 nm. No breakdown of host DNA. Includes λ, χ (chi) and φ80.

See: Francki, R. I. B. *et al.* (1991) Classification and nomenclature of viruses. *Archives of Virology Supplementum*, **2**, 163.

Hayes, W. (1980) Portraits of viruses: bacteriophage lambda. *Intervirology*, **13**, 133—153.

Hendrix, R. W., Roberts, J. W., Stahl, F. W. & Weisberg, R. A. (1983) *Lambda II*. Cold Spring Harbor Laboratory, NY.

Lewin, B. (1977) Phage lambda. In: *Gene expression—3*, pp. 274—535. New York: Wiley. (Infective pathways, pp. 274—411; development, pp. 412—535.)

Family: **Podoviridae** (class I) (coliphage T7 and related phage groups)

Linear double-stranded DNA of M_r 25×10^6. Head 60 nm diameter. Short (20 nm) non-contractile tail. Host DNA breaks down. Includes P22.

See: Francki, R. I. B. *et al.* (1991) Classification and nomenclature of viruses. *Archives of Virology Supplementum*, **2**, 165.

Lewin, B. (1977) Phages T3 and T7. In: *Gene expression—3*, pp. 682—723. New York: Wiley.

Family: **Tectiviridae** (class I) (PRD1 phage group)

Linear double-stranded DNA of M_r 9×10^6 in 63 nm particles. Contains internal lipid. After injection of DNA a tail structure of about 60 nm appears.

See: Francki, R. I. B. *et al.* (1991) Classification and nomenclature of viruses. *Archives of Virology Supplementum*, **2**, 155.

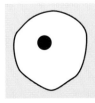

Family: **Plasmaviridae** (class I) (*Mycoplasma* virus type 2 phage group)

Circular double-stranded DNA of 8×10^6 M_r in lipid-containing envelope. 50—125 nm particle with small dense core. Formed by budding. Infects *Mycoplasma*.

See: Francki, R. I. B. *et al.* (1991) Classification and nomenclature of viruses. *Archives of Virology Supplementum*, **2**, 124.

Family: **Corticoviridae** (class I) (PM2 phage group)

Circular double-stranded DNA of M_r 6×10^6, isometric 60 nm particle, no envelope, lipid between protein shells, no tail. Spikes at vertices. Infects *Pseudomonas*.

See: Francki, R. I. B. *et al.* (1991) Classification and nomenclature of viruses. *Archives of Virology Supplementum*, **2**, 157.

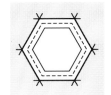

Family: **Microviridae** (class II) (φX174 group)

Circular single-stranded DNA of M_r 1.7×10^6. 27 nm icosahedron with knobs on 12 vertices. Includes G4.

See: Francki, R. I. B. *et al.* (1991) Classification and nomenclature of viruses. *Archives of Virology Supplementum*, **2**, 178.

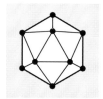

Family: **SSV1-type phages** (class I)

Circular double-stranded DNA of M_r 4.6×10^6 in a lemon-shaped 60×100 nm particle with short spikes at one end. May integrate. Infects *Sulfobulus shibatae*.

See: Francki, R. I. B. *et al.* (1991) Classification and nomenclature of viruses. *Archives of Virology Supplementum*, **2**, 127.

Family: **Lipothrixviridae** (class I)

Linear double-stranded DNA of M_r 10×10^6 in an enveloped, rigid 400×40 nm rod. Infects the archaebacterium. *Thermoproteinus tenax*.

See: Francki, R. I. B. *et al.* (1991) Classification and nomenclature of viruses. *Archives of Virology Supplementum*, **2**, 127.

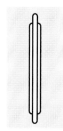

Family: **Inoviridae** (class II)

Rod-shaped phages with a circular single-stranded DNA.

Genera: Inovirus (DNA of M_r $1.9-2.7 \times 10^6$, long flexible filamentous particle up to 1950×6 nm. Host bacteria not lysed. Includes M13 and fd phages)
Plectrovirus (*Mycoplasma* virus type 1 phages. DNA of M_r $2.5-5.2 \times 10^6$ M_r. Short rods of $85-280 \times 14$ nm)

See: Francki, R. I. B. *et al.* (1991) Classification and nomenclature of viruses. *Archives of Virology Supplementum*, **2**, 181.
Zinder, N. D. & Horiuchi, K. (1985) Multiregulatory element of filamentous bacteriophages. *Microbiological Reviews*, **49**, 101–106.

Family: **Cystoviridae** (class III) (phage φ6 group)

Three molecules of linear double-stranded RNA of M_r 2.3, 3.1 and

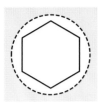

5.0×10^6. Isometric 75 nm particle with lipid envelope. Infects *Pseudomonas*.

See: Mindich, L. (1988) Bacteriophage φ6: a unique virus having a lipid-containing membrane and a genome composed of three dsRNA segments. *Advances in Virus Research*, **35**, 137–176.

Family: **Leviviridae** (class IV) — single-stranded RNA phage group)

Linear single-stranded RNA of M_r 1.2×10^6. 24 nm icosahedron. Includes R17, MS2 and Qβ.

See: Fiers, W. (1979) Structure and function of RNA bacteriophages. *Comprehensive Virology*, **13**, 69–204.
Francki, R. I. B. *et al.* (1991) Classification and nomenclature of viruses. *Archives of Virology Supplementum*, **2**, 306.
Zinder, N. D. (ed.) (1975) *RNA phages.* Cold Spring Harbor Laboratory, NY.

SATELLITE VIRUSES

Satellite viruses have genomes which depend upon coinfection of a host cell with a helper virus. The genome has no homology with that of the helper and hence differs from other types of dependent nucleic acid molecules.

Plant satellite viruses

Most satellite viruses are associated with plant viruses and have one molecule of single-stranded RNA. Some can modulate helper virus replication and alter its effect upon its plant host. Their genomes are classified as:
• large (>0.7 kb) and enclose their own capsid protein;
• large (>0.7 kb) and encode a non-structural protein;
• small (<0.7 kb), encode no protein and never circular;
• small (<0.7 kb), encode no protein and form circular molecules during replication.

Bacterial satellite viruses

P4 is a double-stranded DNA phage dependent upon phage P2.

Animal satellite viruses

Hepatitis delta virus has a large circular single-stranded RNA genome and is dependent upon hepatitis B virus (which has a partly double-stranded circular DNA genome).

Adeno-associated viruses have single-stranded DNA and are dependent upon adenoviruses or herpesviruses.

There is a single-stranded RNA satellite virus dependent upon chronic bee-paralysis virus.

See: Francki, R. I. B. (1985) Plant virus satellites. *Annual Review of Microbiology*, **39**, 151–174.

Maramarosch, K. (ed.) (1991) *Viroids and satellites: molecular parasites at the frontier of life*. Boca Raton, FL: CRC Press.

Roossinck, M. J., Sleat, D. & Palukaitis, P. (1992) Satellite RNAs of plant viruses: structures and biological effects. *Microbiological Reviews*, **56**, 265–279.

Taylor, J. M. (1992) The structure and replication of hepatitis delta virus. *Annual Review of Microbiology*, **42**, 253–275.

VIROIDS

Viroids are very small, circular, single-stranded infectious RNAs (246 to 370 nucleotides) which are never encapsidated and are pathogenic to plants. They have extensive internal base pairing, so that the RNAs resemble double-stranded rods rather than circles. There is no open reading frame and hence they encode no protein. Replication is in the nucleus by the host's RNA polymerase II. Transmission of naked RNA is a problem and viroids have overcome this by being transmitted by vegetative propagation, and some are transmitted in seed, by aphids or through mechanical damage.

See: Maramarosch, K. (ed.) (1991) *Viroids and satellites: molecular parasites at the frontier of life*. Boca Raton, FL: CRC Press.

Symons, R. H. (ed.) (1990) Viroids and related pathogenic RNAs. *Seminars in Virology*, **1**.

FURTHER READING

Francki, R. I. B., Fauquet, C. M., Knudson, D. L. and Brown, F. (1991) Classification and nomenclature of viruses. *Archives of Virology Supplementum*, **2**.

Appendix: 'Inhibitors' and other compounds used in virology

Compound	Activity	Effect on virus or host
Acridine orange	Binds to nucleic acids	dsDNA and dsRNA fluoresce bright green in UV light; ssDNA and ssRNA fluoresce dull red
Actinomycin D	Intercalates in dsDNA or dsRNA between G–C base pairs	Inhibits DNA replication at high concentration and transcription of mRNA and rRNA at low concentration
		Special effect: inhibits influenza virus multiplication by inhibiting synthesis of cell mRNA needed to prime viral mRNA synthesis
Acyclovir	Nucleoside analogue	Inhibits herpes simplex virus types 1 and 2 DNA synthesis; but not cell DNA synthesis as it is phosphorylated only by viral thymidine kinase
α-Amanitin	A cyclic octapeptide which at low concentration binds to DNA-dependent RNA polymerase II	Inhibits cell mRNA synthesis
		Special effect: inhibits influenza virus multiplication (see actinomycin D above)
	At higher concentration binds to DNA-dependent RNA polymerase III (type 1 polymerase is resistant)	Inhibits RNA synthesis
AZT	Nucleoside analogue	Inhibits all retroviruses by preventing DNA chain elongation. Used clinically against HIV
Amantadine	Affinity for endlysosomes and raises local pH	Prevents low-pH-mediated fusion of endocytosed enveloped viruses and uncoating of non-enveloped viruses
	At 100-fold lower concentration	Special effect: inhibits influenza virus multiplication by blocking the virion M2 proton channel
Chloramphenicol	Inhibits peptidyl transferase activity of 50S ribosomal subunit	Inhibits prokaryotic protein synthesis
Cordycepin	3'deoxyadenosine: has no 3' hydroxyl	Blocks RNA synthesis and polyadenylation of mRNA
Cycloheximide	Inhibits elongation phase of translation in eukaryotes	Inhibits eukaryotic protein synthesis
Cyclophosphamide	Of whole animals: prevents mitosis	Loss of immune function
Cytochalasin B	Binds to actin filaments	Nucleus extruded on a stalk and can be snapped off from cell monolayers by centrifugation, giving an enucleated cell

continued on p. 364

Compound	Activity	Effect on virus or host
Erythromycin	Inhibits elongation phase of translation in prokaryotes	Inhibits prokaryote protein synthesis
Ethidium bromide	Binds to nucleic acids	In UV light dsNA fluoresces brightly and ssRNA poorly
Formaldehyde	A polymerized form covalently cross-links free amino groups of bases in nucleic acids	Used to inactivate virus infectivity
Lysosomotropic compounds	High affinity for lysosomes and endosomes raises local pH, e.g. chloroquine, amantidine, NH_4OH	Prevents low-pH-mediated uncoating of viruses
Mitomycin C	Binds to and cross-links dsDNA	Prevents DNA synthesis and inhibits mitosis. Can induce lysogenic phages
Oligo-dT; oligo-U		Attached to insoluble matrix, e.g. sepharose, and used to affinity-purify poly-A-containing mRNA. Anneal in a high-salt and elute with a low-salt wash
Phosphonacetic acid	Pyrophosphate analogue	Inhibits herpes simplex virus types 1 and 2 DNA polymerase
β-Propiolactone	Alkylating agent reacts with amino groups of bases in nucleic acids	Used to inactivate virus infectivity
Protease inhibitors	Prevent cleavage of polypeptides	Non-functional protein in eukaryotes
Protein A	From *S. aureus*; binds with high affinity to Fc region of IgG	Used to affinity-purify IgG or, when labelled, as a probe for IgG
Puromycin	Analogue of aminoacyl portion of tRNA and binds to ribosome	Inhibits prokaryotic and eukaryotic protein synthesis
Streptomycin	Inhibits initiation of translation in prokaryotes	Inhibits prokaryotic protein synthesis
Tetracycline	Binds to 30S ribosomal subunit and inhibits binding of aminoacyl tRNAs	Inhibits prokaryotic protein synthesis
Tunicamycin	Analogue of UDP-*N*-acetylglucosamine; blocks synthesis of core oligosaccharide	Inhibits glycosylation of proteins. Only inhibits N-linked glycosylation of asparagine, the commonest form, not the O-linked glycosylation of serine or threonine which attaches to *N*-acetyl galactosamine
Rifamycin and rifampicin	Binds to β subunit of RNA polymerase	Inhibits initiation of RNA synthesis by blocking binding of DNA-dependent RNA polymerase to template in prokaryotes
UV irradiation	Of DNA converts A→T causing thymidine dimers or in RNA uridine dimers	Inhibits replication and transcription Special effect: inhibits influenza virus multiplication (see actinomycin D)
	Of RNA causes chain breakage of lysogens	Loss of function and infectivity Can induce lysogenic phages

Continued

Compound	Activity	Effect on virus or host
X-irradiation	Of cells causes breaks in nucleic acid chains	Loss of function
	Of whole animals prevents mitosis	Loss of immune function

AZT, 3'azido-2',3'-dideoxythymidine; C, cytosine; DNA, deoxyribonucleic acid; ds, double-stranded; dT, deoxythymidine; G, guanine; HIV, human immunodeficiency virus; IgG, immunoglobulin G; mRNA, messenger RNA; NA, nucleic acid; poly-A, poly-adenine; RNA, ribonucleic acid; rRNA, ribosomal RNA; *S. aureus, Staphylococcus aureus*; ss, single-stranded; tRNA, transfer RNA; U, uridine; UDP, uridine diphosphate; UV, ultraviolet.

Index

Page numbers in *italics* refer to figures and/or tables.

Since the major subjects of this book are viruses, few entries are listed under this keyword and the use of 'viral' as keyword has been kept to a minimum. Readers are advised to seek more specific references.

This index is in letter-by-letter order.

Abbreviations used in subentries: EBV, Epstein–Barr virus; LCMV, lymphocytic choriomeningitis virus; TMV, tobacco mosaic virus.